AF389464

Agricultural Cooperative in the Face of the Challenges of Globalization, Sustainability and Digitalization

Agricultural Cooperative in the Face of the Challenges of Globalization, Sustainability and Digitalization

Editors

Adoración Mozas Moral
Domingo Fernandez Ucles

MDPI • Basel • Beijing • Wuhan • Barcelona • Belgrade • Manchester • Tokyo • Cluj • Tianjin

Editors
Adoración Mozas Moral
University of Jaén
Spain

Domingo Fernandez Ucles
University of Jaen
Spain

Editorial Office
MDPI
St. Alban-Anlage 66
4052 Basel, Switzerland

This is a reprint of articles from the Special Issue published online in the open access journal *Agriculture* (ISSN 2077-0472) (available at: https://www.mdpi.com/journal/agriculture/special_issues/agricultural_cooperative).

For citation purposes, cite each article independently as indicated on the article page online and as indicated below:

LastName, A.A.; LastName, B.B.; LastName, C.C. Article Title. *Journal Name* **Year**, *Volume Number*, Page Range.

ISBN 978-3-0365-3763-4 (Hbk)
ISBN 978-3-0365-3764-1 (PDF)

Contents

About the Editors

Adoración Mozas Moral (Professor) She is Professor of University, Department of Business Organization, Marketing and Sociology. Doctor in Economic and Business Sciences and Extraordinary Doctorate Award from the University of Jaén. She has participated in 50 projects/research contracts (group leader in 14). He has written 18 books, 31 book chapters, 70 scientific articles that have been published in journals of the highest international and national prestige. She has received 9 Research Awards, 1 Teaching Award. She has made 5 stays in foreign centers. She has held many positions related to university management, among which she stands out as Vice-Rector of Students and Labor Insertion. She is currently President of the CIRIEC-Spain and Counselor of Number in the Institute of Giennenses Studies.

Domingo Fernandez Ucles (PhD in Social Sciences) He is Assistant professor in the Business Organization Area of the University of Jaén. He is degree in Business Administration and PhD, with international mention, in Social and Legal Sciences from the University of Jaén. He has focused his research on Information and Communication Technologies, on organic products and on the Social Economy, in the agri-food sector. He has participated in national and international projects and conferences. He is the author of several publications presented at conferences, research articles of the highest international and national prestige and book chapters. http://www4.ujaen.es/~dfucles/. ORCID: 0000-0001-7335-0296.

Editorial

The Agricultural Cooperative in the Face of the Challenges of Globalization, Sustainability and Digitalization

Adoración Mozas Moral and Domingo Fernández Uclés *

Department of Business Organization, Marketing and Sociology, Campus Las Lagunillas s/n, University of Jaén, 23071 Jaén, Spain; amozas@ujaen.es
* Correspondence: dfucles@ujaen.es; Tel.: +34-953-213-624

Citation: Moral, A.M.; Uclés, D.F. The Agricultural Cooperative in the Face of the Challenges of Globalization, Sustainability and Digitalization. *Agriculture* **2022**, *12*, 424. https://doi.org/10.3390/agriculture12030424

Received: 15 March 2022
Accepted: 15 March 2022
Published: 18 March 2022

Publisher's Note: MDPI stays neutral with regard to jurisdictional claims in published maps and institutional affiliations.

The enormous contribution of agricultural cooperative societies to the rural world has not gone unnoticed. This is corroborated by many international entities such as the International Cooperative Alliance, COPA-COGECA, the European Economic and Social Committee (EESC), the United Nations Inter-Agency Working Group on Social and Solidarity Economy and CIRIEC International. The International Cooperative Alliance estimates that 12% of the world's population is linked to one of the 3 million cooperatives that exist worldwide, most of them linked to rural areas. Therefore, cooperative societies are not a marginal phenomenon.

In relation to the role played by agricultural cooperatives in the world, it should be stated that the agricultural cooperative is an enterprise unconditionally and stably linked to the rural environment, to the farmer and the stockbreeder. For this reason, it plays a leading role in the local economy and in the fixation of the population to the territory, thus contributing to the balance and management of the territory, which makes the cooperatives true agents of rural development. On the other hand, cooperative societies have been the guarantors of the structuring of agriculture in rural areas in many countries. These organizations constitute the main structured, organized, professionalized and stable network established throughout the territory, in contact with the rural environment, with the capacity to communicate with and influence farmers and stockbreeders. They directly or indirectly provide much of the employment in the rural world, and cooperative societies by nature develop their activity under cooperative principles and values that make them exponents of socially responsible enterprises. They can therefore be seen as the key to sustainable development as promulgated by the United Nations through the SDGs.

The aim of this Special Issue has been to highlight the importance of agricultural cooperatives in the face of the challenges of globalization, sustainability and digitalization in rural areas. The contributions made to this issue apply to different products, sectors and regions around the world. Below is a summary of these 10 contributions, which are of great interest and topicality.

The efficiency of dairy cooperatives and non-cooperatives in Poland has been evaluated [1]. The results show that, assuming constant returns to scale, dairy cooperatives are technically less efficient than non-cooperatives, while, assuming variable returns to scale, these differences are not statistically significant. Such findings reveal that the technical efficiency of dairy farms in Poland is not differentiated by regional milk production potential. It is recommended to improve the technical efficiency of dairies through the process of consolidation.

Another study develops a micro-meso-macro and territorial evolutionary theoretical framework to study SSE-driven transformation in the sugarcane cluster of Veracruz (Mexico) [2]. The main findings of the article are that the SSE drives the beneficiaries, while the protagonists of the transformation cannot be defined a priori but are shaped by vectors of transformation promoted by the SSE: its values shared by a broad spectrum of actors, the socioeconomic and organizational specificities of the SSE, and its rootedness in the

1

productive system. The fundamental conclusion of the article is the need for a "territorial approach" to SSE impact, as opposed to the dominant "stakeholder-driven approach".

On the other hand, the process of a cooperative merger and its relevant role in the development of these organizations has been investigated [3]. Specifically, the economic, socio-cultural, organizational and process management factors underlying merger processes that fail have been identified: some are aborted at the negotiation stage and others are not approved by the members. The results reveal that, far from being economic factors, defensive localisms, lack of commitment to the merger on the part of partners and directors and communication failures are the most significant factors.

It has also been analyzed whether trust influences the functioning of various forms of collective entrepreneurship in rural Poland [4]. The research shows: the superior role of personal trust over institutional trust in the emergence and functioning of the studied forms of collective entrepreneurship in rural areas; the greater importance of social rather than economic factors in determining the functioning of rural collective entrepreneurship; the positive impact of generalized trust on the trust placed in the forms of entrepreneurship covered by the analysis; the increase of trust over the time of cooperation; and the impact of trust on the functioning of collective entrepreneurship, both in the economic and social dimensions, with a slight advantage of the latter.

Other researchers assess the level of willingness to cooperate among small farmers in Lithuania and elaborate the profiles of small farms that participate and intend to join cooperatives and, conversely, that do not participate in cooperatives and do not intend to do so [5]. The results show that only 8% of the surveyed farms participate in producer groups or cooperatives, while another 8% intend to participate. Small-scale farms in Lithuania have weak market integration, with no bargaining power in input and output markets. The vast majority of small-scale farms are reluctant to participate in cooperative activities in Lithuania. Thus, the main economic factors of farms and social characteristics of managers willing to cooperate are identified.

Another study estimates the factors associated with municipal participation in cooperative membership (MSCM) in Brazil and how the value of production at the municipal level changes with MSCM [6]. The results show that higher education and smaller ownership size are associated with membership in agricultural cooperatives in Brazil. We also estimate how MSCM is associated with agricultural earnings.

The importance of digitalization is also addressed. Specifically, another study identifies which organizational characteristics are directly related to the popularity of Argentine beekeeping organizations in social networks, measured by the number of followers in their accounts [7]. The results show that, beyond the use of Facebook itself, the best organizational practices are associated with factors linked to the cooperative nature of the organization, its localization, environmental sensitivity and its presence on other digital platforms.

Other researchers analyze four rural tourism sites in the suburbs of Chengdu to analyze the influence of farmers' self-identity on their intention to behave responsibly towards the land under multifunctional agricultural perception conditions as variable mediation [8]. The results show that in rural tourism destinations in suburban districts of China farmers' self-identity is an important variable affecting their intention of responsible land behavior. Moreover, the perception of agricultural economic function mediates the relationship between farmers' self-identity and the behavioral intention of land responsibility.

Another study estimates the monetary value of a policy aimed at increasing rural cooperative production in Kazakhstan in order to increase milk production [9]. It analyzes the factors associated with public support for such a policy. In addition, changes in people's WTP before and during the CO-VID-19 pandemic are examined. Among the results obtained, it is shown that psychological factors, i.e., attitude, perceived social pressure and perceived behavioral control, and respondents' awareness of the policy and opinions about the Soviet Union regime are associated with their willingness to pay; sociodemographic factors, namely, age, income and education, are also statistically significant; finally, the effect of COVID-19 fear is negatively associated with respondents' willingness to pay.

To conclude, another study analyzes the level of digitization of the European agri-food cooperative sector based on the construction of a composite synthetic index [10]. The results of the study reveal the existence of a suboptimal and heterogeneous degree of digitization of European agri-food cooperatives, clearly conditioned by their size and the wealth of the country where they operate. The authors recommend promoting public policies that guarantee high-performance digital connectivity, improved training in digital skills and the promotion of cooperative integration processes.

Funding: The funders had no role in the design of the study; in the collection, analyses, or interpretation of data; in the writing of the manuscript, or in the decision to publish the results.

Acknowledgments: We would like to thank the Journal for trusting us with the development of this Special Issue. We would also like to thank all the authors who have submitted their proposals, as well as the reviewers who have participated in the evaluation and improvement of the manuscripts.

Conflicts of Interest: The authors declare no conflict of interest.

References

1. Ziętek-Kwaśniewska, K.; Zuba-Ciszewska, M.; Nucińska, J. Technical Efficiency of Cooperative and Non-Cooperative Dairies in Poland: Toward the First Link of the Supply Chain. *Agriculture* **2022**, *12*, 52. [CrossRef]
2. Gallego-Bono, J.R.; Tapia-Baranda, M. A Territorial-Driven Approach to Capture the Transformative Momentum of the Social Economy Especially from the Agricultural Cooperatives. *Agriculture* **2021**, *11*, 1281. [CrossRef]
3. Meliá-Martí, E.; Lajara-Camilleri, N.; Martínez-García, A.; Juliá-Igual, J.F. Why Do Agricultural Cooperative Mergers Not Cross the Finishing Line. *Agriculture* **2021**, *11*, 1173. [CrossRef]
4. Sieczko, L.; Parzonko, A.J.; Sieczko, A. Trust in Collective Entrepreneurship in the Context of the Development of Rural Areas in Poland. *Agriculture* **2021**, *11*, 1151. [CrossRef]
5. Droždz, J.; Vitunskienė, V.; Novickytė, L. Profile of the Small-Scale Farms Willing to Cooperate—Evidence from Lithuania. *Agriculture* **2021**, *11*, 1071. [CrossRef]
6. Neves, M.D.C.R.; Silva, F.D.F.; Freitas, C.O.D.; Braga, M.J. The Role of Cooperatives in Brazilian Agricultural Production. *Agriculture* **2021**, *11*, 948. [CrossRef]
7. Andrieu, J.; Fernández-Uclés, D.; Mozas-Moral, A.; Bernal-Jurado, E. Popularity in Social Networks. The Case of Argentine Beekeeping Production Entities. *Agriculture* **2021**, *11*, 694. [CrossRef]
8. Cao, X.; Luo, Z.; He, M.; Liu, Y.; Qiu, J. Does the Self-Identity of Chinese Farmers in Rural Tourism Destinations Affect Their Land-Responsibility Behaviour Intention? The Mediating Effect of Multifunction Agriculture Perception. *Agriculture* **2021**, *11*, 649. [CrossRef]
9. Kaliyeva, S.; Areal, F.J.; Gadanakis, Y. Would Kazakh Citizens Support a Milk Co-Operative System? *Agriculture* **2022**, *11*, 642. [CrossRef]
10. Jorge-Vázquez, J.; Chivite-Cebolla, M.; Salinas-Ramos, F. The digitalization of the European agri-food cooperative sector. Determining factors to embrace information and communication technologies. *Agriculture* **2021**, *11*, 514. [CrossRef]

Article

Technical Efficiency of Cooperative and Non-Cooperative Dairies in Poland: Toward the First Link of the Supply Chain

Katarzyna Ziętek-Kwaśniewska *, Maria Zuba-Ciszewska and Joanna Nucińska

The Institute of Economics and Finance, Faculty of Social Sciences, The John Paul II Catholic University of Lublin, Al. Racławickie 14, 20-950 Lublin, Poland; maria.zuba@kul.pl (M.Z.-C.); nucinska@kul.pl (J.N.)
* Correspondence: kwasniewska@kul.pl

Abstract: Several studies conducted in various countries have addressed the technical efficiency of dairies. However, there is a paucity of research on the technical efficiency of dairies in Poland, particularly in relation to their legal form (i.e., cooperatives vs. non-cooperatives). The existing literature also does not provide insights into the technical efficiency of these entities with respect to different regions' milk production capacity. Therefore, this paper aims to: (1) evaluate and compare the technical efficiency of cooperative and non-cooperative dairies in Poland, and (2) examine dairies' technical efficiency due to spatial disparities in milk production potential. We use data envelopment analysis (DEA) to investigate the technical efficiency of 108 dairies in Poland for the year 2019. The milk production capacity of provinces is examined by applying the zero unitarization method. The results show that when assuming constant returns to scale (CRS), dairy cooperatives are less technically efficient than non-cooperatives, whereas when assuming variable returns to scale (VRS), these differences are not statistically significant. For inefficient dairies, we observe the greatest potential for improvement in labor costs and depreciation. Both cooperatives and non-cooperatives operate mostly under decreasing returns to scale. Thus, the potential for enhancing the technical efficiency of dairies through the consolidation process seems to be exploited. Our findings reveal that the technical efficiency of dairies in Poland is not differentiated by regional milk production potential.

Keywords: technical efficiency; cooperatives; dairy processing sector; sustainability; milk production capacity; supply chain; data envelopment analysis

Citation: Ziętek-Kwaśniewska, K.; Zuba-Ciszewska, M.; Nucińska, J. Technical Efficiency of Cooperative and Non-Cooperative Dairies in Poland: Toward the First Link of the Supply Chain. *Agriculture* 2022, 12, 52. https://doi.org/10.3390/agriculture12010052

Academic Editors: Adoración Mozas Moral and Domingo Fernandez Ucles

Received: 19 November 2021
Accepted: 29 December 2021
Published: 1 January 2022

Publisher's Note: MDPI stays neutral with regard to jurisdictional claims in published maps and institutional affiliations.

1. Introduction

The concept of sustainable development is central to political as well as scientific debate. Although definitions of sustainability are varied and fluid depending on different actors' viewpoints [1], this concept has become the cornerstone of global dialogue on the future of humanity [2].

In the presence of limited resources, a growing world population, and climate change, global food security is a major concern [3]. The significance of this problem is strongly emphasized in the United Nations (UN) 2030 Agenda for Sustainable Development [4] by setting 17 Sustainable Development Goals (SDGs), the second of which refers to ending hunger, achieving food security and improved nutrition, and promoting sustainable agriculture (SDG2). In particular, target 2.4 aims to ensure, by 2030, sustainable food production systems and implement resilient agricultural practices [4].

Food systems are extremely diverse and dynamic [5] as well as intrinsically complex, involving many different processes, value chains, actors, and interactions [6]. The concept of a sustainable food system implies sustainability in three dimensions: economic, social, and environmental [7]. The ability to use resources efficiently in production is a prerequisite for the sustainability and competitiveness of the agrifood sector. The significance of food security has been additionally strengthened at the national level by the COVID-19 pandemic [8].

Milk and dairy products are an essential food for human nutrition worldwide [9,10]. Hence, the dairy sector can be considered one of the key building blocks of food systems. The sustainability of the dairy industry can be seen as providing consumers with the nutritional dairy products they demand in an economically viable, environmentally sound, and socially responsible way, now and in the future [11]. The sustainable milk and dairy production life cycle ranges from on-farm milk production, the industrialization and processing of dairy products, all the way to their marketing [10], creating a network structure [12] and a closely knitted process called a supply chain [13].

Performance evaluation has become a significant topic in supply chain management [12], including the dairy sector. Although a number of studies have been conducted on economic sustainability at the farm level (e.g., [14–17]), the discussion cannot be limited to this initial link of the dairy supply chain. Given that milk is a perishable commodity that cannot be stored in its raw form, its processing and transformation are crucial in the dairy sector [18]. For this reason also, the economic sustainability of dairies, which are the next link in the supply chain, should be given equal attention. Nevertheless, research in this area remains scarce. As economic sustainability is considered a complex problem, in this study, and similarly to Popović and Panić [19], we refer to efficiency as its component.

The dairy processing industry belongs to the major subsectors of the food processing industry in the European Union (EU) [8]. An efficient and competitive milk processing industry has been deemed crucial to maintaining sustainable milk production [20].

The relationship between the initial and the intermediate segments of the dairy supply chain has become the rationale for the establishment of cooperatives. Farmers' cooperative ownership has a long tradition in many parts of the world and is the most prevalent form of vertical integration in dairy supply chains [21]. Dairy cooperatives have played an important role in the dairy processing sector in Europe [22]. Poland is a prime example, as more than 70% of its dairies operate as a cooperative compared to about 20% in most EU countries [23]. Poland is one of the leading cow's milk producers and processors in the EU (12.2 million tons cows' milk delivered to dairies in 2019 [24]), characterized by considerable spatial diversity in its milk production capacity [25].

There is a debate concerning the relative efficiency of cooperatives versus explicitly for-profit forms of organization in the dairy processing industry [22,26]. Empirical analyses in this field have employed various methods. Data envelopment analysis (DEA) is a commonly used approach for measuring the relative efficiency and competitiveness of the food and drink industry worldwide (as reviewed in [27]).

Given the importance of the dairy industry's efficiency, the aim of this study is twofold: (1) to evaluate and compare the technical efficiency of cooperative and non-cooperative dairies in Poland, and (2) to examine dairies' technical efficiency due to spatial disparities in milk production potential. We evaluate the technical efficiency of dairies using the DEA approach. By exploring the issue of the dairy sector's efficiency with a focus on the legal form of milk processors, our study contributes to the stream of research on agricultural cooperatives within the context of sustainability.

The article is structured as follows. Section 2 presents the condition of the Polish dairy sector and provides a literature review. Section 3 describes the data and methods. Section 4 presents and discusses the research results. Finally, Section 5 concludes and outlines areas for future research.

2. Theoretical Background

2.1. The Condition of the Dairy Sector in Poland

The agrifood industry is the largest manufacturing sector in the EU, of which the dairy processing industry is a relevant subsector [8]. The EU is the most important supplier of milk and dairy products on the world market [28].

When it comes to dairy products, Poland remains self-sufficient. Indeed, the country's degree of food self-sufficiency in the case of milk and its products, i.e., the ratio of domestic production to domestic consumption [29], has been practically systematically increasing for the past 20 years [30]. Polish dairy cooperatives ensure food security in dairy products, especially the basic ones, in every region of the country [31].

The number of dairy cows in 2019 amounted to 2.16 million, a continuation of a downward trend. In 2019, the number of farms keeping cows was 220,000, but only 118,000 of these supplied milk to the dairy industry, 95% of which were family farms. Despite the increase in the average herd of cows (from 3.9 in 2005 to 11.2 in 2019), most of the farms are still characterized by their low scale of production [32,33]. According to Eurostat data for 2016, the share of farms with more than 30 dairy cows in Poland was only 20%, while for other major milk producers, the figure was as high as 85–97% (specifically in Germany, the United Kingdom, France, and the Netherlands) [34]. However, the process of concentrating cows in large and efficient farms continues, as does the modernization of milk production through improving production technology and the genetics of dairy cattle (farms with small-scale production and low profitability have abandoned dairy cattle breeding). Since Poland's accession to the EU in 2004, the marketability of its milk production has improved significantly (in 2019, over 84% of the volume of raw material produced went to the dairy industry), although it remains below the EU average (94%) [32,33,35].

From the beginning of the market transformation, we have observed a practically systematic downward trend in the number of dairies. Since 1990, when 348 dairies, all of which were cooperative, operated on the market, the share of the cooperative sector both in the number of dairies and in the purchase of milk has declined (to 62.5% and 72.3% in 2017, respectively [31]). However, the dairy industry is currently the only industry in Poland dominated by cooperatives, strengthening the integration of agriculture with the processing industry. Thanks to the concentration and modernization of the dairy industry, which began in the 1990s and continues to the present day, the technical and economic productivity of the average dairies has systematically improved [32,36].

The value of sales has been growing systematically, especially since Poland's accession to the EU (by 84% to PLN 34.7 billion), as well as the share of direct exports in the value of sales (up to 18.4% in 2019). The milk processing sector is characterized by its continued net sales profitability (for over 20 years), and the level thereof has increased from 0.2% in 1999 to 1.4% in 2019, although the share of profitable entities in the industry changed in that period (in 2019 it amounted to 68.1%). The sector also maintains current financial liquidity, albeit the investment rate in 2019 was 1.45, i.e., lower than 20 years earlier. Throughout this period, investment activity increased in years of good economic conditions on the world market along with growing exports of dairy products. Especially since 2011, it has shown a systematic increase (except for 2018, when it decreased, albeit only slightly). This means that in each year, investment expenditures in relation to the annual depreciation increased. These were intended, to the greatest extent, for the purchase of machinery, equipment, and means of transport, i.e., for the modernization of dairies' production potential [32,35–37].

Given that it deals with the industrial processing of collected milk into finished products for consumption or refined raw material for other industries, the dairy industry is closely related to farms. Domestic milk production is characterized by considerable territorial differentiation [25]. Connecting the spatial distribution of processing plants with their raw material base is important due to the territorial dispersion and fragmentation of the production of agricultural raw materials between many farms [38].

In recent years, the phenomenon of the concentration of a fragmented food industry has been clearly noticeable in Poland, because the scale effect is readily apparent, consisting in the dependence of production costs and profits at the scale of production. Food processing thus follows the footsteps of concentration in agriculture [38]. The entity structure of the food industry in Poland is changing, both as a result of Poland's accession to the EU and continued economic globalization. The fact that processes of concentration and consolidation are underway is reflected by the takeovers of Polish enterprises by foreign and domestic investors, as well as the mergers taking place among Polish enterprises [39]. Although the process of concentration of subjective structures in the dairy industry has been faster since 2004 than the average for the entire food industry [40], the industry itself is characterized by its low degree of internationalization [41]. The growing share of large enterprises in the sold production of the dairy sector [42] further evidences the existing trend of industrialization of production in food processing [43]. Currently, the rapid development of large retail chains (often with foreign capital) has changed the balance of power in the national food chain, causing the dominant position of processing companies to decline [44].

2.2. Efficiency of Dairies: Literature Review

The efficient use of resources is an evident driver of economic development [45]. Hence, improvement in this area often becomes one of the sustainability goals of any industry [46]. Enhancing the productivity and efficiency of agriculture input use is regarded as the first step to meeting the challenge of sustainable use of natural resources as well as reducing environmental impacts [47].

The dairy sector's efficiency has been the subject of a number of studies worldwide. While many of them have addressed the question of efficiency at the farm level [28,47–65], the problem of dairy efficiency seems to have received less attention.

The performance of dairy processors has been assessed with various methods. One stream of literature focuses on the financial performance of dairies. Another explores these entities' technical efficiency using DEA or stochastic frontier analysis (SFA) [66]. The SFA approach has been applied to the dairy processing industry by, for instance, Doucouliagos and Hune [67], Soboh et al. [68], Hirsch et al. [23], Čechura and Žáková Kroupová [8], and Beber et al. [69]. The DEA method has also been widely used in studies on the technical efficiency of dairies in various countries. Table A1 in Appendix A summarizes the literature review on the application of DEA to the dairy processing industry.

Not many papers have been published on the efficiency of the dairy processing industry in Poland. Five such studies using the DEA method [9,70–73] are included in the literature review (Table A1). These studies have addressed the following aspects of dairy efficiency: technical efficiency [73], its changes [9,70,71], and scale efficiency [72]. Furthermore, the aforementioned evaluation of technical efficiency has been supplemented by analyses of selected financial ratios for dairies [9,70,71,73]. The data used in these studies cover a specific region [71,73], a whole country [70,72], or more than one country [9].

The aforementioned studies have examined cooperatives, among other forms of dairy processing entities in Poland. Additionally, Špička [9] has emphasized the significance of cooperatives' prevalence in the Polish dairy industry. As the existing body of literature lacks a comprehensive analysis of the technical efficiency of dairy cooperatives in comparison with other organizational forms of dairies in Poland, our study aims to fill this gap.

Previous studies on the efficiency of dairies vary on many dimensions. While the literature does not provide a complete list of sources of efficiency differences, Berger and Mester [74], though in the context of financial institutions, have indicated three of them: (1) the concept of efficiency employed; (2) the methods used to measure efficiency under these concepts; and (3) potential correlates of efficiency. The third source covers at least partially exogenous characteristics that may explain some of the efficiency differences that remain after controlling for conceptual and measurement issues. Correlates of efficiency include, for example, regulatory, market type, or organizational form [74]. Therefore, following this view, the cooperative as a dominant organizational form of dairies is considered a determinant that may have a significant impact on their (in)efficiency [69].

According to Pietrzak [75], farmer cooperatives can pursue a variety of objectives. They show potential to improve the welfare of farmer-members and society as a whole in comparison with profit-maximizing enterprises (investor-owned firms, IOFs) [75]. It is assumed that differences in objectives and organizational structures between IOFs and co-operatives affect their production technology and technical efficiency [68]. On the one hand, cooperatives are less oriented toward efficient input use (especially members' products) and value-added production than on exploiting economies of scale (Hind, 1999, as cited in [22]). On the other hand, their relatively conservative financial structure, low ownership costs, and the homogeneity of member's interests are recognized as factors that make them succeed [69].

Farmer cooperatives are common and significant commercial organizations in many parts of the world [76]. Previous empirical studies on the technical efficiency of dairy cooperatives and IOFs have indicated that the efficiency of both cooperatives and IOFs can be greater depending on the context, the data employed, and the objective of the performance measured [69]. This provides motivation for further research in this area.

3. Data and Methods

3.1. Data

In this paper, we focus on dairies in Poland (NACE Rev. 2 Class 10.51) that were operating in 2019 and were still active as of 20 February 2021 (thus, excluding entities closed and in liquidation). The data used in this paper were obtained from balance sheets and income statements of dairies for the year 2019 retrieved from the Emerging Markets Information Service database (EMIS) [77]. Considering the scope and type of data required, the following criteria guided the selection of entities for the study: (1) availability of financial statements for the year 2019, and (2) presentation of income statement by nature of expense.

Initially, 116 dairies meeting these criteria were selected, i.e., 71% of 163 dairies operating in 2019 [35]. Entities with missing records were then removed. We also eliminated the outliers due to the sensitivity of efficiency scores to their presence: if there is an outlier among the observations, it can result in a significant reduction in the level of technical efficiency of inefficient units [78,79]. The outliers were identified using output to input ratios [66] according to the following procedure. We identified a unit as an outlier if the value of any of the output to input ratios fell outside the interval of the mean plus/minus three standard deviations. Finally, a sample of 108 dairies was used for the empirical investigation. Taking one output and four inputs in our study, this sample size fully satisfied the rule of thumb for determining the appropriate number of decision-making units (DMUs) in DEA, stating that $n \geq \max\{m \times s, 3(m + s)\}$ where n stands for the number of DMUs, m is the number of inputs, and s is the number of outputs [80]. The sample comprised 65 (60.2%) cooperative and 43 (39.8%) non-cooperative dairies. This corresponded to the structure of dairies in Poland by legal form in 2019 (57.5% and 42.5%, respectively) [77].

Table 1 presents the results of a comparative analysis of the financial ratios for cooperative and non-cooperative dairies in the areas of liquidity, profitability, capital structure, and activity. The Mann–Whitney U test was employed for between-group comparisons due to the failure to meet the assumptions of parametric testing. We observed a statistically significant difference between cooperative and non-cooperative dairies in terms of profitability ratios; this applied to ratios based on net profit and operating profit. The profitability ratios were significantly higher in non-cooperative dairies compared to their cooperative counterparts. We also identified statistically significant differences in days receivables outstanding, days payable outstanding, and days inventory outstanding. The Mann–Whitney U test revealed that the cooperative dairies managed their inventories more efficiently, collected receivables more quickly, and also paid off their liabilities faster. Nevertheless, as there was no significant difference in the cash conversion cycle between these groups, the above-mentioned differences ultimately canceled out. We also identified a statistically significant difference in the wage efficiency ratio, i.e., non-cooperative dairies generated significantly higher sales revenue per each PLN paid for the labor factor. In the case of liquidity and capital structure ratios, there were no significant differences between the two groups of dairies. Our findings suggest that for cooperative dairies, maintaining financial security is more important than achieving profitability, and this attitude results from their specificity. This is because dairy cooperatives have a bimodal character, i.e., they involve a community of members and are enterprises that this community has established.

The long-term stability of functioning, and thus the ability to achieve the goals for which the cooperatives were instituted, is more important than short-term profit making [75].

Table 1. A comparative analysis of financial ratios for cooperative and non-cooperative dairies for the year 2019 (authors' calculations based on [70,77,81].

Financial Ratio	Formula	Form	n	Mean	Med	Mann–Whitney			
						Mean Rank	U	Z	p
		liquidity ratios							
current ratio	current assets/current liabilities	cooperative	65	2.419	1.668	58.00	1170.00	−1.428	0.153
		non-cooperative	43	1.659	1.269	49.21			
cash ratio	cash/current liabilities	cooperative	65	1.026	0.356	59.26	1088.00	−1.942	0.052
		non-cooperative	43	0.455	0.054	47.30			
		profitability ratios							
return on sales	net profit/net sales	cooperative	65	−0.039	0.002	46.80	897.00	−3.141	0.002
		non-cooperative	43	−0.003	0.014	66.14			
return on assets	net profit/total assets	cooperative	65	−0.025	0.003	47.29	929.00	−2.940	0.003
		non-cooperative	43	0.049	0.025	65.40			
return on equity	net profit/equity	cooperative	65	−0.086	0.009	46.66	888.00	−3.198	0.001
		non-cooperative	43	0.162	0.065	66.35			
return on sales II	operating profit/net sales	cooperative	65	−0.037	0.000	45.94	841.00	−3.494	<0.001
		non-cooperative	43	0.001	0.017	67.44			
return on assets II	operating profit/total assets	cooperative	65	−0.023	0.001	46.42	872.00	−3.300	<0.001
		non-cooperative	43	0.052	0.029	66.72			
return on equity II	operating profit/equity	cooperative	65	−0.074	0.009	46.65	887.00	−3.206	0.001
		non-cooperative	43	0.165	0.060	66.37			
		capital structure							
equity to assets ratio	equity/total assets	cooperative	65	0.541	0.571	57.71	1189.00	−1.309	0.191
		non-cooperative	43	0.463	0.526	49.65			
long-term debt to assets ratio	long-term debt/total assets	cooperative	65	0.095	0.076	53.22	1314.00	−0.524	0.600
		non-cooperative	43	0.113	0.056	56.44			
short-term debt to assets ratio	short-term debt/total assets	cooperative	65	0.364	0.347	51.98	1234.00	−1.026	0.305
		non-cooperative	43	0.444	0.392	58.30			
equity to fixed assets ratio	equity/fixed assets	cooperative	65	1.727	1.306	58.48	1139.00	−1.622	0.105
		non-cooperative	43	1.336	1.029	48.49			
		activity ratios							
total asset turnover ratio	net sales/total assets	cooperative	65	2.846	2.730	57.75	1186.00	−1.327	0.184
		non-cooperative	43	2.681	2.377	49.58			
fixed asset turnover ratio	net sales/fixed assets	cooperative	65	10.058	6.200	57.09	1229.00	−1.058	0.290
		non-cooperative	43	10.783	5.053	50.58			
equity turnover ratio	net sales/equity	cooperative	65	5.512	5.008	56.06	1296.00	−0.637	0.524
		non-cooperative	43	5.164	3.602	52.14			
wage efficiency ratio	net sales/labor cost	cooperative	65	9.303	7.966	44.18	727.00	−4.208	<0.001
		non-cooperative	43	17.363	11.730	70.09			
raw material efficiency ratio	net sales/raw materials	cooperative	65	2.643	1.357	52.66	1278.00	−0.750	0.453
		non-cooperative	43	3.120	1.446	57.28			
days inventory outstanding (DIO)	(inventory/net sales) × 365	cooperative	65	16.093	12.884	46.45	874.00	−3.286	0.001
		non-cooperative	43	25.211	20.857	66.67			
days receivables outstanding (DRO)	(short-term receivables/ net sales) × 365	cooperative	65	34.865	32.508	45.42	807.00	−3.706	<0.001
		non-cooperative	43	40.816	39.782	68.23			
days payables outstanding (DPO)	(current liabilities/net sales) × 365	cooperative	65	70.344	42.158	47.92	970.00	−2.683	0.007
		non-cooperative	43	77.130	52.799	64.44			
cash conversion cycle (CCC)	DIO+DRO-DPO	cooperative	65	−19.386	2.915	53.91	1359.00	−0.242	0.809
		non-cooperative	43	−11.103	1.658	55.40			

3.2. Method

Our research design consisted of the following phases: (1) assessment of the technical efficiency of cooperative and non-cooperative dairies in Poland; and (2) examination of the technical efficiency of dairies in the context of spatial disparities in milk production potential.

3.2.1. DEA Method

The technical efficiency of cooperative and non-cooperative dairies in Poland was determined by using DEA, a non-parametric approach used in evaluating the performance of DMUs on the basis of multiple inputs and multiple outputs [12]. This method is described, for example, in [82,83].

In order to evaluate the technical efficiency of dairies in Poland, the constant returns to scale (CRS) model was applied first. Whereas the CRS assumption is regarded as appropriate if all units operate at optimal scale [83], when there are differences in the scale of operation of units, the variable returns to scale (VRS) model is considered more suitable [84]. Therefore, in the next step we applied this approach. Using the VRS specification allowed us to calculate the technical efficiency while excluding scale efficiency (SE) effects [82,83]. The SE score is the result of dividing the technical efficiency (TE) obtained under the CRS assumption by the pure technical efficiency (PTE) score from the VRS model. Thus, differences between CRS and VRS technical efficiency scores indicate the presence of scale inefficiency [52,85]. Under the VRS assumption, scale-inefficient DMUs are compared only with efficient ones of similar size [9,61]. The decomposition of the CRS TE score into the PTE and SE allowed us to determine the extent to which the inefficiency of dairies in Poland is related to management issues and an inappropriate scale size (see [86]).

Within the CRS and VRS assumptions, two approaches (i.e., input oriented and output oriented) can be employed. The choice of orientation should take into account "which quantities (inputs or outputs) the managers have most control over" [83], p. 180. For dairy operations, the input-oriented model has been indicated as being more appropriate [72]. This orientation has also been widely adopted in dairy sector efficiency studies [20,66,72,85]. In the present study, therefore, we followed this approach, viewing the dairies, similarly to [66], as cost minimizers.

One output and four input variables were used in the DEA models. The variables were selected on the basis of the literature review (Table A1). The selected output variable was net sales revenue. Given that dairies may offer a variety of products and data on production in physical terms are not presented in their financial statements, the choice of this variable as the output variable seemed appropriate and reasonable. The input variables were:

- Labor costs—due to the lack of data on labor inputs in physical terms, this cost category represents the factor of production in question; it consists of salaries and social security costs;
- Raw material costs—raw materials are of key importance for dairies; by including this cost category, we refer to the involvement of raw materials, mainly milk, in the production process of dairy products;
- Depreciation expense—capital is one of the major factors of production; given that net sales revenue is used as the output variable, for consistency purposes, depreciation expense is adopted as the input of capital factor due to its flow nature; this cost category can be seen as "the financial value of consumption of the long-term assets" [9], p. 177;
- Other operating costs—including other costs related to the production process.

Table 2 presents the descriptive statistics of the output and input variables.

Table 2. Descriptive statistics of output and input variables; values given in thousands PLN (authors' calculations based on [77]).

Form	Variable	Mean	Med	SD	Min	Max	Q1	Q3
cooperative ($n = 65$)				output				
	NS	268,525.49	38,130.18	833,695.80	341.98	5,182,216.01	18,958.07	131,452.70
				inputs				
	LC	18,956.04	5282.10	49,499.37	242.75	336,432.51	2707.10	14,289.93
	RM	207,015.94	26,569.23	654,858.48	53.33	4,060,276.85	9255.83	106,025.00
	DE	5063.03	580.20	15,447.84	7.90	96,878.00	193.45	4247.92
	OC	35,502.32	5031.30	107,198.03	109.01	641,033.23	2540.70	20,564.88
non-cooperative ($n = 43$)				output				
	NS	228,647.14	87,096.48	347,286.35	423.85	1,486,375.50	22,870.88	281,108.00
				inputs				
	LC	13,675.55	6566.71	23,725.76	238.24	131,231.00	2209.23	14,671.11
	RM	144,554.43	48,073.75	224,947.70	95.42	1,192,281.50	14,737.43	175,026.00
	DE	4230.19	1609.79	8957.35	5.46	42,231.00	547.61	3585.27
	OC	56,911.08	12,529.71	116,628.46	140.32	499,853.94	3540.48	51,947.02
total ($n = 108$)				output				
	NS	252,648.00	60,269.39	680,776.08	341.98	5,182,216.01	21,605.88	191,143.70
				inputs				
	LC	16,853.62	5551.52	41,148.91	238.24	336,432.51	2572.14	14,462.88
	RM	182,147.00	34,267.88	526,600.50	53.33	4,060,276.85	10,448.64	143,293.50
	DE	4731.44	890.42	13,205.95	5.46	96,878.00	250.82	3697.86
	OC	44,026.18	6809.61	111,010.84	109.01	641,033.23	2876.41	26,339.29

Abbreviations: n, number of observations; NS, net sales revenue; LC, labor costs; RM, raw material costs; DE, depreciation expense; OC, other operating costs; Med, median; SD, standard deviation; Min, minimum value; Max, maximum value; Q1, lower quartile; Q3, upper quartile.

3.2.2. Regional Analysis

As part of our research, we also wanted to investigate whether the region in which a dairy is located significantly differentiates the technical efficiency of the DMUs studied. As dairies are the second link in the dairy supply chain and are therefore dependent on the operation of dairy farms, and as the milk production capacity at the level of dairy farms varies spatially in Poland, we first determined the milk production potential of each province in Poland. For this purpose, we used the zero unitarization method [87]. Our approach comprised the following steps:

- Identification of a set of potential diagnostic variables substantively related to the phenomenon under study;
- Selection of diagnostic variables meeting the following statistical criteria: coefficient of variation (CV) at least equal to 0.1; max to min ratio at least equal to 2 [88];
- Normalization of diagnostic variables (all selected variables are stimulants) $X_1, X_2, \ldots, X_s$ according to the following formula [87]:

$$z_{ij} = \frac{x_{ij} - \min\limits_{i} x_{ij}}{\max\limits_{i} x_{ij} - \min\limits_{i} x_{ij}}, \left(\begin{array}{l} i = 1, 2, \ldots, r \\ j = 1, 2, \ldots, s \end{array} \right), \tag{1}$$

where: r—number of objects; s—number of diagnostic variables;

- Determination of a synthetic variable Q_i [87]:

$$Q_i = \frac{1}{s} \sum\nolimits_{j=1}^{s} z_{ij} \ (i = 1, 2, \ldots, r) \tag{2}$$

- Division of provinces into three groups (according to the method presented in [87]):

○   Group I—provinces with a high level of milk production capacity:

$$Q_i \in (\max_i Q_i - U, \ \max_i Q_i] \tag{3}$$

○   Group II—provinces with a medium level of milk production capacity:

$$Q_i \in (\max_i Q_i - 2U, \ \max_i Q_i - U] \tag{4}$$

○   Group III—provinces with a low level of milk production capacity:

$$Q_i \in [\min_i Q_i, \ \max_i Q_i - 2U] \tag{5}$$

where:

$$U = \frac{\max_i Q_i - \min_i Q_i}{3} \tag{6}$$

The following variables were selected to assess provinces' milk production capacity: X_1, dairy cow density per 100 ha of agricultural land (in heads); X_2, total cow's milk purchase (in thousands of liters); X_3, share of purchase in milk production (in %); X_4, average milk yield per cow (in liters); X_5, cow's milk production per 1 ha of agricultural land (in liters); X_6, share of cows in farms with more than 50 cows (in %); X_7, average number of cows per farm. All data are for the year 2019 [35,89,90], except for the variable X_7. Due to the lack of more recent data, this variable refers to 2016 [91].

Given the above classification of provinces, each dairy was assigned to an appropriate region (high, medium, or low milk production potential) according to its location. Then, the H Kruskal–Wallis test was used to determine the differences in technical efficiency between these three groups of dairies (as the data did not meet the assumptions for parametric testing).

The DEA was conducted using DEAP Version 2.1 [82]. This program has previously been used by Singh et al. [18,92], Gradziuk [71], Ohlan [85,93], Madau et al. [28], Syp and Osuch [57], Silva et al. [60], and Popović and Panić [19], among others, in studies on the efficiency of the dairy sector (whether dairies or dairy farms). Other calculations were performed using IBM SPSS Statistics, version 27. A p value less than 0.05 was considered statistically significant.

4. Results and Discussion

4.1. Technical Efficiency of Cooperative and Non-Cooperative Dairies

Table 3 presents the summary statistics for the technical efficiency of the examined dairies. Under the CRS assumption, the TE scores ranged from 0.543 to 1. Although the mean TE score of 0.895 exhibited a high degree of technical efficiency of dairies, this result also indicated that there is still scope for improvement in this area. That is, overall, on average, dairies could proportionally reduce their inputs by 10.5% without reducing their output. For the least efficient dairy, this reduction should be as high as 45.7%.

Table 3. Descriptive statistics of TE scores and results of Mann–Whitney U test (authors' calculations based on [77]).

	Form	n	Mean	Med	SD	Min	Max	Q1	Q3	Mean Rank	Mann–Whitney		
											U	Z	p
	cooperative	65	0.879	0.884	0.081	0.543	1.000	0.832	0.927	48.32	995.50	−2.534	0.011
TE	non-cooperative	43	0.920	0.932	0.082	0.747	1.000	0.845	1.000	63.85			
	total	108	0.895	0.899	0.084	0.543	1.000	0.839	0.978				

It is generally assumed that cooperatives are less efficient than other legal forms of plants [18]. Due to the specific nature of cooperatives as presented by Soboh et al. [22], cooperative dairies were expected to have a lower value of input-oriented technical efficiency.

Examination of the TE scores by legal form indeed showed that non-cooperative dairies outperformed their cooperative counterparts. The mean and quartile values of the TE scores for non-cooperative dairies exceeded those for cooperatives. In this regard, a Mann–Whitney U test indicated that technical efficiency in the group of non-cooperatives was statistically significantly higher than in the group of cooperatives (Table 3).

According to the TE scores, 22 dairies (20.4%) were identified as technically efficient (by efficient units, we mean units for which the efficiency score was 1 and all input and output slack values were zero). In their case, it can be stated that the inputs involved were efficiently consumed in the production process. Most of them (14 out of 22) were non-cooperative dairies. The remaining 86 (79.6%) with TE scores below 1 showed inefficiency in input utilization. For them, it is recommended to make efforts to enhance the efficiency of input use.

The percentage of efficient DMUs was lower in the group of cooperative dairies than in the group of non-cooperative ones. Whereas only 12.3% of the former group of dairies was fully efficient, this was true of almost one third of the latter. In order to examine the association between the legal form of dairy and being technically efficient, a chi-square test of independence was performed. The relationship between the above variables was found to be statistically significant (Table 4). Thus, non-cooperative dairies were more likely to be technically efficient than cooperative ones. In this context, however, it should be noted that the examination of the TE scores of only inefficient dairies did not reveal statistically significant differences between cooperatives and non-cooperatives ($U = 707.50$, $Z = -1.087$, $p = 0.277$).

Table 4. Sample structure according to TE scores (authors' calculations based on [77]).

Form	TE		x^2	*df*	*p*
	Efficient	**Inefficient**			
cooperative	8 (12.3%)	57 (87.7%)	6.543	1	0.011
non-cooperative	14 (32.6%)	29 (67.4%)			

Note: row percentages are given in parentheses.

While the analysis of the TE scores provides an insight into the overall technical efficiency, its decomposition into PTE and SE gives us additional valuable information on the efficiency performance of the dairies studied. As shown in Table 5, the PTE scores were at least as high as the TE scores, which is in line with the theory that the VRS frontier is more flexible and envelops the data points more tightly than the CRS frontier [51,86]. The mean PTE score reached 0.935, suggesting that given the scale size, the examined dairies could reduce their input consumption proportionally by 6.5% without altering their output. The least efficient DMU had a PTE score of 0.549, indicating the need for a proportional reduction in inputs of 45.1%.

Table 5. Descriptive statistics of PTE scores and results of Mann–Whitney U test (authors' calculations based on [77]).

	Form	*n*	Mean	Med	SD	Min	Max	Q1	Q3	Mann–Whitney			
										Mean Rank	U	Z	*p*
PTE	cooperative	65	0.926	0.949	0.083	0.549	1.000	0.877	1.000	49.81	1092.50	−1.957	0.050
	non-cooperative	43	0.947	0.994	0.075	0.755	1.000	0.902	1.000	61.59			
	total	108	0.935	0.958	0.080	0.549	1.000	0.881	1.000				

The examination of the PTE scores by legal form did not reveal statistically significant differences between non-cooperative dairies and their cooperative counterparts (Table 5). This result is similar to Singh et al. [18] using the DEA method.

According to the PTE scores, 38 (35.2%) dairies were identified as efficient (PTE = 1) and 70 (64.8%) as inefficient (PTE < 1). There was a significant relationship between being technically efficient (in terms of PTE) and the legal form of the dairy in question. The non-cooperative dairies were more likely to be efficient than the cooperative ones. Whereas in the group of cooperative dairies about one in four units was fully efficient, in the group of non-cooperatives it was almost half (Table 6). It should be added, however, that when comparing the PTE scores of only inefficient dairies by legal form, no statistically significant differences were found between cooperatives and non-cooperatives ($U = 516.00$, $Z = -0.152$, $p = 0.879$).

Table 6. Sample structure according to PTE scores (authors' calculations based on [77]).

Form	PTE		χ^2	df	p
	Efficient	Inefficient			
cooperative	17 (26.2%)	48 (73.8%)	5.84	1	0.016
non-cooperative	21 (48.8%)	22 (51.2%)			

Note: row percentages are given in parentheses.

A more in-depth look at the efficiency reference set (under the VRS assumption), which serves as a benchmark for inefficient DMUs, allowed us to identify the best-practice dairies. Concerning the reference set frequency, the "best performer" (with the highest frequency of 64) was the small dairy from the Wielkopolskie province. The second and third places belonged to the medium dairies from the provinces of Śląskie and Lubelskie (with frequencies of 21 and 20, respectively). All these dairies were non-cooperatives. Of 38 VRS technically efficient dairies, seven (four cooperatives and three non-cooperatives) were never reported as a reference point for inefficient dairies. Another seven (three cooperatives and four non-cooperatives) had a frequency of 1 or 2 in the reference set. Due to a low peer count number, these dairies can hardly be considered best-practice entities.

A more detailed analysis, focusing on the differences between the actual and the target values (under the VRS assumption) of the variables used (taking into account the slacks; slacks represent the remaining inefficiency left after a proportional reduction in inputs or outputs if the DMU cannot achieve the efficiency frontier [94]), was undertaken to reveal the extent to which inefficient dairies should reduce each of the given inputs (no slacks in output were observed) to become efficient. The analysis of inefficiencies in relation to the inputs used in the production process can provide important insights for managers, enabling them to make better decisions [95].

According to the results, the inefficient dairies should reduce their labor costs, raw material costs, depreciation expense, and other operating costs, on average, by 25.0%, 10.1%, 20.8%, and 10.1%, respectively. Therefore, the greatest capacity for improvement can be observed in labor costs and depreciation expense. Given that inefficient dairies could achieve the same output with lower depreciation expense, the above result may suggest that they are not utilizing their fixed assets fully efficiently. Similar to the results of Vlontzos and Theodoridis [20] with regard to the Greek dairy industry, inefficient dairies appear to be overinvested. As Beber et al. [69] have pointed out, it is critical to avoid unplanned overinvestment that could lead to idle capacity. Another possible explanation for this result is that dairies need to maintain spare capacity because of the perishability of their raw materials and products. The issue that seems to be more challenging in the course of business is the reduction of labor costs.

At the aggregate level, that is, considering the total value of inputs consumed by inefficient units, labor costs, raw material costs, depreciation expense, and other operating costs should be reduced by 23.3%, 6.1%, 20.7%, and 6.9%, respectively.

As presented in Table 7, there were no statistically significant differences between cooperatives and non-cooperatives in terms of their potential for input reduction.

Table 7. Potential input reduction in inefficient dairies by legal form (authors' calculations based on [77]).

Input	Form	Potential Input Reduction (%)			Mann–Whitney		
		Mean	Med	Mean Rank	U	Z	p
LC	cooperative	26.1	24.8	34.15	463.00	−0.822	0.411
	non-cooperative	22.4	22.2	38.45			
RM	cooperative	10.0	7.7	35.75	516.00	−0.152	0.879
	non-cooperative	10.3	9.6	34.95			
DE	cooperative	18.7	12.5	36.33	488.00	−0.506	0.613
	non-cooperative	25.5	17.3	33.68			
OC	cooperative	10.0	7.7	35.75	516.00	−0.152	0.879
	non-cooperative	10.3	9.6	34.95			

By comparing the results of the three efficiency scores, i.e., TE (Table 3), PTE (Table 5), and SE (Table 8), it can be observed that the technical inefficiency of the dairies was driven slightly more by managerial inefficiency than by scale inefficiency. This is indicated by the lower mean PTE score accompanied by a higher coefficient of variation. Our results are similar to those of Lima et al. [46] but different from those of Ohlan [85]. As shown in Tables 3, 5 and 8, this observation applies to both cooperative and non-cooperative dairies. In this regard, it should be noted that insufficient knowledge and skills of managers are identified as one of the internal barriers to the development of the dairy processing sector [96,97].

Table 8. Descriptive statistics of SE scores (authors' calculations based on [77]).

	Form	n	Mean	Med	SD	Min	Max	Q1	Q3
	cooperative	65	0.951	0.962	0.048	0.818	1.000	0.923	0.993
SE	non-cooperative	43	0.971	0.989	0.042	0.803	1.000	0.957	1.000
	total	108	0.959	0.977	0.046	0.803	1.000	0.930	0.999

The analysis of returns to scale revealed that the majority of the examined dairies (63.0%) were operating under decreasing returns to scale, implying that these DMUs could enhance their overall technical efficiency by reducing their size (Table 9). The results also indicate that 17 DMUs (15.7%) were experiencing increasing returns to scale, meaning that they were operating below their optimal scale size. Therefore, there is scope for them to improve their technical efficiency by increasing their size. Of the 108 dairies, 23 (21.3%) were operating at optimal scale. Our results in this regard differ from those of Baran [72], according to which, in the years 1999–2010, on average 69% of dairy firms experienced increasing returns to scale, while 22% presented decreasing returns to scale. This may suggest that the possibility of improving technical efficiency through the concentration of the dairy sector and increasing the scale of dairy production in Poland has been exploited. The process of concentration of the milk processing sector in Poland started about 25 years ago [32]. In addition, Poland's accession to the EU intensified competition on the milk market [96], which was a driving force for further concentration. This process was initiated mainly by large dairies, which took over smaller units, thus increasing their territorial range and the amount of milk processed. Large dairies began to specialize their plants in the production of technologically similar products [96]. In this regard, it is worth noting that specialization of dairies may result in a decrease in the number of products they offer [31]. In light of the structural changes in the milk processing sector, small dairies need to seek their market niche by, for example, producing regional products [96].

Table 9. Types of returns to scale by legal form of dairy (authors' calculations based on [77]).

Form	drs	crs	irs	χ^2	df	p
cooperative	46 (70.8%)	8 (12.3%)	11 (16.9%)	7.92	2	0.019
non-cooperative	22 (51.2%)	15 (34.9%)	6 (14.0%)			
total	68 (63.0%)	23 (21.3%)	17 (15.7%)			

Abbreviations: drs, decreasing returns to scale; crs, constant returns to scale; irs, increasing returns to scale. Note: row percentages are given in parentheses.

As can be observed from Table 9, non-cooperative dairies were considerably more likely to be scale efficient, presenting a higher frequency of constant returns to scale than their cooperative counterparts. While the majority of both cooperatives and non-cooperatives showed decreasing returns to scale, the proportion of such units was higher for cooperatives. The relationship between these variables was statistically significant.

To summarize our findings, similarly to Mahajan et al. [98], we grouped the DMUs according to their technical, pure technical, and scale efficiency scores (Table 10).

Table 10. Classification of dairies according to TE, PTE, and SE scores (authors' elaboration).

Case	TE = 1	TE < 1		
		PTE = 1 SE < 1	PTE < 1 SE = 1	PTE < 1 SE < 1
total	22 (20.4%)	16 (14.8%)	1 (0.9%)	69 (63.9%)
cooperatives	8 (7.4%)	9 (8.3%)	0 (0.0%)	48 (44.4%)
non-cooperatives	14 (13.0%)	7 (6.5%)	1 (0.9%)	21 (19.4%)
Recommendation	no action required	adjustment in the scale of operations	improvement in managerial performance	both adjustment in the scale of operations and improvement in managerial performance

Note: percentages for the whole sample are given in parentheses.

4.2. Technical Efficiency of Cooperative and Non-Cooperative Dairies: The Spatial Perspective

In order to ascertain whether the technical efficiency of dairies is spatially differentiated, we divided the provinces of Poland into three groups, i.e., provinces with high, medium, or low milk production capacity. For this purpose, we first assessed the milk production capacity of each province using the zero unitarization method on the basis of the following diagnostic variables: X_1, dairy cow density per 100 ha of agricultural land (in heads); X_2, total cow's milk purchase (in thousands of liters); X_3, share of purchase in milk production (in %); X_4, average milk yield per cow (in liters); X_5, cow's milk production per 1 ha of agricultural land (in liters); X_6, share of cows in farms with more than 50 cows (in %); X_7, average number of cows per farm. Table 11 presents the descriptive statistics for selected diagnostic variables.

Table 11. Descriptive statistics of diagnostic variables (authors' calculations based on [35,89–91]).

Variable	Mean	Med	SD	CV	Min	Max	Max/Min
X_1	12.83	10.85	9.25	0.72	2.80	40.20	14.36
X_2	739,243.69	294,841.00	826,913.82	1.12	77,853.00	2,604,942.00	33.46
X_3	82.80	85.55	11.73	0.14	47.36	94.02	1.99
X_4	5129.50	5416.50	1200.12	0.23	2678.00	6760.00	2.52
X_5	796.63	546.00	613.09	0.77	190.00	2579.00	13.57
X_6	28.03	29.28	13.91	0.50	6.60	53.14	8.05
X_7	0.18	0.13	0.14	0.77	0.04	0.46	12.87

Note: X_3 variable had a max/min ratio of slightly less than 2; however, due to the substantive importance of this variable, we decided to include it in the set of diagnostic variables.

The values of the synthetic variable of milk production capacity and the resulting classification of provinces are given in Table 12 and Figure 1. Based on these results, we assigned cooperative and non-cooperative dairies to distinguished groups of provinces according to their location (i.e., dairies located in a region of low, medium, or high milk production capacity).

Table 12. Values of the synthetic variable and classification of provinces by milk production capacity (authors' calculations based on [35,89–91]).

Rank	Province	Q_i	Group
1	Podlaskie	0.7541	I: high milk production capacity
2	Mazowieckie	0.6077	
3	Wielkopolskie	0.5964	
4	Opolskie	0.5131	II: medium milk production capacity
5	Dolnośląskie	0.4592	
6	Kujawsko-Pomorskie	0.4582	
7	Warmińsko-Mazurskie	0.4581	
8	Śląskie	0.4180	
9	Łódzkie	0.4095	
10	Zachodniopomorskie	0.4016	
11	Lubuskie	0.3764	
12	Pomorskie	0.3627	
13	Lubelskie	0.3083	III: low milk production capacity
14	Świętokrzyskie	0.2550	
15	Podkarpackie	0.2300	
16	Małopolskie	0.1169	

Group I : $Q_i \in (0.5417; 0.7541]$
Group II : $Q_i \in (0.3293; 0.5417]$
Group III : $Q_i \in [0.1169; 0.3293]$

Figure 1. Milk production capacity by province (authors' elaboration based on data from Table 12).

The *H* Kruskal–Wallis test showed that there was no statistically significant difference in technical efficiency scores (in terms of PTE)—this conclusion also held under the CRS assumption—between these three groups of dairies (Table 13). This finding held for both cooperative and non-cooperative DMUs, suggesting that the environmental factor of a region's capacity to produce milk does not significantly differentiate the efficiency of dairies in converting inputs into output. It has been recognized that milk plants prefer milk surplus areas that have higher milk production than their respective milk demand [18], hence we argue that the environmental conditions of a region may influence the location of the dairy processing industry [99]. Moreover, they may affect issues such as the marketability of milk production [25], the organization of raw material transportation, and thus transportation costs and milk prices [100], and the technological quality of milk purchased by dairies [101]. In summary, while the environmental factor may influence the above-mentioned aspects of dairy processors' functioning, we found no evidence of significant difference in PTE scores between groups of dairies distinguished by their location (i.e., dairies located in a region of low, medium, or high milk production capacity).

Table 13. Results of PTE scores analysis for cooperative and non-cooperative dairies by region of location (authors' calculations based on [35,77,89–91]).

	Milk Production Capacity of Region	*n*	Mean	Med	Kruskal–Wallis			
					Mean Rank	*H*	*df*	*p*
cooperative	low	9	0.919	0.924	27.06	1.051	2	0.591
	medium	30	0.930	0.951	33.93			
	high	26	0.924	0.955	33.98			
non-cooperative	low	5	0.990	1.000	29.40	2.482	2	0.289
	medium	19	0.944	0.987	20.05			
	high	19	0.940	1.000	22.00			

5. Conclusions

This paper has examined the technical efficiency of dairies in Poland on the basis of data for the year 2019. Given that the Polish dairy processing industry is predominated by cooperatives, our research has focused on comparing their technical efficiency with that of dairies of other legal forms, thereby contributing to the scientific debate on this issue. Due to the inherent link between dairy operations and their access to raw materials, this study has additionally explored the technical efficiency of dairies in the context of spatial disparities in milk production potential. Thus, we have provided insights into the technical efficiency of dairies from a supply chain perspective. To our knowledge, such an analysis has not previously been conducted.

We have investigated the technical efficiency of dairies using the DEA method by taking net sales revenue as the output and labor costs, raw material costs, depreciation expense, and other operating costs as the input variables. The estimates of efficiency scores were obtained under the CRS and VRS assumptions. We have also identified the types of returns to scale of the given dairies. In order to examine the technical efficiency of the dairies in relation to the milk production capacity of the region in which they are located, we have used the zero unitarization method, dividing the provinces of Poland into three groups: provinces with high, medium, or low milk production capacity.

The results indicate that, assuming CRS, the level of technical efficiency of the dairy processing sector in Poland was on average 0.895. In this regard, we found that non-cooperative dairies were significantly more efficient than cooperatives.

In the search for sources of inefficiencies, in the next step we examined the results for pure technical efficiency and scale efficiency. The PTE score was on average 0.935. On this point, the differences between non-cooperatives and cooperatives were not statistically significant. Thus, when referring to managerial performance in converting inputs into output, we found no evidence of lower efficiency in dairy cooperatives. For both the TE and PTE scores, dairy non-cooperatives revealed a greater proportion of units identified as efficient. However, it is important to note the limitations of such a zero-one approach, as an entity may not be fully efficient (with an efficiency score of 1) and yet still exhibit high efficiency.

According to our results, inefficient dairies presented the greatest potential for reducing labor costs and depreciation expense. This implies that dairies in Poland could reduce these costs while maintaining the same level of output. We did not identify significant differences in potential for input reduction between cooperatives and non-cooperatives.

About one in five of all the dairies studied were scale efficient. For non-cooperative dairies, the proportion of such units was nearly three times that of cooperatives. Most dairies showed decreasing returns to scale, meaning that they were too large relative to their optimal scale. This observation applied to both cooperative and non-cooperative dairies; however, the former group showed a higher proportion of units operating under decreasing returns to scale. The prevalence of dairies exhibiting decreasing returns to scale may be a result of the intensification of the concentration process in the milk processing sector, caused in particular by Poland's accession to the EU. Our results can be perceived as a sign of the saturation of the dairy sector in Poland with the consolidation process. Therefore, in light of the above, the continuation of this process does not seem to be recommended as far as technical efficiency is concerned. Due to structural changes in the milk processing sector, small dairies need to find a market niche if they want to compete with large units.

The examination of the PTE scores taking into account the spatial disparities in milk production potential did not provide evidence for the claim that the technical efficiency of dairies was affected by the milk production capacity of their location region. This finding indicates that although the availability and abundance of raw milk may affect the density of and the competition among dairies, they do not significantly differentiate their technical efficiency.

The results suggest some directions for further research. Given that only one year of data was used in this study, it would be valuable to examine the technical efficiency of cooperative and non-cooperative dairies in Poland over a longer period of time. Moreover, in further research it would be beneficial to investigate technical efficiency from a supply chain perspective more broadly, i.e., taking into account the distribution conditions as the next link in the supply chain. Such a perspective would provide a more holistic view regarding the efficiency of the dairy sector in Poland. Another possible stream of research would consider the economic sustainability of the Polish dairy supply chain in association with its environmental and social dimensions.

Author Contributions: Conceptualization, K.Z.-K., M.Z.-C. and J.N.; methodology, K.Z.-K., M.Z.-C. and J.N.; software, K.Z.-K.; validation, K.Z.-K. and J.N.; formal analysis, K.Z.-K.; investigation, K.Z.-K., M.Z.-C. and J.N.; resources, K.Z.-K., M.Z.-C. and J.N.; data curation, K.Z.-K., M.Z.-C. and J.N.; writing—original draft preparation, K.Z.-K., M.Z.-C. and J.N.; writing—review and editing, K.Z.-K., M.Z.-C. and J.N.; visualization, K.Z.-K.; project administration, K.Z.-K.; funding acquisition, K.Z.-K. All authors have read and agreed to the published version of the manuscript.

Funding: The APC was funded by The John Paul II Catholic University of Lublin, grant number 1/6-20-21-05-0501-0002-0593.

Institutional Review Board Statement: Not applicable.

Informed Consent Statement: Not applicable.

Data Availability Statement: EMIS data: restrictions apply to the availability of these data; data obtained from EMIS are available from the authors with the permission of EMIS. Publicly available datasets were also analyzed in this study. These data can be found here: https://stat.gov.pl/ (accessed on 6 November 2021); https://ec.europa.eu/eurostat/data/database (accessed on 2 November 2021).

Conflicts of Interest: The authors declare no conflict of interest.

Appendix A

Table A1. Review of literature (authors' elaboration based on source publications).

Source Publication	DEA Model	Output Variables	Input Variables	Sample	Year/Period	Country
Singh et al. (2001) [18]	VRS, IO	(1) aggregate dairy products' variable	(1) raw material (mainly raw milk); (2) labor; (3) capital—depreciation, repairs, maintenance, and interests of the machinery and building; (4) other inputs (administration, fuel, power, insurance, etc.)	13 cooperative and 10 private dairy plants from Haryana and Punjab states	1992/93 and 1996/97	India
Baran and Kołyska (2009) [70]	M	(1) net sales revenue	(1) number of staff; (2) fixed assets	205–248 dairy processing firms, including cooperatives	1998–2005	Poland
Gradziuk (2009) [71]	CRS, VRS, OO; M	(1) net sales revenue	(1) sum of depreciation, material and energy consumption, and contracted services costs; (2) labor costs	12 large dairy processing companies from the Mazowieckie province	2001–2007	Poland
Soboh et al. (2012) [22]	VRS, IO	(1) total turnover	(1) fixed assets; (2) material costs; (3) labor costs	133 dairy processing companies: 90 investor-owned firms and 43 cooperatives	2004	Belgium, Denmark, France, Germany, Ireland, the Netherlands
Baran (2013) [72]	CRS, VRS, IO	(1) net sales revenue	(1) labor costs; (2) costs of material and energy consumption; (3) fixed assets	743 observations of dairy processing firms, including cooperatives	1999–2010	Poland
Ohlan (2013) [85]	CRS, VRS, IO	(1) net value added	(1) fixed capital; (2) working capital; (3) labor; (4) raw materials; (5) fuel	Data obtained from Annual Survey of Industry, Ministry of Commerce and Industry, Government of India	1980–2008	India
Kapelko and Oude Lansink (2013) [66]	VRS, IO	(1) turnover	(1) employee costs; (2) material costs; (3) fixed assets	Unbalanced panel of 3509 observations of 264–380 dairy processing firms	2000–2009	Spain
Vlontzos and Theodoridis (2013) [20]	CRS, VRS, IO, M	(1) revenue; (2) mixed profit	(1) overall depreciation; (2) costs of sold products; (3) shared capital; (4) value of stock; (5) short-term liabilities	29 dairy companies, 20% of them cooperatives	2006–2007 for CRS, VRS, IO; 2003–2007 for M	Greece

Table A1. *Cont.*

Source Publication	DEA Model	Output Variables	Input Variables	Sample	Year/Period	Country
Domańska et al. (2015) [73]	VRS, IO	(1) net sales revenue	(1) fixed assets; (2) number of staff	12 dairy processing companies from the Lubelskie province, including 10 cooperatives	2010–2012	Poland
Špička (2015) [9]	VRS, IO, M	(1) sales revenue	(1) material and energy costs; (2) staff costs; (3) depreciation and amortization	130 dairy processors	2008–2013	Czech Republic, Poland, Slovakia
Lima et al. (2018) [46]	CRS, VRS, IO, MS	(1) revenue	(1) payroll; (2) processed milk volume; (3) boiler, fuel, and electricity costs	40 dairy establishments, of which 85% were private and 15% were cooperatives	2014/2015	Brazil
Popović and Panić (2019) [19]	VRS, IO, MS	(1) sales revenue	(1) costs of material (mainly raw milk); (2) labor costs; (3) energy costs; (4) other costs (depreciation, costs of purchased commodities, contracted services, non-material costs, and interest paid)	79 non-cooperative dairy processing companies	2016	Serbia
Ruales Guzmán et al. (2021) [102]	VRS, IO, OO	(1) revenue (2) profit	(1) current assets; (2) property, plant, and equipment; (3) non-current liabilities; (4) equity	19 dairy industry companies	2017	Colombia

Note: CRS, constant returns to scale; VRS, variable returns to scale; IO, input-oriented; OO, output-oriented; M, Malmquist; MS, multi-stage model.

References

1. Hooks, T.; Macken-Walsh, Á.; McCarthy, O.; Power, C. Farm-Level Viability, Sustainability and Resilience: A Focus on Cooperative Action and Values-Based Supply Chains. *Stud. Agric. Econ.* **2017**, *119*, 123–129. [CrossRef]
2. Orr, D.W. Four Challenges of Sustainability. *Conserv. Biol.* **2002**, *16*, 1457–1460.
3. Augustin, M.A.; Udabage, P.; Juliano, P.; Clarke, P.T. Towards a More Sustainable Dairy Industry: Integration across the Farm–Factory Interface and the Dairy Factory of the Future. *Int. Dairy J.* **2013**, *31*, 2–11. [CrossRef]
4. United Nations. Transforming Our World: The 2030 Agenda for Sustainable Development. Available online: https://sustainabledevelopment.un.org/content/documents/21252030%20Agenda%20for%20Sustainable%20Development%20web.pdf (accessed on 6 November 2021).
5. Bricas, N. The Scope of the Analysis: Food Systems. In *Food Systems at Risk. New Trends and Challenges*; Dury, S., Bendjebbar, P., Hainzelin, E., Giordano, T., Bricas, N., Eds.; FAO: Rome, Italy; CIRAD: Montpellier, France; European Commission: Brussels, Belgium, 2019; pp. 15–18, ISBN 978-92-5-131732-7.
6. Tendall, D.M.; Joerin, J.; Kopainsky, B.; Edwards, P.; Shreck, A.; Le, Q.B.; Kruetli, P.; Grant, M.; Six, J. Food System Resilience: Defining the Concept. *Glob. Food Sec.* **2015**, *6*, 17–23. [CrossRef]
7. Nguyen, H. Sustainable Food Systems: Concept and Framework. Available online: https://www.fao.org/3/ca2079en/CA2079EN.pdf (accessed on 6 November 2021).
8. Čechura, L.; Žáková Kroupová, Z. Technical Efficiency in the European Dairy Industry: Can We Observe Systematic Failures in the Efficiency of Input Use? *Sustainability* **2021**, *13*, 1830. [CrossRef]
9. Špička, J. The Efficiency Improvement of Central European Corporate Milk Processors in 2008–2013. *Agris -Online Pap. Econ. Inform.* **2015**, *7*, 175–188. [CrossRef]
10. Feil, A.A.; Schreiber, D.; Haetinger, C.; Haberkamp, Â.M.; Kist, J.I.; Rempel, C.; Maehler, A.E.; Gomes, M.C.; da Silva, G.R. Sustainability in the Dairy Industry: A Systematic Literature Review. *Environ. Sci. Pollut. Res.* **2020**, *27*, 33527–33542. [CrossRef]
11. Miller, G.D.; Auestad, N. Towards a Sustainable Dairy Sector: Leadership in Sustainable Nutrition. *Int. J. Dairy Technol.* **2013**, *66*, 307–316. [CrossRef]
12. Ang, S.; Zhu, Y.; Yang, F. Efficiency Evaluation and Ranking of Supply Chains Based on Stochastic Multicriteria Acceptability Analysis and Data Envelopment Analysis. *Int. Trans. Oper. Res.* **2021**, *28*, 3190–3219. [CrossRef]
13. Rao, K.H.; Raju, P.N.; Reddy, G.P.; Hussain, S.A. Public-Private Partnership and Value Addition: A Two-Pronged Approach for Sustainable Dairy Supply Chain Management. *IUP J. Supply Chain. Manag.* **2013**, *10*, 15–25.
14. Zorn, A.; Esteves, M.; Baur, I.; Lips, M. Financial Ratios as Indicators of Economic Sustainability: A Quantitative Analysis for Swiss Dairy Farms. *Sustainability* **2018**, *10*, 2942. [CrossRef]
15. Wilczyński, A. Farm Economic Sustainability—Financial Ratio Analysis. *Res. Pap. Wroc. Univ. Econ. Bus.* **2020**, *64*, 120–131. [CrossRef]
16. Bórawski, P.; Pawlewicz, A.; Mickiewicz, B.; Pawlewicz, K.; Bełdycka-Bórawska, A.; Holden, L.; Brelik, A. Economic Sustainability of Dairy Farms in the EU. *Eur. Res. Stud. J.* **2020**, *23*, 955–978. [CrossRef]
17. Bánkuti, F.I.; Prizon, R.C.; Damasceno, J.C.; Brito, M.M.D.; Pozza, M.S.S.; Lima, P.G.L. Farmers' Actions toward Sustainability: A Typology of Dairy Farms According to Sustainability Indicators. *Animal* **2020**, *14*, s417–s423. [CrossRef]
18. Singh, S.; Coelli, T.; Fleming, E. Performance of Dairy Plants in the Cooperative and Private Sectors in India. *Ann. Public Coop. Econ.* **2001**, *72*, 453–479. [CrossRef]
19. Popović, R.; Panić, D. Economic Sustainability of Dairy Processing Sector in Serbia. In *Sustainable Agriculture and Rural Development in Terms of the Republic of Serbia Strategic Goals Realization within the Danube Region. Sustainability and Multifunctionality*; Subić, J., Jeločnik, M., Kuzman, B., Andrei, J.V., Eds.; Institute of Agricultural Economics: Belgrade, Serbia, 2019; pp. 686–700, ISBN 978-86-6269-068-5.
20. Vlontzos, G.; Theodoridis, A. Efficiency and Productivity Change in the Greek Dairy Industry. *Agric. Econ. Rev.* **2013**, *14*, 14–28.
21. Jansik, C.; Irz, X.; Kuosmanen, N. *Competitiveness of Northern European Dairy Chains*; MTT Economic Research, Agrifood Research Finland: Helsinki, Finland, 2014; ISBN 978-951-687-177-9.
22. Soboh, R.; Oude Lansink, A.; Van Dijk, G. Efficiency of Cooperatives and Investor Owned Firms Revisited. *J. Agric. Econ.* **2012**, *63*, 142–157. [CrossRef]
23. Hirsch, S.; Mishra, A.; Möhring, N.; Finger, R. Revisiting Firm Flexibility and Efficiency: Evidence from the EU Dairy Processing Industry. *Eur. Rev. Agric. Econ.* **2020**, *47*, 971–1008. [CrossRef]
24. Eurostat. Cows' Milk Collection and Products Obtained-Annual Data. Available online: https://appsso.eurostat.ec.europa.eu/nui/show.do?dataset=apro_mk_cola&lang=en (accessed on 2 November 2021).
25. Parzonko, A. *Globalne i Lokalne Uwarunkowania Rozwoju Produkcji Mleka*; Wydawnictwo SGGW: Warszawa, Poland, 2013; ISBN 978-83-7583-461-1.
26. Boyle, G.E. The Economic Efficiency of Irish Dairy Marketing Co-Operatives. *Agribusiness* **2004**, *20*, 143–153. [CrossRef]
27. Gardijan Kedžo, M.; Lukač, Z. The Financial Efficiency of Small Food and Drink Producers across Selected European Union Countries Using Data Envelopment Analysis. *Eur. J. Oper. Res.* **2021**, *291*, 586–600. [CrossRef]
28. Madau, F.A.; Furesi, R.; Pulina, P. Technical Efficiency and Total Factor Productivity Changes in European Dairy Farm Sectors. *Agric. Food Econ.* **2017**, *5*, 17:1–17:14. [CrossRef]

29. Thomson, A.; Metz, M. *Implications of Economic Policy for Food Security: A Training Manual*; FAO: Rome, Italy, 1999; ISBN 92-5-104379-5.
30. Zuba-Ciszewska, M. Structural Changes in the Milk Production Sector and Food Security—The Case of Poland. *Ann. Pol. Assoc. Agric. Agribus. Econ.* **2019**, *21*, 318–327. [CrossRef]
31. Zuba-Ciszewska, M. Rola spółdzielni w zapewnieniu dostępności żywności w Polsce − na przykładzie produktów mleczarskich. *Wieś Rolnictwo* **2020**, *186*, 93–119. [CrossRef]
32. Seremak-Bulge, J.; Hryszko, K.; Pieniążek, K.; Rembeza, J.; Szajner, P.; Świetlik, K. *Rozwój Rynku Mleczarskiego i Zmiany Jego Funkcjonowania w Latach 1990–2005*; Instytut Ekonomiki Rolnictwa i Gospodarki Żywnościowej—Państwowy Instytut Badawczy: Warszawa, Poland, 2005; ISBN 83-89666-34-0.
33. Szajner, P. (Ed.) *Rynek Mleka. Stan i Perspektywy*; Instytut Ekonomiki Rolnictwa i Gospodarki Żywnościowej: Warszawa, Poland, 2019; Volume 57.
34. Eurostat. Bovine Animals by NUTS 2 Regions—Annual Data. Available online: https://appsso.eurostat.ec.europa.eu/nui/show.do?dataset=ef_lsk_bovine&lang=en (accessed on 2 November 2021).
35. Szajner, P. (Ed.) *Rynek Mleka. Stan i Perspektywy*; Instytut Ekonomiki Rolnictwa i Gospodarki Żywnościowej: Warszawa, Poland, 2020; Volume 58.
36. Zuba-Ciszewska, M. Structural Changes in the Dairy Industry and Their Impact on the Efficiency of Dairies—A Polish Example. In Proceedings of the 2018 International Scientific Conference 'Economic Sciences for Agribusiness and Rural Economy', Warsaw, Poland, 7 June 2018; Volume 2, pp. 116–123. [CrossRef]
37. Szczepaniak, I.; Tereszczuk, M. Confronting the Polish Dairy Industry with the International Competition in the EU Food Market. *Rev. Socio Econ. Perspect.* **2017**, *2*, 31–51. [CrossRef]
38. Urban, S. Trzeci agregat agrobiznesu—Przetwórstwo surowców rolniczych żywnościowych i nieżywnościowych. In *Agrobiznes i Biobiznes. Teoria i Praktyka*; Urban, S., Ed.; Wydawnictwo Uniwersytetu Ekonomicznego we Wrocławiu: Wrocław, Poland, 2014; pp. 46–58, ISBN 978-83-7695-408-0.
39. Kraciuk, J. Koncentracja produkcji w polskim przemyśle spożywczym. *Zeszyty Naukowe SGGW w Warszawie. Problemy Rolnictwa Światowego* **2008**, *5*, 33–41.
40. Chechelski, P. Ocena procesów koncentracji struktur podmiotowych w branżach przetwórstwa produktów pochodzenia zwierzęcego w Polsce. *Ann. Pol. Assoc. Agric. Agribus. Econ.* **2017**, *19*, 62–67. [CrossRef]
41. Baran, J. *Intensywność i Zasięg Geograficzny Internacjonalizacji Sektora Przetwórstwa Mleka*; Wydawnictwo SGGW: Warszawa, Poland, 2019; ISBN 978-83-7583-848-0.
42. Zuba-Ciszewska, M. Rola przemysłu spożywczego w gospodarce Polski. *Nierówności Społeczne a Wzrost Gospodarczy* **2020**, *64*, 69–86. [CrossRef]
43. Mroczek, R. Pozycja przemysłu spożywczego w łańcuchu żywnościowym w Polsce na przełomie XX/XXI wieku. *Zeszyty Naukowe SGGW w Warszawie. Problemy Rolnictwa Światowego* **2018**, *18*, 23–37. [CrossRef]
44. Chechelski, P. Ewolucja łańcucha żywnościowego. In *Przemysł Spożywczy—Makrootoczenie, Inwestycje, Ekspansja Zagraniczna*; Firlej, K., Szczepaniak, I., Eds.; Fundacja Uniwersytetu Ekonomicznego w Krakowie: Kraków, Poland, 2015; pp. 45–64, ISBN 978-83-65173-07-2.
45. Nazarko, J.; Chodakowska, E. Labour Efficiency in Construction Industry in Europe Based on Frontier Methods: Data Envelopment Analysis and Stochastic Frontier Analysis. *J. Civ. Eng. Manag.* **2017**, *23*, 787–795. [CrossRef]
46. Lima, L.P.; Ribeiro, G.B.D.; Silva, C.A.B.; Perez, R. An Analysis of the Brazilian Dairy Industry Efficiency Level. *Int. Food Res. J.* **2018**, *25*, 2478–2485.
47. Alem, H. The Role of Technical Efficiency Achieving Sustainable Development: A Dynamic Analysis of Norwegian Dairy Farms. *Sustainability* **2021**, *13*, 1841. [CrossRef]
48. Mbaga, M.D.; Romain, R.; Larue, B.; Lebel, L. Assessing Technical Efficiency of Québec Dairy Farms. *Can. J. Agric. Econ.* **2003**, *51*, 121–137. [CrossRef]
49. Le, S.; Jeffrey, S.; An, H. Greenhouse Gas Emissions and Technical Efficiency in Alberta Dairy Production: What Are the Trade-Offs? *J. Agric. Appl. Econ.* **2020**, *52*, 177–193. [CrossRef]
50. Skevas, I.; Emvalomatis, G.; Brümmer, B. Heterogeneity of Long-Run Technical Efficiency of German Dairy Farms: A Bayesian Approach*. *J. Agric. Econ.* **2018**, *69*, 58–75. [CrossRef]
51. Theodoridis, A.M.; Psychoudakis, A. Efficiency Measurement in Greek Dairy Farms: Stochastic Frontier vs. Data Envelopment Analysis. *Int. J. Econ. Sci. Appl. Res.* **2008**, *1*, 53–67.
52. Kelly, E.; Shalloo, L.; Geary, U.; Kinsella, A.; Wallace, M. Application of Data Envelopment Analysis to Measure Technical Efficiency on a Sample of Irish Dairy Farms. *Ir. J. Agric. Food Res.* **2012**, *51*, 63–77.
53. Mohd Suhaimi, N.A.B.; de Mey, Y.; Oude Lansink, A. Measuring and Explaining Multi-Directional Inefficiency in the Malaysian Dairy Industry. *Br. Food J.* **2017**, *119*, 2788–2803. [CrossRef] [PubMed]
54. Skevas, I.; Oude Lansink, A. Dynamic Inefficiency and Spatial Spillovers in Dutch Dairy Farming. *J. Agric. Econ.* **2020**, *71*, 742–759. [CrossRef]
55. Jaforullah, M.; Whiteman, J. Scale Efficiency in the New Zealand Dairy Industry: A Non-Parametric Approach. *Aust. J. Agric. Resour. Econ.* **1999**, *43*, 523–541. [CrossRef]

56. Soliman, T.; Djanibekov, U. Assessing Dairy Farming Eco-Efficiency in New Zealand: A Two-Stage Data Envelopment Analysis. *N. Z. J. Agric. Res.* **2021**, *64*, 411–428. [CrossRef]
57. Syp, A.; Osuch, D. Zmiany efektywności i produktywności gospodarstw polowych i mlecznych w województwie lubelskim w latach 2014–2016. *Ann. Pol. Assoc. Agric. Agribus. Econ.* **2018**, *20*, 137–142. [CrossRef]
58. Wilczyński, A.; Kołoszycz, E.; Świtłyk, M. Technical Efficiency of Dairy Farms: An Empirical Study of Producers in Poland. *Eur. Res. Stud. J.* **2020**, *23*, 117–127. [CrossRef]
59. Świtłyk, M.; Sompolska-Rzechuła, A.; Kurdyś-Kujawska, A. Measurement and Evaluation of the Efficiency and Total Productivity of Dairy Farms in Poland. *Agronomy* **2021**, *11*, 2095. [CrossRef]
60. Silva, E.; Almeida, B.; Marta-Costa, A.A. Efficiency of the Dairy Farms: A Study from Azores (Portugal). *Eur. Countrys.* **2018**, *10*, 725–734. [CrossRef]
61. Barnes, A.P. Does Multi-Functionality Affect Technical Efficiency? A Non-Parametric Analysis of the Scottish Dairy Industry. *J. Environ. Manage.* **2006**, *80*, 287–294. [CrossRef] [PubMed]
62. Demircan, V.; Binici, T.; Zulauf, C.R. Assessing Pure Technical Efficiency of Dairy Farms in Turkey. *Agric. Econ. Czech* **2010**, *56*, 141–148. [CrossRef]
63. Tauer, L.W. Productivity of New York Dairy Farms Measured by Nonparametric Malmquist Indices. *J. Agric. Econ.* **1998**, *49*, 234–249. [CrossRef]
64. Stokes, J.R.; Tozer, P.R.; Hyde, J. Identifying Efficient Dairy Producers Using Data Envelopment Analysis. *J. Dairy Sci.* **2007**, *90*, 2555–2562. [CrossRef] [PubMed]
65. Mugera, A.W. Measuring Technical Efficiency of Dairy Farms with Imprecise Data: A Fuzzy Data Envelopment Analysis Approach. *Aust. J. Agric. Resour. Econ.* **2013**, *57*, 501–520. [CrossRef]
66. Kapelko, M.; Oude Lansink, A. Technical Efficiency of the Spanish Dairy Processing Industry: Do Size and Exporting Matter? In *Efficiency Measures in the Agricultural Sector: With Applications*; Mendes, A.B., Soares da Silva, E.L.D.G., Azevedo Santos, J.M., Eds.; Springer: Dordrecht, The Netherlands, 2013; pp. 93–106, ISBN 978-94-007-5739-4.
67. Doucouliagos, H.; Hone, P. The Efficiency of the Australian Dairy Processing Industry. *Aust. J. Agric. Resour. Econ.* **2000**, *44*, 423–438. [CrossRef]
68. Soboh, R.A.M.E.; Oude Lansink, A.; Van Dijk, G. Efficiency of European Dairy Processing Firms. *NJAS Wagen. J. Life Sc.* **2014**, *70–71*, 53–59. [CrossRef]
69. Beber, C.L.; Lakner, S.; Skevas, I. Organizational Forms and Technical Efficiency of the Dairy Processing Industry in Southern Brazil. *Agric. Food Econ.* **2021**, *9*, 23:1–23:22. [CrossRef]
70. Baran, J.; Kołyska, J. Porównanie wykorzystania zasobów małych, średnich i dużych przedsiębiorstw przemysłu mleczarskiego w latach 1998–2005. *EQUIL* **2009**, *2*, 159–169. [CrossRef]
71. Gradziuk, K. Efektywność przedsiębiorstw przemysłu spożywczego na przykładzie branży mleczarskiej. *Ann. Pol. Assoc. Agric. Agribus. Econ.* **2009**, *11*, 117–122.
72. Baran, J. Efficiency of the Production Scale of Polish Dairy Companies Based on Data Envelopment Analysis. *Acta Sci. Pol., Oecon.* **2013**, *12*, 5–13.
73. Domańska, K.; Kijek, T.; Tomczyńska-Mleko, M. Efektywność przedsiębiorstw przemysłu mleczarskiego w województwie lubelskim. *Ann. Pol. Assoc. Agric. Agribus. Econ.* **2015**, *17*, 29–34.
74. Berger, A.N.; Mester, L.J. Inside the Black Box: What Explains Differences in the Efficiencies of Financial Institutions? *J. Bank. Financ.* **1997**, *21*, 895–947. [CrossRef]
75. Pietrzak, M. *Fenomen Spółdzielni Rolników. Pomiędzy Rynkiem, Hierarchią i Klanem*; CeDeWu: Warszawa, Poland, 2019; ISBN 978-83-8102-191-3.
76. Grashuis, J.; Su, Y. A Review of the Empirical Literature on Farmer Cooperatives: Performance, Ownership and Governance, Finance, and Member Attitude. *Ann. Public Coop. Econ.* **2019**, *90*, 77–102. [CrossRef]
77. Emerging Markets Information Service (EMIS). Available online: https://www.emis.com/ (accessed on 6 November 2021).
78. Guzik, B. *Podstawowe Modele DEA w Badaniu Efektywności Gospodarczej i Społecznej*; Wydawnictwo Uniwersytetu Ekonomicznego w Poznaniu: Poznań, Poland, 2009; ISBN 978-83-7417-368-1.
79. Prędki, A. *Modelowanie Zmienności Danych w Ramach Metody DEA*; Wydawnictwo Uniwersytetu Ekonomicznego w Krakowie: Kraków, Poland, 2016; ISBN 978-83-7252-725-7.
80. Cooper, W.W.; Li, S.; Seiford, L.M.; Zhu, J. Sensitivity Analysis in DEA. In *Handbook on Data Envelopment Analysis*; Cooper, W.W., Seiford, L.M., Zhu, J., Eds.; International Series in Operations Research & Management Science; Springer: New York, NY, USA, 2011; pp. 71–91, ISBN 978-1-4419-6151-8.
81. Nowak, E. *Analiza Sprawozdań Finansowych*, 4th ed.; Polskie Wydawnictwo Ekonomiczne: Warszawa, Poland, 2017; ISBN 978-83-208-2256-4.
82. Coelli, T. A Guide to DEAP Version 2.1: A Data Envelopment Analysis (Computer) Program. *CEPA Work. Pap.* **1996**, *96/08*, 1–49.
83. Coelli, T.J.; Rao, D.S.P.; O'Donnell, C.J.; Battese, G.E. *An Introduction to Efficiency and Productivity Analysis*, 2nd ed.; Springer: New York, NY, USA, 2005; ISBN 978-0-387-24266-8.
84. Pai, P.; Khan, B.M.; Kachwala, T. Data Envelopment Analysis—Is BCC Model Better than CCR Model? Case of Indian Life Insurance Companies. *NMIMS Manag. Rev.* **2020**, *38*, 17–35.
85. Ohlan, R. Efficiency and Total Factor Productivity Growth in Indian Dairy Sector. *Q. J. Int. Agric.* **2013**, *52*, 51–77.

86. Kumar, S.; Gulati, R. An Examination of Technical, Pure Technical, and Scale Efficiencies in Indian Public Sector Banks Using Data Envelopment Analysis. *Eurasian J. Bus. Econ.* **2008**, *1*, 33–69.
87. Jędrzejczyk, Z.; Kukuła, K.; Skrzypek, J.; Walkosz, A. *Badania Operacyjne w Przykładach i Zadaniach*, 6th ed.; Wydawnictwo Naukowe PWN: Warszawa, Poland, 2011; ISBN 978-83-01-16483-6.
88. Kukuła, K.; Luty, L. O wyborze metody porządkowania liniowego do oceny gospodarki odpadami w Polsce w ujęciu przestrzennym. *Zeszyty Naukowe SGGW w Warszawie. Problemy Rolnictwa Światowego* **2018**, *18*, 183–192. [CrossRef]
89. GUS [Statistics Poland]. Fizyczne Rozmiary Produkcji Zwierzęcej w 2019 roku. Available online: https://stat.gov.pl/obszary-tematyczne/rolnictwo-lesnictwo/produkcja-zwierzeca-zwierzeta-gospodarskie/fizyczne-rozmiary-produkcji-zwierzecej-w-2019-roku,3,15.html (accessed on 6 November 2021).
90. GUS [Statistics Poland]. Zwierzęta Gospodarskie w 2019 Roku. Available online: https://stat.gov.pl/obszary-tematyczne/rolnictwo-lesnictwo/produkcja-zwierzeca-zwierzeta-gospodarskie/zwierzeta-gospodarskie-w-2019-roku,6,20.html (accessed on 6 November 2021).
91. GUS [Statistics Poland]. Charakterystyka Gospodarstw Rolnych w 2016 r. Available online: https://stat.gov.pl/obszary-tematyczne/rolnictwo-lesnictwo/rolnictwo/charakterystyka-gospodarstw-rolnych-w-2016-r-,5,5.html (accessed on 6 November 2021).
92. Singh, S.; Fleming, E.; Coelli, T. Efficiency and Productivity Analysis of Cooperative Dairy Plants in Haryana and Punjab States of India. *Work. Pap. Ser. Agric. Resour. Econ.* **2000**, *2000-2*, 1–18.
93. Ohlan, R. Productivity and Efficiency Analysis of Haryana's Dairy Industry. *Productivity* **2011**, *52*, 42–50.
94. Ozcan, Y.A. *Health Care Benchmarking and Performance Evaluation. An Assessment Using Data Envelopment Analysis (DEA)*, 2nd ed.; Springer: New York, NY, USA, 2014; ISBN 978-1-4899-7472-3.
95. Kapelko, M. Measuring Inefficiency for Specific Inputs Using Data Envelopment Analysis: Evidence from Construction Industry in Spain and Portugal. *Cent. Eur. J. Oper. Res.* **2018**, *26*, 43–66. [CrossRef] [PubMed]
96. Malak-Rawlikowska, A.; Milczarek-Andrzejewska, D.; Fałkowski, J. Restrukturyzacja sektora mleczarskiego w Polsce—Przyczyny i skutki. *Roczniki Nauk Rolniczych, Seria G* **2007**, *94*, 95–108.
97. Brodziński, M.G. *Oblicza Polskiej Spółdzielczości Wiejskiej. Geneza-Rozwój-Przyszłość*; Wydawnictwo FREL: Warszawa, Poland, 2014; ISBN 978-83-64691-04-1.
98. Mahajan, V.; Nauriyal, D.K.; Singh, S.P. Technical Efficiency Analysis of the Indian Drug and Pharmaceutical Industry: A Non-Parametric Approach. *Benchmarking Int. J.* **2014**, *21*, 734–755. [CrossRef]
99. Nowak, M.M. Wpływ spółdzielni mleczarskich na przemiany przestrzenne, ekonomiczne i środowiskowe we współczesnej gospodarce. *Prace Naukowe Uniwersytetu Ekonomicznego we Wrocławiu* **2013**, *296*, 251–260.
100. Roman, M. *Uwarunkowania i Kierunki Zmian Zasięgu Geograficznego Rynku Mleka Surowego w Polsce*; Wydawnictwo SGGW: Warszawa, Poland, 2017; ISBN 978-83-7583-756-8.
101. Rudziński, R. Organizacja logistyki w zakładach przetwórstwa mleka. *Zeszyty Naukowe Uniwersytetu Przyrodniczo-Humanistycznego w Siedlcach, Seria: Administracja i Zarządzanie* **2010**, *87*, 157–177.
102. Ruales Guzmán, B.V.; Rodríguez Lozano, G.I.; Castellanos Domínguez, O.F. Measuring Productivity of Dairy Industry Companies: An Approach with Data Envelopment Analysis. *J. Agribus. Dev. Emerg. Econ.* **2021**, *11*, 160–177. [CrossRef]

Article

A Territorial-Driven Approach to Capture the Transformative Momentum of the Social Economy Especially from the Agricultural Cooperatives

Juan Ramón Gallego-Bono [1,*] and MariaR Tapia-Baranda [2,*]

[1] Department of Applied Economics, University of Valencia, Avda. de Tarongers s/n, 46022 Valencia, Spain
[2] Instituto Tecnológico de Estudios Superiores de Monterrey, Avda. Eugenio Garza Sada, 2501 Sur, Monterrrey 64849, Mexico
* Correspondence: Juan.R.Gallego@uv.es (J.R.G.-B.); mariar.tapia@tec.mx (M.R.T.-B.)

Abstract: In the last few lustrums, the literature has searched for more precise methods to assess the socio-economic importance of the Social and Solidarity Economy (SSE). On that basis, this article offers a new way of assessing the SSE impact, enhancing the understanding of the SSE potential for socio-economic transformation. An evolutionary micro–meso–macro and territorial theoretical framework is developed, utilizing, along with the assistance of a qualitative methodology, studies on the transformation promoted by the SSE on the sugar cane cluster of Veracruz (Mexico). The main results of the article are that the SSE boost beneficiaries, while the protagonists of the transformation cannot be defined a priori, but are rather conformed by transformation vectors promoted by the SSE: their values shared by a wide spectrum of actors, the SSE socio-economic and organizational specificities, and their rooting in the productive system. The fundamental conclusion of the article is the need for a "territorial-driven approach" of the SSE's impact, compared to the dominant "stakeholder-driven approach". The main limitations (and suggestions for future studies) are the empirical investigation of a single case, and the need to develop a qualitative and quantitative system of indicators of the transformative drive of SSE.

Keywords: social and solidarity economy; evolutionary approach; territorial-driven approach; agricultural cooperatives

Citation: Gallego-Bono, J.R.; Tapia-Baranda, M.R. A Territorial-Driven Approach to Capture the Transformative Momentum of the Social Economy Especially from the Agricultural Cooperatives. *Agriculture* **2021**, *11*, 1281. https://doi.org/10.3390/agriculture11121281

Academic Editors: Adoración Mozas Moral and Domingo Fernandez Ucles

Received: 5 November 2021
Accepted: 13 December 2021
Published: 16 December 2021

Publisher's Note: MDPI stays neutral with regard to jurisdictional claims in published maps and institutional affiliations.

1. Introduction

In recent decades, an extensive literature has been developed to explain how to value and measure the socioeconomic importance of SSE (Social and Solidarity Economy) [1–7]. A part of this literature quantified the sets of SSE activities with a series of variables, such as production, employment, etc. This exercise has also been carried out at different levels on the reality of the SSE, covering individual initiatives (micro-level), initiatives in different sectors or regions (meso level), or the economic level as a whole (macro-level) [2,3,5,6]. The analysis underlying the SSE, measured in these terms, is very important, especially when its evolution over time is analyzed, showing some highly relevant dynamic aspects such as its greater resilience than capitalist companies when faced with crises, measured, for example, by their greater ability to maintain employment or production [8,9].

Nowadays, this way of measuring the significance of the SSE is only an indication that does not allow us to capture other aspects that acquire great importance in the projection of the SSE within the whole socio-economic system.

We will argue that SSE constitutes an essential transformation instrument for socioeconomic systems. However, this role goes beyond the scope of the SSE itself. Thus, to capture and measure this transformation potential, we need to understand how it contributes to, and drives, change on the path of socio-economic development.

This article will try to demonstrate the transformative potential in SSE by showing its ability to promote a more socially inclusive and environmentally friendly development path.

We have considered that analyzing the potential for transformation that SSE may possess, is not separate from the socio-economic structures in a social formation. Therefore, it is convenient to delimit the type of realities under study.

In this sense, although the analysis contained in this work could have a more general field of validity, the article will focus on the problems of Latin America and, more specifically, on a set of countries and regions that face important obstacles to their development. Such is the case of the different behaviors of SSE entities in the sugarcane cluster in Veracruz, Mexico.

Indeed, the article will pay special attention to a type of society in which the entities that are created following the legal forms of the SSE, especially cooperatives, are captured since their formation by the network of dominant political and economic actors.

Recently, however, this has not prevented some of these initiatives from developing in a spontaneous, endogenous, and territorial-rooted process, generating a new transformative path promoted by the SSE values. This process will be connected with other local movements and civil society, which have shown a huge transformation capacity [10–13]. This is the case, in particular, of a series of regions and clusters in Latin America that show a certain capacity to exhibit a more inclusive and environmentally sustainable development process [14].

In line with the latter, is important to clarify the different types of SSE entities that exist in the Veracruz sugarcane cluster and whose essential role will be analyzed in this article. In the first place, it must be said that in the sugar cane of Veracruz we have not found companies that had the legal form of cooperatives, surely for the reasons indicated above. However, we do find a new type of SMEs (small and medium-sized companies) organized almost strictly following cooperative principles, although without presenting the cooperative legal form. This is specifically true regarding firms such as Mastevia or Balandra Foundry both in Cordoba, Veracruz, México. Along with these firms, we also have two types of associations. On the one hand, the associations of sugarcane producers or sugarcane growers. While some of them are part of the politico-economic sugarcane system dominated by multinationals, other alternative associations have recently appeared that behave according to the principles of the SSE and that have developed a different strategy. Among the associations also stand out Asociación de Cañeros Independientes (ACI), and the UPV (Union of Piloncilleros de Veracruz); this last one has behavior similar to that of a second grade agri-food cooperative.

In this sense, the theoretical–empirical originality of the article lies in showing how, compared to the business model of the sugarcane cluster organized and dominated by multinationals, which has high participation in the sugar mills (sugarcane producing companies and derived from it), based on a standardized product and a plot that minimizes relationships and marginalizes local knowledge and actors, another alternative organization is possible. Indeed, another trajectory has emerged that tries to develop a differentiated product based on local relationships and knowledge.

In this context, this article offers in the following section a conceptual framework that will show the SSE capacity to promote a process of the clusters transformation in developing countries (particularly in Latin America).

To address this conceptual challenge, and this is the main theoretical contribution of the article, a micro–meso–macro evolutionary approach (initially conceptualized by Dopfer, Foster, and Potts [15,16] will be developed, in which these three different levels are defined in dynamic and structural terms, with meanings in some cases different from how they are usually understood. In a third section, we will use this framework to illustrate the case of the Veracruz sugarcane cluster, where the SSE is managing to promote the creation of a new path of inclusive and environmentally sustainable development. The leadership that a series of SSE entities exercises over a broader set of actors will be evidenced, thanks to the existence of a series of world views and values shared by these actors, which are largely those of the SSE: cooperation, trust, acknowledgement, and transparency relationships. The article will end with a discussion and conclusions section.

The theoretical contribution of this article develops an evolutionary approach that articulates the micro and meso levels, allowing the derivation of a series of vectors for explaining an essentially spontaneous process of endogenous base transformation driven by SSE entities.

In the practical field, the capacity for self-organization of civil society led by the SSE is evident. Furthermore, this process may indeed have limitations in the absence of support from public administrations. However, it is no less true that this process of self-organization shows a series of behaviors and relationships that could constitute safe references (because they are rooted in the territory and the local actors themselves) for the definition of an alternative policy to promote an endogenous trajectory and development of the territories and clusters of Latin America.

2. Conceptual Framework and Methods

2.1. Moving from a Stakeholder-Driven Approach to a Territorial-Driven Approach

2.1.1. Evolutionary Roots of the Shift from the SSE Impact

We will start with the thought that, "the evaluation of the social economy reflects the role the social economy is expected to play in the development model and its transformation", Bouchard [2]. Insisting on this same idea, Richez-Battesti [14,17] emphasizes that, "what is at stake with the evaluation is also—and above all the definition of the field of the SSE and its modes of regulation."

Bouchard argues that diverse evaluation approaches are associated with different underlying theoretical approaches or paradigms. She connects the managerial and strategic perspective, the neo-institutionalist economics, and the institutionalist sociology perspective with different approaches to assessing the importance of the SSE [2]. In this sense, the emphasis of [2,17] on the transformation and modes of regulation that the SSE is capable of printing, is part of the latter institutionalist approach. Now, based on our perspective, it is key to complete this institutionalist sociology approach with an evolutionary approach, where retaining the sociological importance of power relations, and in the search for a more democratic and inclusive economic model, the contribution of the SSE to a process of change and transformation is inserted, where new actors, competencies, relationships, and values appear.

Both at the theoretical and empirical levels, the article focuses on the problem of clusters and territories. By cluster, we understand a network of companies connected by their link to the same value chain that participates in the production of a good, as well as a set, of activities of actors and support activities (universities, technology centers, service companies, etc.) within a more or less wide geographic space [18]. By territory, we understand a space socially constructed by the actors through their individual and collective interactions, and their interaction with the environment [19–22]. A key hypothesis of this article is that in the same movement, SSE entities enhance relations and territorial knowledge and could also contribute to developing the clusters that a territory hosts.

The SSE actors can promote a transformation in the clusters and territories in which they operate because, by involving a plurality of players with partially different competencies, relationships, and values, they are capable of promoting (through interaction with these other actors) the shift (territorial) from the micro to the meso level, and this shift is of vital importance for the effective institutionalization of individual SSE initiatives [2,23,24].

In this sense, an important literature advancement to evaluate the incidence of SSE is the work of Ebrahim and Rangan [4]. As they synthetically point out, "outputs don't necessarily translate into outcomes, and outcomes don't necessarily translate into impact". What this means is that the true impact of SSE entities lies in the meso-sphere of the diffusion and institutionalization of individual (micro) initiatives. Now, to appreciate the true impact of the SSE on this meso level, we have to conceptualize how this meso impact goes beyond the SSE entities themselves, and how this process is linked to the transformation mechanism associated with the SSE. For that purpose, first, we are going to formulate this general principle of transformation in the remainder of this section. Then, in the next section, it will be shown how the SSE gives direction and how this transformation process is supported in its productive territorial rooting, which precisely confers to the SSE its scope in terms of generating a path towards more inclusive and environmentally sustainable development.

We will begin by advancing the general element of the evolutionary foundation of the need for a territorial-driven approach to the impact of SSE. A characteristic that Saïd et al. [5] attribute to some quantitative approaches that try to assess the impact of SSE, is that of stakeholder-driven approaches. In other words, a static approach that seeks to assess the impact of the SSE by measuring the effects of its entities on the stakeholders (interest groups) [5]. With this approach, stakeholders are defined a priori. Now, from an evolutionary perspective, SSE initiatives generate ideas and innovations that in turn connect with other ideas and innovations, which likewise stimulate the development of new ones [25]. An emerging process of transformation and structural change is generated, which cannot be anticipated a priori [25–27]. The territory is expected to constitute a privileged space within which these chains of ideas, innovations, and connections are produced. Especially when it is reasonable to expect that the actors are going to be favored by interacting with other actors on these same terms, and which are also very close concerning their expectations, worldviews, and values [28]. Hence, in this article, we support the hypothesis that there is a need to assess the impact of SSE as territorial driven. Because, in coherence, with the open nature of the interaction and change in complex systems [26], the stakeholders will define themselves by the process of structural transformation itself, while they will also give it feedback. This is what we are going to conceptualize in more detail below with the help of the micro–meso–macro approach, but connecting it with the territorial roots of the SSE.

2.1.2. The Values of the SSE: Motor of Transformation and Basis of the Development Trajectory and Its Rooting in the Production System
The Micro–Meso Articulation

We are defending the hypothesis that in the context of clusters in developing countries, SSE may be the engine of a change in the dominant development model in the territory.

This development model consists of the emergence of new economic activities and a new development path in the territory [28]. To support this hypothesis, we developed an evolutionary approach that emphasizes the generation of micro-variety (actors, competencies, relationships, and values) at the heart of change [25,27,29–32]. Now, in the micro–meso–macro evolutionary approach [15,16] being defended, innovations and changes do not occur when they are generated by an entity or organization (micro), but through a meso-trajectory of generation, adoption, diffusion, and institutionalization of these innovations (meso-rule) in the set of entities or organizations that define a population. These meso-rules are new routines in the behavioral, cognitive, technological, and socio-organizational fields [15,16].

The macro-order is generated through the adjustment between different meso-trajectories. Therefore, the deployment of a meso-trajectory can destabilize it, which frequently does not happen because there is a set of institutions or meta-institutions that operate at a high level of abstraction (values, beliefs, etc.), which allow coordination behaviors, limiting the variation margins of the meso-trajectories, so that they are compatible within a macro-order [16,33,34].

From an evolutionary perspective, the macro-order supposes the existence of a behavior pattern and this can only happen when there is a structural coherence between the activities that make up this order [25]. This reasoning is very important to appreciate the macro level of incidence of SSE from a new perspective. The core of the approach that concerns us constitutes the passage from the micro to the meso. We will anticipate that it is the internal heterogeneity of the aforementioned population, which can promote both the institutionalization of change and the transformation process itself.

In this sense, emphasis is placed on the need to focus on the capabilities of the SSE to reach the underlying problems (of inequality, discrimination, etc.) [4] or, as previously advanced, the need to institutionalize individual SSE initiatives [23]. In Latin America, Coraggio has pointed out that, "there is a clash between: (1) The urgent survival needs of the impoverished and excluded sectors, as well as the targeted public programs of individuals or small groups related to self-employment designed to deal with this urgent situation; and (2) The longer time frames required to give proper consideration to the possibilities of building a system of SSE and to allow for the cultural changes it entails." Both levels are needed. At the very least it is necessary to intervene at the first level while keeping the second level in mind, to shift from a micro to a meso-level perspective (promoting articulation, complementarity, in territories, and communities)" (Coraggio) [24].

Our perspective also places the micro–meso articulation as being of vital importance for the SSE, but placing more emphasis on the dynamic and all-encompassing virtues of the SSE. In effect, we argue that the actors, competencies, relationships, and values of the SSE can promote a transformation in clusters and territories by promoting interaction between actors that are at the same time similar as well as different. This capacity of the SSE will be supported by two key elements. First, the ability of SSE values to bring together different players who share a series of world views and essential values in the face of the status quo, represented by the dominant actors [10,13,14,32,35]. These values are supported by a set of principles "involved in the institutionalization of new economic activities" [24]. Principles that go from the market or the in-depth use of power relations, which are usually used by the dominant actors [14,24,36], to other values more typical of the SSE, such as reciprocity, fair trade, redistribution, non-exploitation of work, non-extractive development, responsible consumption, the transformation of property relations [24], and the recognition of knowledge from other actors, transparency, trust, and non-discrimination [13]. Second, linking the SSE to the productive system gives it a potential capacity to transform the development model. Thus, for example, Coraggio [24] highlights that policies to stimulate the development of new forms of production should be seen as a necessary complement to redistributive policies.

The combination of both elements gives the SSE entities the ability (so far dismissed) to have an impact on the territory. It has been argued that the principles and values of the SSE lead them to commit to the needs of the community in which they operate. This can favor both a commitment to local development initiatives (which favors the inclusion of groups with difficulties/disadvantaged groups), as well as a predisposition to address environmental problems and to attract people committed to both, inclusiveness and equity, as well as to environmental sustainability [37,38]. Now, the connection with the productive systems in the framework of clusters in Latin America, means that there is an impact at a deeper level, from a structural perspective, in which the SSE and its integration in the territory can promote a process of change and transformation through a new development model. From this perspective, on the issue of inclusiveness, it is not only about showing the capacity of the SSE to develop programs and activities that incorporate people and groups with difficulties into employment and services. It is about showing that the SSE, because of its values, socio-economic, and organizational characteristics, promotes a shift towards a development model based on previously marginalized resources, actors, capacities, and relationships.

In the same way, these values and worldview, are what make a certain type of institutions (rules of the game) and coordination mechanisms between actors based on reciprocity, trust, recognition, and acceptance of grass-root knowledge (for example, a tacit character and based on the experience of the others) [13,36]. This then causes a type of embeddedness [39] and an organized (cognitive, organizational, and institutional) proximity between actors [20,21,40], which constitutes an essential condition for cooperation between actors so they can introduce innovations in industrial ecology (the use of waste from some productive activities as inputs to others, as occurs in natural ecosystems) [13,41,42].

In this section, we have shown that the essential drivers of micro–meso articulation and the basis of the role of the SSE to drive this process are: the generation of microdiversity that SSE entities can set in motion and the capacity of the SSE itself to bring together various actors, linked in turn, to the nature of their values and the roots of the SSE in the production system. Both engines define the transformative capacity of the SSE, which is at the core of this article. Now, these two engines operate concretely through their articulation with the organizational characteristics of the SSE, and this will allow us to deduce a series of transformation channels promoted by the SSE.

The Channels of Transformation Carried by the SSE in Terms of Territorial Networks

It should be noted, on the one hand, the link between values and principles of the SSE are shared by actors who do not always belong to the SSE, but who are marginalized (in various ways) by the status quo, and, on the other hand, the development of SSE's innovative initiatives from the production system itself. These two elements have another essential implication from the (meso) perspective of the institutionalization process of the SSE's innovations [23]. It is about the SSE capacity to promote the development of a whole series of practice, epistemic, and political communities among different actors capable to spread transformative impulses from the micro to the meso level, creating a transformative territorial network [13,25]. This ability of the SSE proceeds in three essential ways. First, the organization of the SSE into second- and higher-degree entities gives them a great capacity to spread the new routines among all the other SSE entities that are related to each other [43,44]. This is an area in which agricultural cooperatives have played an essential role because they constitute the paradigm of commitment to the community and the territory highlighted above. This commitment leads a cooperative, concerning its members, or in a second-degree cooperative concerning its base cooperatives, to exercise its *voice* in the sense of Hirschman [45]. For example, when the production of the members or the base cooperatives, respectively do not meet the quality standards required for the successful commercialization of agricultural production, the cooperatives do not suspend these members, or these base cooperatives (*exit*), nor do they maintain the relationship as if there was no problem (*loyalty*). On the contrary, what cooperatives do is express (*voice*) their dissatisfaction with the situation and systematically demand, as well as aid, members to improve their routines and achievements. This enables the cooperatives and grass-root cooperatives to raise the skills and capacity for innovation of their members [43]. However, in the framework of social and political realities where cooperatives are systematically captured and denatured by the dominant political-economic powers, other collective entities such as associations could play a similar role.

Second, through the specific resources and know-how of the territory. The commitment of cooperatives and other entities of the SSE with the community and the territory, places them in excellent condition to mobilize a set of resources that are the result of collective-learning processes developed by local actors when dealing with productive, organizational, and commercial problems [19,46], as well as the in-house know-how developed by the actors who actively participate in this process. It is a process in which the cooperatives and other entities of the SSE reinforce the ties of their organizations with the community and the territory through a whole web of social relations and social capital linked overall to the productive system [47].

Third, due to the richness of inter-sectoral relationships that innovative actors frequently carry in clusters of developing countries. This is usually associated with the development of a great diversity of venture activities in the territory by the most proactive and critical actors with the status quo, as a way to face uncertainty and improve their resilience. The growth of this productive diversity makes the same (more dynamic) local actors belong to and vitalize different practice and epistemic communities, which operate with different values, organizations, and knowledge. This process is going to have a special scope in terms of creating a space of structural coherence when the activities of these pioneering actors develop simultaneously, not only in different links of the sugar value chain but also in various value chain activities (of clusters and economic sectors) [48].

Succinctly, we understand from the above, the need to move from a stakeholder driven to a territorial-driven evaluation logic. We mean that the territory itself channels an essentially "meso-generating process" of a whole set of connections and interactions between actors that diffuse and amplify in complexity the innovative SSE initiatives, toward: (a) other SSE entities; (b) other entities that behave with the same principles and values or (c) other entities that only share certain world views and values; and (d) diffusion of routines or innovations by exaptation, this is to serve a different objective for which it was originally developed [49]. However, we should not underestimate "the capacity of appropriation and co-optation of any forms of innovation by corporate power in the dominant food system" (Rossi et al.) [2,50], and that could completely neutralize this transformative power of innovation by exaptation carried out by entities other than SSEs. All this makes it difficult to define a priori the stakeholders that ultimately benefit from the impact of the SSE.

The Characterization of the Generation Process of New Connections

Now, this creative process of generating new connections is a process with fewer features of indeterminacy than those contemplated in the general explanation of Metcalfe [20]. This is because in our framework, as in the reality that it is intended to represent, there is polarization and fragmentation (social, political, and economic) that causes the actors to only selectively engage with certain actors. The shared values then are those that define the top priority when selecting the actors with whom they will be preferentially linked.

This process of expansion of innovations in the territory, only partially indeterminate, allows us to see the meaning of the macro-order in a new light. Indeed, we have highlighted above that the deployment of meso-rules by meso-trajectories generates tensions, but also evolutionary opportunities [44], in other meso-rules and meso-trajectories and can destabilize the macro-order that links the different meso-rules [16]. We have also seen the importance of a series of institutions that operate at a high level of abstraction (meta-institutions) to regulate the macro-order, thanks to the establishment of certain limits to meso changes so that they do not destabilize the macro-order [16]. "The [macro] order produces a pattern" [7,25], in such a way that there must be a certain structural fit between the meso-rules so that a minimum coherence is generated, which is often very complex. This is where the question of intermediary actors that can generate a space of confluence between different development paths arises [28], that in regions and clusters of developing regions are often the reflection of different meso-rules and meso-trajectories. This is also a crucial aspect from the perspective of shaping spaces for overcoming socio-economic fragmentation, capable of leading to the shaping of more integrated and less polarized realities, and therefore more coherent from a structural perspective.

This space capable of expanding to the macro-order, understood in a structural sense, opens a new field of research for the assessment of the impact of the SSE. For example, given the predominance of extractive behaviors [24], and little concern for efficiency in the clusters of Latin America [32], the attitudes that seek better use of resources through the industrial ecology could generate a broad cross-sectorial space. This process could stimulate a movement shared by various actors (such as companies, technicians, researchers, and public officials) at the system's margin, those who are uncomfortable with the status quo and are willing to change it [14]. This movement could contribute to the territorial extension of some of the new routines and innovations driven by the SSE values and defended by SSE entities and other leading actors of the network, and the progressive shaping of discourse in favor of a transformative frame [51], more extensive, with more options to sustainably transform the cluster. Nonetheless, the synchronous coexistence of different discourses or even of different socio-technical micro-systems is highly probable.

2.2. Methodology and Information Sources

The empirical objective of this article is to explain the transformation process of the Veracruz sugarcane cluster and identify the actors, competencies, and relationships that drive this process. The central hypothesis to be shown is that the scope of the SSE goes far beyond the importance and presence of the entities with a formal legal structure of SSE. Consequently, a reality in which the dominant (public and private) actors capture some lawful entities of the SSE, defines an exceptional case to contrast this hypothesis. The research is based on a qualitative methodology with field visits and 110 in-depth personal interviews conducted between 2017 and 2021. A closed questionnaire was used for 43 key informants (23 experts, 8 local politicians, and 12 researchers) and an open questionnaire for 67 stakeholders (28 farmers, 4 farmer associations, 9 sugar mills, 16 piloncillo makers and 10 entrepreneurs). A pilot study was conducted to check the relevance of the questions by interviewing five reliable actors. In general, it was sought to delve into the origin of the business, their markets, the internal and external relations, and the nature of their networks and innovations.

The sample intentionally includes all the sugarcane players. However, considering the objective of this research, the sample gives special importance to the actors that belong to the SSE (Union of Piloncilleros de Veracruz—UPV- and Asociación de Cañeros Independientes—ACI-) and to the emerging business initiatives that belong to the ESS, or that are organized following the principles of the ESS. The group of experts, based on their reliability and leadership, are businessmen, professionals, public servants, researchers, and university professors.

The sample was made using the networking or snowball technique that increased with the contacts of the interviewees until the saturation of the information was achieved.

The interviews were audio-recorded and transcribed to obtain the results by making a systematic language analysis. It was sought to clarify: who are the actors that lead the generation and diffusion of innovations (social, environmental); the type of link (direct or indirect) that they have with the SSE, and their involvement in the innovation processes in the territory. Special attention was also paid to the processes of generation of organized proximity and shared identities between actors, in so far as they can constitute the basis for the creation of networks of diverse nature and scope between them. The main blocks of questions directed towards the different actors in the value chain were the following. Origin, motivation, and form of organization of the entrepreneurial initiative. Main products and markets. Leading innovations developed in recent years and actors participating in them. Main forms of cooperation, main networks with other companies and institutions, main principles and values that guide business activity, and main elements that help bring them closer to or distance them from other actors in the territory. Role of insertion in markets and external value chains as a mechanism for consolidating and/or changing existing routines in the sugarcane cluster.

3. Results

3.1. SSE as a Driver of Structural Change in the Sugarcane Cluster in Veracruz (Mexico)

3.1.1. Historic Roots

Historically, the high mountain region in Veracruz, Mexico, has stood out since colonial times for its economic importance since the sugarcane plantation was favored, which is a sector that has been the economic and social engine of this territory. The VC (value chain) of sugarcane extends to consumers in the countries that make up the United States–Mexico–Canada Agreement (USMCA), but sugar production in this region is sold as a commodity, without adding more value to the GVC (global value chain). Currently, it has a crowded rural-urban territory with 65 urban centers, abundant water, and biodiversity [13,52].

The purpose of the USMCA, the extension of the North American Free Trade Agreement (NAFTA), is to give the region advantages to compete with the Asian giants, incorporating some axes of the Circular Economy and the Sustainable Development Goals (ODS).

There is also the challenge of stopping corruption. The relevance of the economic actors in this territory to face these challenges is set out below.

3.1.2. Internal and External Cluster Organization: The Formation of Two Polar Networks

The (external) insertion of the cluster as a whole in USMCA has contributed both to the consolidation of a *conservative* network and to the emergence of a transformative network [13]. Now, we can better understand these networks if we take into account that they diverge both in their organization within the Veracruz cluster and in the organization of their relations with the outside world.

External Dependency and Internal Hierarchical Organization: The Conservative Network

The integration of the Mexican sugar sector into NAFTA in 1990 has simultaneously defined two radically different socio-economic realities. On the one hand, the vertical integration to the GVCs, driven by powerful multinational actors, "has had a strong negative impact on agricultural products, on the quality of employment and on the environment" [53], accelerating migration and fragmenting knowledge that drives innovations. These large multinational groups operate in many other activities, including the food and beverage industry, and have taken important positions in the traditional sugarcane production mills (with an increase in control over agricultural production) and maintain close ties with the public powers. They also have solid connections with universities and national research centers linked to sugarcane cultivation, forming a *conservative network*. This network focuses its innovations on the field of agricultural production and, within it, on scientific–technological aspects. In this way, local knowledge based on experience is largely neglected, marginalizing the actors who possess it, and making it difficult for these local actors to have access to new scientific–technological knowledge [13].

On the other hand, aided by globalization, different Mexican SMEs have appeared, benefiting from the territory's competitive advantage to compete on the international market with more specialized products and services. This way, a *transformative network* will be formed around this local resource that is opposed to the status quo represented by the conservative network. As a consequence of the greater circulation of local resources, the value chain expands within the territory [13]. From 1990, the Mexican sugar sector was "cartelized", guarantee prices for sugar cane and exchange quotas for sweeteners (cane sugar and corn syrup) were set between countries [54]. Government actions (laws and policies) are shaped by the interests of different networks and levels of power. The conservative network is made up of the large corporate groups highly integrated into globalization (Cornbelt, sugar industry, and US oil refineries). National business groups and organized groups with great social and political power are also incorporated into this network, such as farmers' associations (Confederación Nacional Campesina—CNC- and Confederación Nacional de Propietarios Rurales—CNPR-), sometimes infiltrated by drug trafficking. This power dynamic in the conservative network shows little commitment to health (healthy food) and fair trade, key dimensions for socio-technical change. By being isolated from these objectives (socio-technical and environmental), and by ignoring the inclusion of new actors and knowledge, the opportunity of insertion into the territory and in other economic sectors of the sugar cane cultivation and the sugar mills is reduced [13]. In the last two years, the CNC has managed to free itself from its traditional ties to the PRI (Partido Revolucionario Institucional) and now acts with notable autonomy and independence from the political parties.

The opportunities to promote inclusion through cooperative initiatives (especially agricultural ones), and other forms of SSE are captured since their formation by some actors of this conservative network, to divert public money and for electoral purposes: "federal deputies ask us for 10% or up to 50% of the money destined to the formation of agricultural cooperatives" (interviews with the cane, coffee, and papaya producers 2016–2020).

In terms of Industrial Ecology initiatives within this conservative network, they are based on inter-industrial relationships, such as the use of biomass to produce electricity for self-consumption and the usage of some by-products as inputs in other establishments to produce feed and fertilizers [55]. However, these inter-industrial relationships occur within the same company [13]. This together with the vertical integration of the VC favors the concentration of public spending on R + D + i in a national scientific–technological–political system, which restricts its scope of research to a very limited field. The network of actors that make up this system, includes the actors aforementioned and the national public research centers, public universities, main political parties, and government officials. So, as the sugarcane crop is basically destined for sweetener production, research in this only line is carried out, but its creation potential of other products is not recognized (biodegradable containers, paper, building, fabrics, medicines, solvents, etc.) that may include opportunities for specialization and entrepreneurship supported by the resources (human and natural) within the territory. This limits the inter-enterprise exchanges that stimulate socio-organizational innovations, restricts the opportunities to the now marginalized actors, and the scope of environmental protection [13,36] (see Table 1).

Table 1. Business behavior according to subsectors of the value chain in the conservative and transformative networks (results expressed in %).

Different Areas of Business Behavior		Total Firms in Absolute Values 54	Group of Conservative Actors AV 32					Group of Innovative and Intermediate Actors AV 22		
			Agricultural Suppliers AV: 1	Small Farmers *Ejidatarios* AV 21	Sugar Mills AV 6	Sugar Mills Services AV 4	Sugar Services & Trade AV 2	Medium Farmers AV 8	Sugar Mills AV 2	Piloncillo Producers AV 10
Kind of products and services	Generic	77.8	100.0	100.0	100.0	100.0	0.0	100.0	100	0.0
	Specialised	40.7	0.0	0.0	0.0	0.0	100.0	100.0	100	100.0
Origin of companies	Local	81.5	0.0	100.0	0.0	25.0	100.0	100.0	100	100.0
	National	14.8	100.0	0.0	66.6	75.0	0.0	0.0	0.0	0.0
	International	3.7	0.0	0.0	33.4	0.0	0.0	0.0	0.0	0.0
Sales markets for products and services	National	100.0	100.0	100.0	100.0	100.0	0.0	100.0	100.0	100.0
	NAFTA	50.0	0.0	0.0	100.0	0.0	50.0	100.0	100.0	100.0
	Resto of the world	13.0	0.0	0.0	100.0	0.0	50.0	0.0	0	0.0
Types of innovation	On process or product	50.0	0.0	0.0	15.0	100.0	100.0	100.0	100.0	100.0
	On Market	37.03	0.0	0.0	0.0	0.0	100.0	100.0	0.0	100.0
	Ecological of technological kind	90.9	100.0	100.0	100.0	0.0	100.0	100.0	100.0	100.0
	Ecological of social and organizational kind	25.9	0.0	0.0	15.0	0.0	100.0	0.0	100.0	100.0

Source: Own made, based on interviews with enterprises (2017–2020) and Gallego and Tapia 2021.

The Spur to Change from a Transformative Network by the Actors That Drive the SSE Values

Actors and Values

Dissimilar to the conservative network dynamics by the large and powerful political and business groups, a territorial transformative network emerges [13,14] The diffusion effect of this transformative net is vital because, unlike the powerful lobbies of the conservative network, they do not depend so much on political–economic negotiation power, as on their ability to recruit new actors and locally rooted knowledge, and from their relationships with each other. The actors that form this transformative network are small and medium farmers, local associations, and local SMEs (suppliers of sugarcane, biomass, raw material and packaging, biofuel, healthy and/or traditional foods, export services, administration, recycling, IT, and logistics for resident companies), as well as regional universities that favor the integration of local knowledge, preparing professions ingrained in territorial needs, with a more technical profile. This group of actors spearheads specialized services and products as well as innovation activities ecological of social and organizational type (see Table 1). Although many of these SMEs are not formally entities of the SSE; in fact, they follow the principles of internal work organization of the SSE (see Table 2). These ventures assume horizontal and non-hierarchical forms in their structure, so then stimulate the internal participation of all members (transparency and cooperation), placing them sociologically very close to the SSE (SMEs Interviews 2017–2020).

Table 2. Different democratic and SSE values appreciation.

	Cooperation	Trust	Capacities Acknowl-edgement	Transparency	Influence Re-lationships
Actors total on absolute values (AV): 45	15.6	17.8	17.8	55.6	24.4
(A) Government actors AV: 4					
CONADESUCA CIDCA SEP City officials	25.0	0.0	25.0	0.0	0.0
(B) Group of conservative actors AV: 34					
Agricultural supplier producers and services AV 28	0.0	14.2	7.1	35.7	21.42
Sugar mills AV: 6	0.0	0.0	0.0	33.3	50.0
(C) Group of Innovative actors AV: 7					
Services, trade & piloncillo makers	85.7	57.1	71.4	71.4	28.6

Source: Own made, based on interviews with actors (2017–2020).

The relationships between the sugar cane cluster's stakeholders make up two different networks: the dominant conservative network includes transnational and national sugar groups that are in charge of lobbying political actors and main associations (CNIAA, ejidatarios and Confederación Nacional de Propietarios Rurales—CNPR-), and promoting agricultural research with R&D national system (see Table 3).

The actors in the transformative network are cross-linked with other economic sectors, by affinity mainly with groups more committed to egalitarian values (cooperation, trust, acknowledgement of capacities and transparency). They are the Asociación de Cañeros Independientes (ACI), Unión de Piloncilleros de Veracruz (UPV), local SMEs, medium farmers, some practice and epistemic communities, and local researchers. In this second network, the proximity (geographic, social, organizational, and ethical) generates inclusive innovations, with actors and knowledge rooted in the territory and mechanisms of agglutination around environment protection [13].

A group of intermediate actors work with both networks, they are the independent sugar mills, SMEs, universities, and researchers, all locally rooted (see Figure 1).

Table 3. Relations of R&D system with the conservative, intermediate and innovative network actors.

Actors in the Scientific and Technological System	Group of Conservative Actors				Group of Innovative and Intermediate Actors			
	Agricultural Suppliers	Small Farmers Ejidatarios	National Sugar Mills and CNIAA	Sugar Mills Services	Sugar Services and Trade	Medium Farmers	Local Sugar Mills	Piloncillo Producers
		Farmers associations CNC Y CNPR						
✗ no relationships					✓ with relationships			
NATIONAL GOVERNMENTAL SCIENTIFIC AND TECHNOLOGICAL INSTITUTIONS								
CONADESUCA (*)	✗	✗	✓	✗	✗	✗	✓	✗
CIDCA (*)	✗	✓	✓	✗	✗	✓	✓	✗
INIFAP (*)	✗	✗	✓	✗	✗	✗	✓	✗
COLPOS (*)	✓	✗	✓	✗	✗	✓	✓	✗
UNIVERSITIES AND COLLEGES (REGIONAL)								
VERACRUZ UNIVERSITY Agricultural Sciences college	✗	✗	✓	✓	✗	✓	✓	✗
VERACRUZ UNIVERSITY Chemical sciences college	✗	✗	✗	✗	✓	✓	✓	✓
ITESM (*)	✗	✗	✗	✓	✗	✓	✓	✗
UTCV (*)	✗	✗	✗	✓	✓	✓	✓	✓
* CNIAA	Cámara Nacional de las Industrias Azucarera y Alcoholera National Chamber of the Sugar and Alcohol Industries							
* CONADESUCA	Cámara Nacional de las Industrias Azucarera y Alcoholera National Chamber of the Sugar and Alcohol Industries							
* CIDCA	Centro de Investigación y Desarrollo de la Caña de Azúcar Sugarcane Research and Development Center							
* INIFAP	Instituto Nacional de Investigaciones Forestales, Agrícolas y Pecuarias National Institute of Forestry, Agricultural and Livestock Research							
* COLPOS	Colegio de Postgraduados en ciencias agrícolas Postgraduate College of Agricultural Sciences							
* ITESM	Instituto Tecnológico de Estudios Superiores de Monterrey Technological Institute of Higher Studies of Monterrey							
* UTCV	Universidad Tecnológica del Centro de Veracruz Technological University of the Center of Veracruz							

Source: Own made, based on interviews with actors (2017–2020). * refers to the name of the institution.

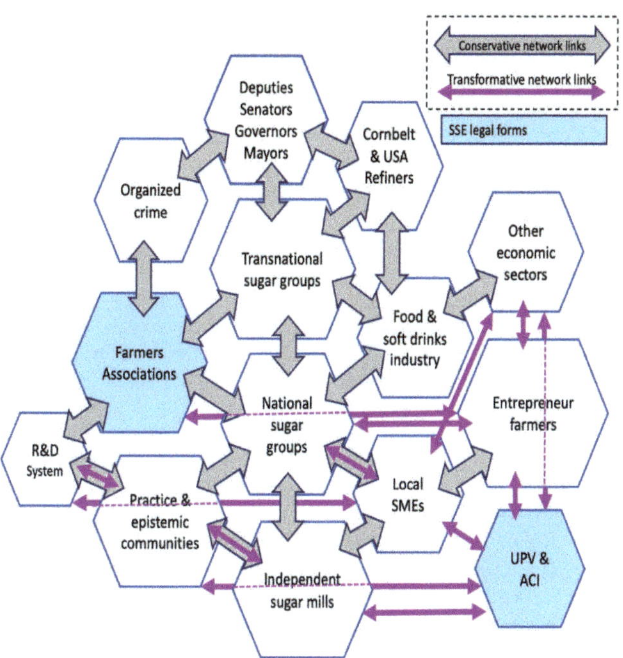

Figure 1. The limited scope of SSE legal form in the conservative and transformative networks.

Figure 2 shows the same network map, with the links of meso rules diffusion through the different mechanisms (values and organizational principles) of the SSE. It also shows the differences between the two groups of SSE entities (one made up with farmers associations belonging to the conservative network, and the other with UPV and ACI belonging to the transformative network) (see Figure 2).

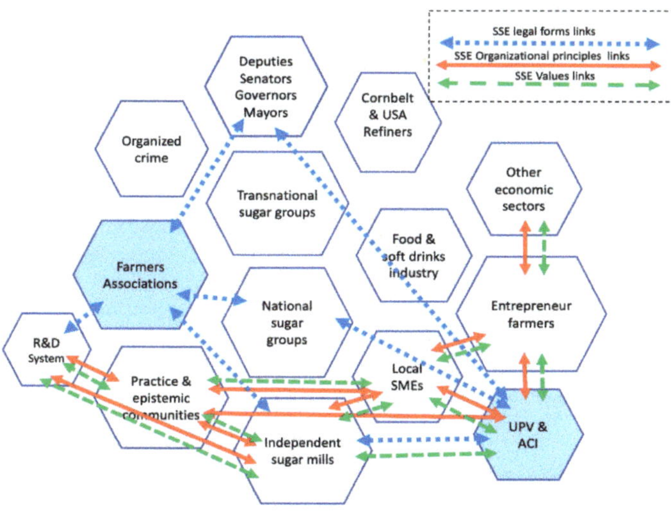

Figure 2. Links of meso rules diffusion through the different mechanisms: the expanded scope of the SEE entities. Source: Own made.

The Emergence of the Shorter Supply Chain

Linked by this vision on proximity and trust, members of SSE local associations, (such as the Unión de Piloncilleros de Veracruz (UPV) and the Asociación de Cañeros Independientes (ACI) and other actors who are not part of the SSE, but who share some of its principles and values, are connected. The actors meet, organize, and offer their goods and services among themselves, creating a shorter supply chain and a more circular sustainable (socially, environmentally, and economically) development process within the territory.

The UPV was created in 2005 with approximately 200 medium-sized cane producers/owners from the municipalities of Huatusco and Zentla Veracruz. They are artisan manufacturers of piloncillo (artisan sweet) that saw the need to rely on a quality standard/certification for their manufacture. This standard/certification requires that 100% pure cane juice is used to obtain a culturally recognized product of high nutritional quality, using renewable energy, thereby creating a cycle that protects the ecosystem. "To leave a better planet for our children" (interview with a piloncillo maker, 2018). This statement shows the commitment with the territory as a key to the cultural and institutional proximity among local actors. The associates are professionals and entrepreneurs who grow sugar cane and, unlike the small owners (ejidatarios), obtain better yields in their crops, and hire harvest groups (workers who organize informally). In this way, they not only disseminate the demands for improvement and the obligation to comply with a series of norms and standards (meso-rules) for all members of the association following the operating model of agricultural cooperatives, but they also push to improve the groups of workers who run as wage earners for a significant number of farmers (inclusiveness). As we will see below, the piloncillo makers also operate and improve other areas of the production, marketing, and the sugar cane value chain and its derivatives. Jointly they diversify their yields with other crops, livestock, and trees to serve specialized markets that demand quality and natural products (environmental innovation). This association enabled them to take advantage of their personal relationships (neighbors, relatives, and colleagues) and cultural identity (most are descendants of Italian migrants or their neighbors), coordinating their knowledge and vision of development to solve the problems they are faced with (intermediaries, health, environment, transparency, employment, security, etc.). The partners strengthen the density of their relationships by including some small farmers (ejidatarios) who share this vision, and expand into a series of activities (commerce, transportation, leisure, and services) that broaden the value chain within the territory. At the same time, through the USMCA, they are engaged with markets, practice and epistemic communities in the USA and the rest of the world, updating their formal and empirical know-how. Driven by the same values (trust, cooperation, and transparency) and vision that brings together actors of the transforming network, and based on proximity (geographical, social, organizational, and ethical), small and medium-sized sugarcane producers, tired of the corruption and the few benefits that they obtain from being forced to depend on one of the main producer associations (Comisión Nacional Campesina—CNC- y CNPR, which belongs to the Institutional Revolutionary Party (PRI)), join independent associations [13].

For example, in the municipalities of Omealca Veracruz and its surroundings, the ACI was created with the support of another political party. The purpose of this association was to provide greater transparency over the management of the resources that are used to improve the harvest yields (reduction in the waiting time of the trucks that transport the cane from the fields to the batey to unload them, detection of pests, the price of implements and products for till and harvest, shared load transport, etc.) [13]. The ACI also coordinates health campaigns, scholarships for students, sports tournaments, and participates in the local religious festivals.

It must also be taken into account that Mexico's insertion into NAFTA and later into the USMCA, makes the actors from the other two more developed countries of the agreement (especially the USA) monitor the behaviors of the Mexican actors with whom they cooperate. Thus, recently the car manufacturers of the USA filed a complaint against the non-transparent practices of Mexican unions in this sector, which has led to a radical change in union representation in this sector. Everything points in the direction that a similar process could be taking place in the sugarcane cluster. They are, therefore, facing an external exercise of the *voice* in the presence of the behaviors considered inappropriate by the Mexican partners belonging to the same network. By forcing a change in the routines of some actors (associations) of the conservative network of the cluster, they can also change the framework of the relationships within the transforming network as well as the relationships between both networks. In effect, on the one hand, they represent support for the change in routines already initiated by some actors within the transforming network. We are thinking, for example, of the emergence of the ACI, which, as we have seen, came about due to the dissatisfaction of many producers concerning the other traditional associations integrated into the conservative network. On the other hand, it can turn these newly emerged associations into references of their peers in the conservative network. Thus, this process can bring about, unintentionally, external actors as vectors of change and external intermediaries that contribute to the creation of new meeting space (transparent and trustworthy) between different networks and, consequently, to the expansion of a structural coherent space meso–macro.

The ambivalent role of the insertion into the GVC and, in particular, into NAFTA and USMCA, based on the dynamics of change in the sugarcane cluster derived from all of the above. The insertion into groups with interests in other sectors (fuel, food, beverages, etc.) and the encouragement of vertical integration around the mills generates inertia, because it minimizes the interaction with other actors and with local forms of knowledge and reinforces the actors and dominant routines of the conservative network. However, as some of these routines of its Mexican partners go against their economic interests, the external actors react by becoming determined agents of change and with a reach that can transcend the conservative network itself.

The Role of Intermediate Actors

The transformative net integration includes actors that can be considered intermediaries between the two polar networks (local independent mills—San José de Abajo and Motzorongo-, R&D system, and SMEs from other regions and other sectors). The double actors that make up this intermediate network are organized crosswise and horizontally, they create relationships with other communities of practice and epistemic communities and markets that sometimes coincide, but often are different from the relationships that the other two networks have. By including local knowledge, they carry out their research, with local universities, with their suppliers, other professionals, and with other sectors and countries. The processes of these actors are also more transparent and specialized, using and developing local resources (human and material), which include a greater diversity of activities and, as a result, the products are more specialized and comply with international quality standards (SMEs interviews 2018). Although the innovations that are produced here have an impact on the territory, very slowly they can reach diffusion/transmission (meso) in the context of violence and corruption that prevails in the territory. That is why, these actors strengthen their relationships in safe spaces, such as in local events (sports, religious, and cultural), school, or family, "where the attendees are well known and for a long time" (interview with a manager of independent sugar mill, 2018). Along with this intermediation work that some actors can often carry out on an individual basis, it is also important to highlight the special capacity to take on this role of intermediaries between polar networks played by some actors integrated into collective entities.

In this sense, along with the leading piloncillo makers who play a key role in the dissemination of norms and standards among other piloncillo makers, but who are much more advanced than the vast majority of small owners (ejidatarios), there are other piloncillo makers who are more followers than leaders who can also play an important role. In effect, this condition of followers makes them suitable references for some restless ejidatarios who would like to modernize their productive and commercial practices but in a non-disruptive way. This discussion shows that in this case, the conditions are optimal for a perfect combination to be made between the degree of homophily (similarity) and heterophily (difference) among actors. This is key so that the actors involved are similar enough to be able to imitate one another but are also sufficiently different so that information and new routines can flow between them [56]. Some non-leading piloncillo makers would be imitated by some proactive ejidatarios, which would actually provide positive feedback on an important kind of actor in the conservative network. Now, in addition to this combination of similarity/difference between the actors, which encourages the dissemination of innovations of the two networks, there is another element that is vital in the ability of the actors to bridge between both network groups especially linked to agricultural production. It is that agricultural production best represents the structural bases of the productive system, where the key aspects of inequality, discrimination, and environmental problems will be defined, as well as the terrain where to most effectively attack all these imbalances.

4. Discussion and Conclusions

This article is part of recent works that have defended the SSE's capacity to transform the economic model [2] and the institutions of socio-economic regulation [17] as a starting point and nucleus of assessment of the importance of the SSE. In this case, our article presents itself as a novelty, in that it adopts an evolutionary and territorial perspective. This allows us to see in a new light this transformation capacity of the SSE and, consequently, the way to value and measure it. Thus, it is a question of deepening the research line opened by the works that have emphasized the importance of assessing the SSE's incidence beyond the direct weight that their entities have [4], the analyses that highlight the importance of the transition from the micro to the meso [4,5,24], or even the key role of its institutionalization [23].

This discussion section and conclusions will focus on this transformation capacity of the SSE territory, highlighting three dimensions of the process: (a) the territorial-driven approach; (b) the role of meso dynamic; and (c) the economic policy perspective.

Regarding the first aspect, what differentiates our article from those works is that it defends and provides evidence in favor of the hypothesis that the SSE transformation capacity goes beyond the importance of its entities, but its path cannot be limited a priori to stakeholders (interest groups), this being the "stakeholder-driven approach" [5]. In effect, we advocate a "territorial-driven approach" because we consider (as was theoretically and empirically evidenced) that the SSE effective field of influence (and, consequently, the stakeholders) cannot be defined a priori but are the result of a creative and open process of interaction of ideas, innovations, and connections [25] driven by the territory itself. However, because it is guided by the territory, it is not an indeterminate process, as Metcalfe [25] argues, but only partially indeterminate. Indeed, it is shown that the SSE values are what allow it to expand its sphere of action, through a "transformative network" creation, which encompasses many more actors than the formal SSE entities, but it is selective. In other words, the cumulative sequences of innovations and connections are essentially limited to actors who share worldviews and values, although these too are transformed by the process. This transformative network demonstrates the capacity of local actors to challenge the capture of SSE entities by the established political–economic network [10,32,36,57]. This process is possible because SSE's entities and other entities that do not have the legal SSE form, but follow their routines and principles of organization, are capable of generating creative processes around themselves. This includes not only innovations generation but also their diffusion among a wide spectrum of actors in the territory. These actors are progressively enrolled in the new transformative network [13]. This enrollment occurs through a set of communities of practice, epistemic, and political, which are based on (and contribute to extend) shared expectations, worldview, and values. In addition, this process supports the embedding of the SSE in the production system. This means that inclusive and environmental innovations (for example, industrial ecology) have a greater scope than has been considered so far in the literature. Indeed, as the Veracruz sugarcane case shows, inclusive innovation allows actors to emerge and enroll in the transformative network, whose resources, knowledge, and relationships have been marginalized from the productive system until now. A transformation that goes far beyond redistributive processes in the form of helping disadvantaged actors [24]. This transformative network that was formed in Veracruz also promotes the adoption of industrial ecology innovations through inter-sectoral cooperation between actors based on organized proximity [13]. Ultimately, it is the endogenous linkage of the SSE that gives it this enormous, and at the same time, open potential for territorial transformation. However, the development of a micro–meso–macro approach allows us to advance our understanding of the (dynamic) forces that guide this process.

In fact, and as a second highlight of the article, the micro–meso–macro approach that we have followed, and whose development we are trying to contribute in this article, has argued that changes occur in the meso domain; that is, while a new rule (routine or innovation) is diffused among the entities of the same nature that make up a population [15,16]. Now, based on our article, it is possible to defend the idea that the population of entities cannot be defined a priori, but rather is defined with the process of generation and diffusion of innovation (meso-rule). Hence, the need for a territorial-driven approach to understanding the ultimate reach of the SSE, and also, the possibility of building a typology of mechanisms for the diffusion of innovation from the SSE that could expand the number of actors (population) who adopt the innovations even beyond the area of reach of the SSE entities. In this sense, it has been possible to demonstrate the strategic role of several agricultural associations and other entities that follow the principles of organization of the SSE in the dissemination of new routines, and the formation of a broad transformative network that houses a large number of actors that end up being in the orbit of the ESS. Therefore, we can conclude that as the meso-rule that originated in the SSE spreads to increasingly different actors, the greater, is also the structural (macro), scope of the diffusion and institutionalization of the meso-rule and, therefore, its impact is also more comprehensive. Following this reasoning, we expose an inclusion effect associated with: (1) the promotion of activities based on new organizational principles; (2) the mobilization of knowledge, which by itself, is marginalized by the dominant network; (3) the use of proximity relationships between actors; and (4) a key aspect that has not been very prominent until now, and that causes our work to the surface, is that through the simultaneous affiliation of the same entrepreneurs to several very different communities of practice, can generate a cross-sectional diffusion of some meso-rules. While this process could destabilize the macro-order [16], and generate opportunities in other areas [44], it could also broaden the scope of the transformative process and extend along with it the macro-order from a structural perspective. Indeed, through the dissemination of the meso-rule, new spaces of structural coherence are opened between productive activities that are part of the same value chain or different value chains that intersect at some points.

In the following Table 4, we present some basic aspects of this theoretical framework, connecting the specific features and actors of sugarcane with a typology of mechanisms of diffusion, proximity, and transmission of the meso rules from the entities of the SSE. Explaining the possible transformative scope in the specific case of the Veracruz sugar cane cluster.

The article offers some clues to advance future work with more accurate measurements of the transforming scope of the ESS. This constitutes a contribution to the development of a micro–meso–macro approach. The most important issue is that the article allows us to derive from a double conceptual and empirical perspective, a series of guiding criteria for the selection and definition of a system of indicators. At the micro level, it is about putting the focus on the diversity and heterogeneity of the actors (new and already established) as key factors in generating innovations. The SSE stand out in this regard for their innovative social and environmental dynamism, their commitment to the territory and their link to the productive system. At the meso level, the indicator selection criteria must capture the various mechanisms for the diffusion of micro innovations. The organizational advantages of some entities of the SSE, the predicament of its principles and values outside its own legal field, especially in contexts of violence, inequality, and fragmentation of knowledge, as well as the role of the practice and epistemic communities in the diffusion of innovations among the entities of the same population, can be essential. Third, at the level of macro indicators of structural coherence, two key criteria emerge from our article. It is about, on the one hand, the detection of intermediary actors with the capacity to pave the way for the formation of new broad-based evolutionary trajectories by "merging" various networks of actors that until then would follow differentiated paths. On the other hand, associated with the propensity of the innovative actors of the territory to develop activities in different sectors, very significant possibilities of macro (structural) change are generated following the possibilities of interrelation between different practice and epistemic communities and, the possibility of generating intersectoral innovations that these local innovators convey. Finally, and in a general way for the three micro, meso, and macro levels, it follows from the article that this system of indicators should not focus only, or fundamentally, on the innovations generated by the SSE, but rather on the amplifying effect on territorial innovations that exercises the entry into the action of the entities of the SSE.

Table 4. Meso-rules diffusion mechanisms by SSE.

Proximity Sources (Shared Elements)	Operational Mechanisms (Instruments)	The Transformative Reach of the Territory	Examples in Veracruz Sugar Cane Cluster
Type A: -Legal form -Organizational principles -Values	-1st- and 2nd-degree cooperatives, -Associations -Other SSE collective forms	Medium	Dissemination of new regulations and commercial-production practices in UPV (Unión de Piloncilleros de Veracruz), and ACI (Asociación de Cañeros Independientes)
Type B: -Organizational principles -Values	-Communities and business networks -USMCA -SDG-UN -Communities and political networks	High	Interaction between piloncillo makers and SMEs that are organized according to the principles and values of the SSE
Type C: -Values	-Practice communities, -Epistemic communities, -Communities and business networks -Communities and political networks -Social movements -Activists	Very high	Intersectoral interactions of piloncillo makers and SMEs, with small owners (ejidatarios), traders, transporters, professionals, including universities and local/regional research centers
Type D: -Adoption of the meso rules by exaptation. -Without there being any affinity	-Communities of practice, -Epistemic communities, -Communities and business networks -Communities and political networks	Very high	In this case, these have not been detected.

Source: Own made.

This last result is complementary to those obtained recently by some other research focused on the study of certain alternative structures of agrarian organizations, such as *Community Supported Agriculture* [58]. Some of these studies conclude that the attractiveness of these alternative structures increases when these structures establish relationships of trust with external actors [58].

From an economic policy perspective, the article has important implications. First, it has shown the important self-organizing capacity of SSE actors, especially in cooperation with actors who share their organizational principles and values. However, even if these transformation processes have largely taken place outside the public authorities, their role should not be underestimated. Indeed, the actors in the transforming network sometimes rely on public actors occupying peripheral positions, generating a symbiotic relationship based on mutual respect and recognition, which is very useful for the performance of the specific functions of both. Secondly, these processes are an example of the transformative capacity associated with individual and collective actors (e.g., associations) that are acting as entities that have a great deal of autonomy from the public authorities, especially the mainstream political parties. This contrasts with the traditional strong linkage of individual and collective actors in the conservative network with the public authorities, and in particular the political parties. Third, and connected with the contrast we have just seen, the possible demonstration effect of some minority actors of the transforming network on some majority actors of the conservative network should not be underestimated. Consequently, the transforming network's status as a lever of territorial change should make it the focus of any public policy to stimulate democracy, the fight against corruption and the commitment to an endogenous development model fully inserted in the international economy. In this sense, after we conducted the bulk of the interviews that served as the basis for this research, there has been a process of growing autonomy or independence by one of the large associations of sugarcane producers with respect to the major political parties. Trying to deepen the connections that may exist between the latter process and the development and consolidation of a transformative network appears as an interesting field of research for the future that can help to understand the complex dynamic interrelationship between the formal entities of the SSE (agricultural associations and cooperatives, etc.) and other entities that share its organizational principles and values.

Even though we believe that the article opens up some interesting lines of research, to make them effective, it is necessary to address some of the limitations of the article. Firstly, the article has focused on the study of the Veracruz sugarcane cluster (Mexico). No doubt studying the cases of other regions (developing and developed) and comparing them with each other would help to refine the theoretical framework and to fine-tune the proposed system of indicators. Secondly, the methodology used in the empirical analysis is essentially qualitative. Now, it would be necessary to develop some quantitative tools to try to measure the different defining categories of proximity between actors and the diffusion of the meso-rules identified in the article. This could make an important contribution to the development of an operational methodology for the measurement of the SSE that integrates the various methods that revolve strictly around the entities of the SSE and the qualitative method developed here to capture the ability of the SSE to drive a process of broader (territorial) transformation.

Author Contributions: Conceptualization, J.R.G.-B.; methodology, M.R.T.-B. and J.R.G.-B.; validation, M.R.T.-B. and J.R.G.-B.; formal analysis, J.R.G.-B. and M.R.T.-B.; investigation, J.R.G.-B. and M.R.T.-B.; resources, J.R.G.-B. and M.R.T.-B.; data curation, M.R.T.-B.; writing—original draft preparation and writing—review and edition, J.R.G.-B. and M.R.T.-B.; funding acquisition, J.R.G.-B. and M.R.T.-B. All authors have read and agreed to the published version of the manuscript.

Funding: This work was supported by the Spanish Ministry of Science, Innovation, and Universities (GOBEFTER II Project, CSO2016-78169-R) and by the Mexican National Council of Science and Technology (CONACYT) (grant number CVU 442382).

Acknowledgments: This article is a revised version of the communication presented in the 8th CIRIEC International Research Conference on Social Economy "Solidarity and Social Economy and The Agenda 2030: Inclusive and Sustainable Development through Innovative Social Practices", San José (Costa Rica), 8–10 September 2021. The authors appreciate the suggestions of the participants, especially Roberto Cañedo Villarreal. The authors also appreciate the interesting and detailed suggestions of three anonymous reviewers who have helped us many to improve the quality of the article, although only the authors are responsible for the final work.

Conflicts of Interest: The authors declare that they have no conflict of interest concerning the research reported in this manuscript.

References

1. Defourny, J.; Monzon, J.L. Économie Sociale. Entre Economie Capitaliste et Economie Publique/The Third Sector. In *Cooperative, Mutual and Nonprofit Organizations*; De Boeck Université—CIRIEC: Brussels, Belgium, 1992.
2. Bouchard, M.-J. Methods and Indicators for Evaluating the Social Economy. In *The Worth of the Social Economy: An International Perspective*; Bouchard, M.-J., Ed.; Peter Lang: Brussels, Belgium, 2009; pp. 19–34.
3. Bouchard, M.-J.; Rousselière, D. *The Weight of the Social Economy: An International Perspective*; International Centre of Research and Information on the Public, Social and Cooperative Economy: Liège, Belgium, 2015; p. 333.
4. Ebrahim, A.S.; Rangan, V.K. What Impact? A Framework for Measuring the Scale and Scope of Social Performace. *Calif. Manag. Rev.* **2014**, *56*, 118–141. [CrossRef]
5. Saïd, I.; Ladd, P.; Yi, I. *Measuring the Scale and Impact of Social and Solidarity Economy*; UNRISD: Geneva, Switzerland, 2018; pp. 1–6. Available online: https://www.unrisd.org/80256B3C005BCCF9/(httpAuxPages)/B35B595F32BF48C6C12582F900363EC4/$file/IB9%20-%20Measurement-SSE.pdf (accessed on 25 January 2021).
6. Barea, J.; Monzon, J.L. *Manual for Drawing Up Cooperative and Mutual Society Satellite Accounts*; European Commission, D.G. for Enterprise and Industry: Brussels, Belgium, 2008; p. 194.
7. Salathé-Beaulieu, G.; Bouchard, M.J.; Mendell, M. *Sustainable Development Impact Indicators for Social and Solidarity Economy. State of the Art*; UNRISD (United Nations Research Institute for Social Development): Geneva, Switzerland, 2019; pp. 1–39. Available online: https://www.unrisd.org/80256B3C005BCCF9/(httpPublications)/5A72131687B4AE478025848D003847E4?OpenDocument (accessed on 5 October 2020).
8. Chaves-Ávila, R. Crisis del Covid-19: Impacto y respuestas de la economía social. *Not. de la Econ. Pública Soc. Y Coop.* **2020**, *63*, 28–43.
9. Bretos, I.; Bouchard, M.J.; Zevi, A. Institutional and Organizational Trajectories in Social Economy Enterprises: Resilience, Transformation and Regeneration. *Ann. Public Coop. Econ.* **2020**, *91*, 351–358. [CrossRef]
10. Develtere, P. Cooperative Movements in the Developing Countries: Old and New Orientations. *Ann. Public Coop. Econ.* **1993**, *64*, 179–208. [CrossRef]
11. Tomás Carpi, J.A. The prospects for a Social Economy in a changing world. *Ann. Public Cooperative Econ.* **1997**, *68*, 247–279. [CrossRef]
12. Defourny, J.; Develtere, P. The Social Economy: The Worldwide Making of a Third Sector. In *L'économie Sociale au Nord et au Sud*; De Boeck Université: Paris, France, 1999; pp. 1–35. Available online: https://www.researchgate.net/publication/240335888 (accessed on 2 February 2021).
13. Gallego-Bono, J.R.; Tapia-Baranda, M.R. Industrial ecology and sustainable change: Inertia and transformation in Mexican agro-industrial sugarcane clusters. *Eur. Plan. Stud.* **2021**, 1–21. [CrossRef]
14. Berdegué, J.A.; Bebbington, A.; Escobal, J. Conceptualizing Spatial Diversity in Latin American Rural Development: Structures, Institutions and Coalitions. *World Dev.* **2015**, *73*, 1–10. [CrossRef]
15. Dopfer, K.; Foster, J.; Potts, J. Micro-meso-macro. *J. Evol. Econ.* **2004**, *14*, 263–279. [CrossRef]
16. Dopfer, K.; Potts, J. *The General Theory of Economic Evolution*; Routledge: London, UK, 2008; p. 134.
17. Richez-Battesti, N.; Trouvé, H.; Rousseau, F.; Eme, B.; Fraisse, L. Evaluation the Social and Solidarity-Based Economy in France. In *The Worth of the Social Economy: An International Perspective*; Bouchard, M.-J., Ed.; Peter Lang: Brussels, Belgium, 2009; pp. 87–110.
18. Porter, M.E. Clusters and the new economics of competition. *Harv. Bus. Rev.* **1998**, *76*, 77–90.
19. Colletis, G.; Pecqueur, B. Intégration des espaces et quasi-intégration des firmes: Vers de nouvelles rencontres productives? *Rev. Econ. Reg. Urb.* **1993**, *3*, 489–508.
20. Torre, A.; Rallet, A. Proximity and localization. *Reg. Stud.* **2005**, *39*, 47–59. [CrossRef]
21. Torre, A. Territorial development and proximity relationships. In *Handbook of Regional and Development Theories*; Edward Elgar Publishers: Cheltenham, UK, 2019; p. 674.
22. Gallego-Bono, J.R.; Nácher-Escriche, J.M. *Elementos Básicos de Economía*; Un enfoque Institucional, Tirant lo Blanch: Valencia, Spain, 2001; p. 525.
23. Spear, R. Social Accounting and Social Audit in the UK. In *The Worth of the Social Economy: An International Perspective*; Bouchard, M.-J., Ed.; Peter Lang: Brussels, Belgium, 2009; pp. 133–150.
24. Coraggio, J.L. Institutionalising the social and solidarity economy in Latin America. In *Social and Solidarity Economy: Beyond the Fringe*; Utting, P., Ed.; Zed Books Ltd.: London, UK, 2015; pp. 130–149.
25. Metcalfe, J.S. Knowledge of growth and the growth of knowledge. *J. Evol. Econ.* **2002**, *12*, 3–15. [CrossRef]
26. Potts, J. Complexity, economics, and innovation policy: How two kinds of science lead to two kinds of economics and two kinds of policy. *Complex. Gov. Netw. Spec. Issue: Complex. Innov. Policy* **2017**, *3*, 22–34. [CrossRef]
27. Nelson, R.R. A perspective on the evolution of evolutionary economics. *Ind. Corp. Chang.* **2020**, *29*, 1101–1118. [CrossRef]
28. Hassink, R.; Isaksen, A.; Trippl, M. Towards a comprehensive understanding of new regional industrial path development. *Reg. Stud.* **2019**, *53*, 1636–1645. [CrossRef]
29. Nelson, R.R.; Winter, S.G. *An Evolutionary Theory of Economic Change*; Harvard University Press: Cambridge, MA, USA, 1982; p. 437.
30. Nooteboom, B. *A Cognitive Theory of the Firm. Learning, Governance and Dynamic Capabilities*; Edward Elgar: Cheltenham, UK, 2009; p. 280.

31. Castells, M.; Banet-Weiser, S.; Hlebik, S.; Kallis, G.; Pink, S.; Seale, K.; Servor, L.J.; Swartz, L.; Varvarousis, A. *Another Economy Is Possible*; Polity Press: Cambridge, UK, 2017; p. 224.
32. Tomás Carpi, J.A.; Sánchez Andrés, A. *Conflictos de Mercado y de Estado en la Política Económica*; Tirant lo Blanch: Valencia, Spain, 2017; p. 160.
33. Hayek, F.A. *The Fatal Conceit: The Errors of Socialism*; University of Chicago Press: Chicago, IL, USA, 1988; p. 194.
34. Dopfer, K. The origins of meso economics. *J. Evol. Econ.* **2012**, *22*, 133–160. [CrossRef]
35. Magrini, M.B.; Martin, G.; Magne, M.A.; Duru, M.; Couix, N.; Hazard, L.; Plumecocq, G. Agroecological transition from farms to Territorialised agri-food systems: Issues and Drivers. In *Agroecological Transitions: From Theory to Practice in Local Participatory Design*; Berget, J.-E., Audouin, E., Therond, O., Eds.; Springer: Cham, Switzerland, 2019; pp. 69–98. [CrossRef]
36. Gallego-Bono, J.R.; Tapia-Baranda, M.R. Los valores de la Economía Social como impulsores del cambio en clústeres con fuerte fragmentación del conocimiento: El caso de la caña de azúcar de Veracruz (México). *CIRIEC España Rev. Econ. Pública Soc. Coop.* **2019**, *97*, 75–109. [CrossRef]
37. Chataway, J.; Hanlin, R.; Kaplinsky, R. Inclusive innovation: An architecture for policy development. *Innov. Dev.* **2014**, *4*, 33–54. [CrossRef]
38. Millstone, C. Can Social and Solidarity Economy organisations complement or replace publicly traded companies. In *Social and Solidarity Economy: Beyond the Fringe*; Utting, P., Ed.; Zed Books: London, UK, 2015; pp. 86–99.
39. Boons, F.; Howard-Grenville, J.A. *The Social Embeddedness of Industrial Ecology*; Edward Elgar Publishing: Cheltenham, UK, 2009; p. 284.
40. Boschma, R.A. Proximity and innovation: A critical assessment. *Reg. Stud.* **2005**, *9*, 61–74. [CrossRef]
41. Velenturf, A.; Jensen, P.D. Promoting Industrial Symbiosis: Using the Concept of Proximity to Explore Social Network Development. *J. Ind. Ecol.* **2015**, *20*, 700–709. [CrossRef]
42. Niang, A.; Torre, A.; Bourdin, S. Territorial governance and actors' coordination in a local project of anaerobic digestion. A social network analysis. *Eur. Plan. Stud.* **2021**. [CrossRef]
43. Gallego-Bono, J.R.; Chaves-Ávila, R. El modelo cooperativo de sistemas agroalimentarios de innovación: El caso ANECOOP y el sistema citrícola valenciano. *ITEA Inf. Tec. Econ. Agrar.* **2015**, *111*, 366–383. [CrossRef]
44. Gallego-Bono, J.R.; Chaves-Avila, R. Innovation cooperative systems and structural change. An evolutionary analysis of ANECOOP and Mondragón cases. *J. Bus. Res.* **2016**, *69*, 4907–4911. [CrossRef]
45. Hirschman, A.O. *Exit, Voice, and Loyalty: Responses to Decline in Firms, Organizations and States*; Harvard University Press: Cambridge, MA, USA, 1970; p. 176.
46. Gallego-Bono, J.R.; Lamanthe, A. Relations de pouvoir et regulations extra-locales dans l'adaptation des systèmes agroalimentaires au contexte de mondialisation. Une étude de cas France/Espagne. *J. Food Agric. Environ.* **2009**, *90*, 185–213.
47. Mozas-Moral, A.; Bernal-Jurado, E. Desarrollo territorial y economía social. *CIRIEC España Rev. Econ. Pública Soc. Coop.* **2006**, *55*, 125–140.
48. Gadille, M.; Gallego-Bono, J.R. Rebuilding a Cluster While Protecting Knowledge within Low-Medium-Tech Supplier SMEs: A Spanish and French Comparison. *Sustainability* **2021**, *13*, 11313. [CrossRef]
49. Gallego-Bono, J.R.; Chaves-Avila, R. How to boost clusters and regional change through cooperative social innovation. *Econ. Res. -Ekon. Istraživanja* **2020**, *33*, 3108–3124. [CrossRef]
50. Rossi, A.; Coscarello, M.; Biolghini, D. (Re)Commoning Food and Food Systems. The Contribution of Social Innovation from Solidarity Economy. *Agriculture* **2021**, *11*, 548. [CrossRef]
51. Schot, J.; Steinmueller, W.E. Three frames for innovation policy: R&D, systems of innovation and transformative change. *Res Policy* **2018**, *47*, 1554–1567. [CrossRef]
52. GEV (Gobierno del Estado de Veracruz). *Programas Regionales Veracruzanos. Programa Región de las Montañas. 2013–2016*; Secretaría de Finanzas y Planeación: Xalapa, Mexico, 2013; p. 11. Available online: http://www.veracruz.gob.mx/wp-content/uploads/sites/2/2014/04/tf07-pr-montana.pdf (accessed on 2 February 2021).
53. Riquelme, R. Libre Comercio de América del Norte. 11 Datos Para Comprender el TLCAN, el Acuerdo Que Trump Quiere Negociar. *El Economista*, 27 August 2018. Available online: https://www.eleconomista.com.mx/internacionales/Que-es-el-Tratado-de-Libre-Comercio-de-America-del-Norte-20161123-0111.html (accessed on 4 July 2019).
54. Morales, R. Refresqueras, con intereses cruzados en disputa de azúcar. *El Economista*, 14 May 2017. Available online: https://www.eleconmista.com.mx/empresas/refresqueras-cn-intereses-cruzados-en-disputa-de-azucar-20170515.html (accessed on 4 July 2018).
55. Aguilar-Rivera, N. A Framework for the Analysis of Socioeconomic and Geographic Sugarcane Agro Industry Sustainability. *Socio-Econ. Plan. Sci.* **2019**, *66*, 149–160. [CrossRef]
56. Rogers, E.M. *Diffusion of Innovations*; Free Press: Nueva York, NY, USA, 1983; p. 453.
57. Lévesque, B. Le potential d'innovation et de transformation de l'économie social: Quelques élements de problématique. *Interações* **2008**, *9*, 191–216. [CrossRef]
58. Gugerell, C.; Sato, T.; Hvitsand, C.; Toriyama, D.; Suzuki, N.; Penker, M. Know the Farmer That Feeds You: A Cross-Country Analysis of Spatial-Relational Proximities and the Attractiveness of Community Supported Agriculture. *Agriculture* **2021**, *11*, 1006. [CrossRef]

agriculture

Article

Why Do Agricultural Cooperative Mergers Not Cross the Finishing Line?

Elena Meliá-Martí [1,*], Natalia Lajara-Camilleri [1], Ana Martínez-García [2] and Juan F. Juliá-Igual [1]

[1] Centro de Investigación en Gestión de Empresas, CEGEA, Universitat Politècnica de València, 46071 Valencia, Spain; nalade@cegea.upv.es (N.L.-C.); jfjulia@cegea.upv.es (J.F.J.-I.)

[2] ALIMER, Alimentos del Mediterráneo, 30815 Murcia, Spain; anamartinez527@hotmail.com

* Correspondence: emelia@cegea.upv.es

Abstract: Mergers have played a relevant role in the business development of many agri-food cooperatives and have led to the consolidation of large cooperative groups which are leaders in their respective business sectors. However, many of the merger processes undertaken fail: some are aborted at the negotiation stage, and others are not approved by members. These failures entail financial and social costs due to frustrated expectations and the time invested in the negotiation process. The objective of this paper is to establish the economic, socio-cultural, organisational and process management factors that underlie this outcome. A survey was conducted among the directors and administrators of a sample of Spanish agri-food cooperatives that had participated in merger processes which were aborted at the negotiation stage or were not approved by their members. Factor and discriminant analyses established the aspects which had the greatest impact on the failure of the merger processes. Far from being economic factors, these analyses reveal that defensive localisms, a lack of commitment to the merger on the part of members and directors, and communication failures were more significant.

Keywords: mergers; agri-food cooperatives; failure; integration; approval; negotiations

Citation: Meliá-Martí, E.; Lajara-Camilleri, N.; Martínez-García, A.; Juliá-Igual, J.F. Why Do Agricultural Cooperative Mergers Not Cross the Finishing Line? *Agriculture* **2021**, *11*, 1173. https://doi.org/10.3390/agriculture11111173

Academic Editors: Adoración Mozas Moral and Domingo Fernandez Ucles

Received: 22 October 2021
Accepted: 18 November 2021
Published: 20 November 2021

Publisher's Note: MDPI stays neutral with regard to jurisdictional claims in published maps and institutional affiliations.

1. Introduction

The agri-food sector is subject to many sources of uncertainty. Collaborative networks have emerged to mitigate them and ensure competitive advantages. They also aim to shorten social and physical distances between consumers and producers and reduce the number of intermediaries in the food supply chain [1,2]. This process of shortening complex agri-food supply chains is engendering new market relationships which are built around new forms of association and collaborative organizational structures [3], which can improve competitiveness [4]. These processes include diversification into new activities, increasing the value added to farm products, e.g., through an ecological or regional identity, and involving new forms of cost reduction [5].

Integration processes, including mergers, are one of the ways used by agri-food companies to reduce costs and become more competitive.

Mergers and integration processes are highly complex operations in which numerous interests are at stake. The outcome of these processes is significantly affected by the same factors that influence and shape agri-food collaborative networks: governance, behaviour (trust, transparency), performance assessment, intensity of collaboration, strategy and operations management, amongst others [5]. They are also shaped by various factors which stakeholders often find difficult to control or predict, thus hindering their success [6]. Much of the research into mergers and acquisitions has looked at strategic and financial aspects without addressing socio-cultural, organisational and managerial issues. Even so, there is a growing consensus among researchers that these variables are at least as important for value creation in an integration process as financial ones [7–10]. Cooperatives are mutual undertakings. This means the success or failure of their mergers is not just

49

influenced or determined by factors which might be considered standard in an integration process. This occurs because of the special relationship between cooperatives and their members [11], who are both its owners and users and may also play the role of proprietor, employee, consumer or supplier [12]. Moreover, cooperative mergers are a result of the application of one of the Principles of the International Cooperative Alliance: cooperation among cooperatives. This states that cooperatives serve their members most effectively and strengthen the cooperative movement by working together through local, national, regional and international structures. Indeed, the issues which may drive cooperative members to back a merger are not simply the traditional expectation of generating value for the shareholder, which is characteristic of limited liability companies. In these cases, other operational and social aspects take on greater importance. Thus, member concerns about a merger depend on the ability of the process to address questions such as: Will the services the cooperative delivers to its members get better after the merger? Will the importance and power of each member or group of members in the resulting cooperative be lessened? Will the merger lead to the relocation of the cooperative's facilities? Will the cooperative's job offering in the municipality or region in which it is located be affected?

Meanwhile, studies about the factors underlying the failure of merger processes tend to focus on ex-post analyses of the merger [6,13–15]. They examine the company's evolution during the following years to unpack the impact which various aspects, decisions or actions taken by the cooperative may have had on this unsatisfactory performance.

Nonetheless, a large number of merger processes fail to cross the finishing line. In some cases, they are shelved due to disagreements in the negotiation stage itself, while in others, General Meetings do not endorse them after being submitted for approval. Although rarely analysed, these processes may also be considered failures as they entail considerable costs. These include the time spent by directors and administrators, the financial outlay of the studies and experts hired, and what might well be even more serious, the opportunity cost of not carrying out the merger or the value of not choosing the best option.

1.1. Objectives and Structure

This study has sought to add to knowledge of these processes in their earliest stages as these have not been explored in the cooperative arena. The general purpose of this paper was to establish the cultural, organisational, social (related to both members and employees) and operational (tied to the implementation of the process) grounds which shape the termination or failure of some of the merger processes undertaken by cooperatives and to determine their degree of impact on the rejection of the merger. To this end, a survey was sent to directors and administrators of Spanish agri-food cooperatives which took part in a failed merger project in the period 2005–2015 in the aforementioned terms.

The study was conducted in Spain as this is one of the European Union countries with the greatest problem of fragmentation in its agri-food cooperatives [16–20]. Numerous merger processes between cooperatives had been instigated over the study period, which enabled the build-up of a sufficient population of aborted processes.

Firstly, the mergers that yielded this outcome were described (type of process undertaken, number of merging cooperatives, existence of collaborative relationships prior to the merger, whether they are in the same or a different subsector of activity, geographical proximity, etc.). Secondly, the extent of rejection or abandonment of the merger plan in each of the stages was determined by establishing in which ones the cooperative merger process was halted.

Thirdly, the factors or groups of factors which are triggers for this kind of outcome were pinpointed together with the ones that influenced or caused the merger process to be called off in one or other of the identified stages.

Finally, the study aimed to validate a hypothesis that has been much discussed in the literature on cooperative mergers, albeit not confirmed to date by other research, which is that "emotional factors, especially ones related to ties with the territory, colloquially known as localisms, are more responsible for the failure of many merger processes than financial ones" [21,22].

The study's main difficulty lay in finding a sufficient population of cooperatives that had been involved in a merger process which was not successfully completed, since unlike mergers approved by members, this type of outcome is not usually publicised.

This paper is organised as follows: first, the existing literature about qualitative factors shaping the outcome of corporate merger processes and specifically in cooperatives is reviewed. This is followed by the methodology used, the results derived and a discussion of them. Finally, the conclusions of the study are presented.

1.2. Theoretical Framework

Numerous studies and research projects have addressed the factors shaping the success or failure of merger processes using a variety of approaches and standpoints which basically fall into two main categories [23]:

(A) Quantitative or hard factor analysis. These are largely anchored in the analysis of tangible and therefore measurable factors such as economic and financial aspects [24–27]. Briefly, since they are not the subject of this paper, the most recurrent factors examined include (1) the initial size of the participating undertakings [25] and the difference in size between the merging firms [9,28] and the impact of size on merger outcomes; (2) pre-merger economic and financial situation [29,30]; (3) financial health of the acquired firm and how the deal was financed [31,32]; and (4) restructuring and adjustments made in merger processes [33,34].
(B) Qualitative or soft factor analysis. These papers look at intangible factors by examining socio-cultural, organisational and managerial variables [35–38].

Studies of mergers and acquisitions tend to be much more detailed in their analysis of hard factors, in many cases sidelining the impact of socio-cultural, organisational and managerial aspects even though these variables are crucial to the success or failure of the process. Hence, as Papadakis points out [9], studies encompassing the widest range of factors are consequently much needed in order to gain a more comprehensive understanding of the success or failure of these processes.

A review of the literature on soft factors identified and classified those which had the greatest impact on the outcome of a merger process (Table 1). They were grouped into two blocks: the first includes factors which may be considered common to corporate merger processes, insofar as they affect companies regardless of their legal form, while the second consists of factors more specific to cooperatives.

Table 1. Qualitative factors which are relevant in a merger.

Common Qualitative Factors	
Stakeholder management	BenDaniel and Rosenbloom, 1998; Child et al., 2001.
Commitment to the process/resistance to change	Child et al., 2001; Cho et al., 2017; Gomes et al., 2017.
Communication	BenDaniel and Rosenbloom, 1998; Denisi and Shin, 2005, in Stahl and Mendenhall, 2005; Daber, 2013; Davenport and Barrow, 2017.
Existence of a merger team	BenDaniel and Rosenbloom, 1998; Rodríguez-Sánchez et al., 2019.
Previous experience	Schuler and Jackson, 2001; Hopkins, 2008; Cuypers et al., 2017.
Leadership ability	Teerikangas, Véry and Pisano 2011; Daber, 2013; Rodríguez-Sánchez et al., 2019.
Merger speed	Budden et al., 2002; Stahl and Mendenhall, 2005; Oh and Johnston, 2020.
Cultural factors	Buono et al., 1985; BenDanieland Rosenbloom, 1998; Budden et al., 2002; Cartwright and McCarthy, 2005; Sarala et al., 2019.
Qualitative factors specific to cooperatives	
Human relations	Swanson, 1985; Reynolds, 1997; Senise, 2007; Saisset et al., 2017.
Local level characteristics and actors	Giagnocavo and Vargas-Vasserot, 2012

1.2.1. Common Qualitative Factors (Not Specific to Cooperatives) Which Have a Major Influence on the Success of a Merger

Qualitative factors that have an impact on the outcome of a merger include the following:

Handling the interests of the stakeholders in a merger (members of the board of directors (BD), managing directors, managers, members and employees): Some studies suggest that it is not uncommon for several of these groups, especially the members of the BD or managing directors, to put their own interests before those of the shareholders or members, often driven by the expectation of seeing their salaries increase, earning higher profits or improving promotion opportunities due to the larger size of the company resulting from the merger [39]. Being the managing director in a larger firm is also often associated with higher social status than a similar post in a smaller firm [35].

Commitment to the process, professional competence and resistance to change are undisputed determinants of merger outcomes. Regardless of the way in which the merger has been conducted, the chances of success are greatly diminished if human resources are not involved and committed to it as the engagement and motivation of the team are essential success factors [35,40–42].

Managing communication in the process and controlling information about it among employees are crucial. Meynerts-Stiller et al. [43] and Davenport and Barrow [44] argue that having a communication plan in place is a decisive factor for the success of the process. Hence, information should be clear and timely to prevent some employees or members from feeling left out. Furthermore, communication can lessen stress levels among employees and thus increase the chances of a successful merger [45]. Anticipating changes and problems is pivotal so that they can be controlled or forestalled [40].

A further factor for success in the view of BenDaniel and Rosenbloom [40] and Rodríguez-Sánchez et al. [46] is the existence of a merger team that handles the process. Setting up this type of a team builds confidence in the company's commitment to the process while providing assurance of the utmost thoroughness and professionalism, an assessment which spreads across different strata of the company and engenders feedback.

In relation to previous experience, Schuler and Jackson [47] maintain that companies with previous merger experience achieve better results, suggesting that there is a learning curve whereby firms that have dealt with previous mergers are more likely to deliver more successful integrations. Here, Cuypers et al. [48] argue that experience can lead to the creation of a capability that helps with another important M&A issue, that is, value capturing, plausibly through improved bargaining and negotiation skills. In addition, they highlight the importance of considering differential experience between both parties as the source of experience advantage. Conversely, Hopkins [49] asserts that too much experience can be as detrimental as no experience at all since it can lead to arrogance or overconfidence. However, Greenberg et al., in Stahl and Mendenhall [50], qualify the contribution this factor may make to success by pointing out that experience helps to some extent as each situation and agreement is unique and different.

Leadership and good management: Leadership is a crucial organisational requirement as it provides a mechanism for control, efficiency, commitment and understanding in a company and has been shown to be a key factor in a successful merger [6]. Teerikangas et al. define the "integration manager" as the project manager appointed by the company to coordinate all integration-related activities [51].

Integration speed and time horizon: The speed of the integration is another influential variable. Budden et al. contend that the faster the integration, the better the outcomes [36]. Decisions should be made promptly, and they suggest a timeframe of about 90 days as the period in which it is feasible to consolidate the process. Oh and Johnson look at post-merger integration duration and suggest that slower integration minimises conflicts between merger partners, enhances trust-building and reduces the disruption of existing resources and processes in both firms, which may benefit M&As [52]. Reynolds et al. stress that all sides need sufficient time to grasp the impact of the process and emphasise that excessive delays or continuous postponements often have negative consequences, thus increasing the likelihood of failure [53].

Cultural factors: Cartwright and McCarthy argue that corporate culture is the most important facet of organisational adjustment in a merger [54]. Buono et al. note that each organisation has a unique, individual corporate culture and that unexpected problems and conflicts often arise when two or more corporate cultures meet [55]. The prevailing view is that the greater the differences between them, the greater the conflicts and problems for the resulting company [40]. Hence, scholars such as Budden et al. recommend that these processes should take place between companies in the same industry, thereby leveraging any strategic similarities between them [36].

1.2.2. Specific Factors Influencing the Outcome of a Merger of Cooperatives

Numerous factors in a cooperative may have a positive or negative impact on members and other stakeholders when it comes to backing or rejecting a merger. Del Real Sánchez-Flor notes that in any cooperative integration process, it is essential to work in two areas in lockstep: firstly, on human relations, and secondly on technical aspects (legal, financial, labour, etc.) [56], which although in principle are the most important part of the process, in the cooperative industry are often overshadowed by personal issues.

Human relations (members, employees and managing directors): One distinctive element that sets cooperatives apart from other types of firms is the dual condition of their members as both proprietors and suppliers or clients [12]. Within this framework, authors such as Reynolds et al. point out that a major challenge in merging cooperatives is bringing together different cultures and standards for decision-making [53]. They suggest that the negotiation process is an initial test for cooperatives to overcome their differences in order to reach an agreement. They also advocate the appointment of an impartial, fair facilitator to assist in the process. In interpersonal dynamics, Vandeburg et al. specify trust, communication and commitment coupled with the role of the managing director and the rest of the cooperative's staff as key factors for the success of mergers [57]. These conclusions are endorsed by other papers such as the ones by Van Duren et al. [58] and Fulton et al. [59].

At the negotiation stage, it is critical to avoid digressions about who is contributing the most to the process and also to keep an open mind in shaping the new unified cooperative undertaking. It is incumbent upon the organisations involved to be prepared and able to make difficult concessions and even sacrifices, as many negotiations break down when an issue stalls and cannot be resolved [53]. Others point out that managing and handling stakeholders who might oppose the merger transaction is particularly relevant [56,60] and who may be: (1) board members (fearing the loss of prestige and authority), (2) employees (who fear the loss of their employment status in their cooperative) and (3) members, either because of a lack of confidence that the process will yield appropriate results or fear that it may, in some way, alter their operations in the cooperative, especially in terms of the services provided to them, or sometimes because of mere antagonism to the board's management.

Negotiating the makeup of the first board of the ensuing cooperative is often a bone of contention, frequently related to the number of representatives of each cooperative and the choice of chair. When choosing the chair and the members of the board of the resulting cooperative, it is advisable to put a premium on competence and experience rather than the mere fact that they come from one cooperative or another, bearing in mind that in many cooperatives, there is still high demand for professional management and competent members to occupy positions in different bodies [61]. The same principles apply to the choice of managing director.

Local level characteristics and actors: Another recurrent problem and cause of failure in many cooperative merger processes is localism. This refers to the preference of members and staff for the cooperative to remain tied to the municipality, town, region and even the country in which it is domiciled and their refusal to see this situation change after the merger. Regional identity, related to agri-food products, is a key element in the configuration of agri-food networks, which has given rise to different designations of origin (protected designation of origin, protected geographical indication, etc.) [3]. This sense of identity, which goes beyond the fact that the cooperative has a differentiated product with one of these designations, can become a problem when it overrides economic criteria and the interests of the member, the cooperatives' main asset. Giagnocavo and Vargas-Vasserot cite overcoming resistance to change and localism as one of the main challenges facing cooperatives in Spain [18]. Montero [62] argues that the solution lies in having open-minded managing directors who are capable of overcoming individualist approaches, localism, grandstanding, politicisation and vested interests. Likewise, Swanson pinpoints other problematic strategic factors such as choosing the new undertaking's name, mapping out its general objectives and approach, unifying operations and deciding which facilities should be closed or sold after the merger and which should be kept running [60].

Opportunism, meaning the shrewd pursuit of self-interest [63]: This kind of behaviour is found in inter-organisational and internal relationships and has very harmful consequences. Wathne and Heide point to passive opportunistic behaviour as a reluctance to adapt [64]. As Sánchez-Navarro et al. noted, this behaviour can lead to long-term negative effects, including the abandonment of the project in the case of a merger process [63].

Equity valuation: Performance assessment is a key element in sustainable agri-food networks. Its accurate quantification of efficiency, effectiveness and relationships between firms according to social, environmental and economic perspectives is essential [65]. A common problem in merger negotiations comes from the financial valuation of the cooperatives' equity and setting the share capital to be recognised in the resultant cooperative for the members of the cooperatives involved. Wadsworth and Chesnick note that managing fairness in members' equity is perhaps the trickiest financial issue in a merger and advocate respecting and protecting members' contributions [66], while members should understand that their financial stake in the merged cooperative is significant. Likewise, Swanson points out that the evaluation of assets and liabilities and the transfer of capital during cooperative mergers are thorny issues which have to be handled fairly or in a mutually acceptable way by putting in place a plan that includes alternative methods and implications (impacts whether positive or negative) for combining the assets, liabilities and equity of the cooperatives involved [60]. Scholars suggest that the capital to be recognised for each member in the resulting cooperative should be calculated in real terms along with the monetary compensation (or merger fee) which the members of a cooperative may have to pay in order to offset equity imbalances. Similarly, García Sanz argues that the value of the cooperative as established in the negotiations between the boards of directors should be used to calculate the exchange ratio rather than the book value [67].

Strategy-related aspects: The study "Support for Farmer Cooperatives" points out that an inherent problem in many merger processes is that the merger is not the result of a purposeful, considered decision, and consequently, its strategy is not sufficiently discussed [61]. Indeed, Saisset et al., following an analysis of 14 wine cooperative mergers in Languedoc-Roussillon, found that some were done without any backup strategy, others responded to local policies, and finally, in other cases, mergers were based on real corporate strategic projects leading to synergies between the involved cooperatives [68]. Thus, in many cases, the decision to undertake a merger is often made in response to several years of losses, either due to poor management or periods of crisis. Similarly, Reynolds et al. contend that not all combinations of cooperatives are economically profitable and that at times, some members may be reluctant to merge with a much larger cooperative as they think that it might weaken control over the quality and type of services provided [53]. Meanwhile, Martínez Morillo-Velarde argues that cooperatives have traditionally been wary of gaining size, expanding and internationalising. In a nutshell, they are afraid of losing control [69].

Communication in the process: Duft and Zagelow [70] and Montero [62] stress the negative effect of failing to properly explain to members and directors the benefits of the merger as well as the features and operations of the resulting cooperative, very much in line with what has been noted above in the general factors. They point to disinformation as a generator of uncertainty and as a propitiator of rumours or fake information, while honesty, integrity and effective communication are the only factors proven to be useful in addressing the problems arising from a lack of information. Likewise, Del Real Sánchez-Flor contends that good management of internal information to members averts rumours, information distortion and unjustified misgivings while building a climate of trust and transparency which, in the long run, will help to achieve the expected goal and to sound out the views of members and anticipate any problems that may come up [56]. It is additionally essential to ensure the cohesion of the largest number of individuals on each board and draw up an agreed message to members and employees stressing the benefits of the integration process.

Team ability: A firm's staff, with their knowledge, skills, experience and motivation, is considered one of the resources with the greatest potential for generating sustainable competitive advantages [71,72]. However, as noted by a good part of the social economy literature, cooperative firms have difficulties in attracting and retaining executives who are both valuable and committed to cooperative values [72–75]. Accordingly, it is essential in a merger process to bring in professional management with the appropriate training, experience and skills as this is one of the most influential and decisive factors in the success of the cooperative resulting from the integration process [76]. The aim, therefore, is to eschew monistic management models, in which both the cooperative's democratic structure (distribution of decision-making power) and business structure (relations between staff and management) are controlled by the actual members, who may behave in an exploitative or opportunistic manner.

The above review of the literature is now followed by an empirical comparison of the main factors determining the failure of mergers of agri-food cooperatives in the equity valuation, negotiation and General Meeting approval stages.

2. Materials and Methods

2.1. Study Population

There are no databases or sources which include failed merger processes (called off at the negotiation or approval stage), which meant that the information had to be compiled by asking the federations (representative organisations) of agri-food co-operatives in Spain's 17 autonomous regions. Information was gathered from all of them bar three (Canary Islands, Navarre and the Basque Country).

The population identified after making enquiries to the cooperative federations was made up of 104 agri-food cooperatives which had been involved in 36 unsuccessful or aborted mergers between 1995 and 2015 (Table 2).

For the purposes of this research, a failed merger was defined as a merger that did not succeed due to one of the following reasons:

(a) During the initial analysis or assessment stage, the boards of directors or negotiating teams decided not to press ahead with the negotiation and did not submit it to the members for approval at the General Meeting.

(b) The merger plan was submitted to the General Meetings for approval and was turned down by the members.

There are 3669 agri-food cooperatives in Spain [77]. Therefore, the cooperatives which had taken part in a failed merger process accounted for around 2.7% of the total in the period under study. The difficulty of getting hold of this population should be stressed. Although merger processes have to be announced in newspapers and official journals once they have been approved by the General Meetings, information on mergers which were not completed is not available except through the cooperatives themselves and, in this case, their federations. This problem is compounded by the fact that cooperatives often have an overt interest in not publicising these processes because they have not been successfully concluded.

2.2. Sample Selection

The reliability of the conclusions drawn in a study depends to a large extent on the way in which the sample is selected [78]. In this case, the questionnaire was addressed to the 104 agri-food co-operatives identified by the cooperative federations as participants in 36 failed or aborted merger processes. The response rate was 42%, which means the analysed sample was made up of 44 cooperatives and was thus representative of the analysed population at 94% (Table 2).

Table 2. Survey factsheet.

Confidence Interval	94%
Margin of error for a sample proportion	6%
Population	104 (finite)
Sample size	44
Fieldwork	29/01/2015 to 01/03/2016
Type of interview	Electronic survey via email
Scope of action	National (Spain)

2.3. Survey

The data collection methodology used was a survey (see Supplementary Materials), which is one of the most commonly employed approaches in social science research to empirically study the characteristics and interrelationships of socio-economic and organisational variables [79,80].

The surveys were conducted via email and addressed to the managing directors and administrators of the cooperatives.

2.4. Conducting the Questionnaire

The first part of the questionnaire consisted of general information about the cooperatives, such as their type, their geographical location and the sector they operated in. The second part investigated the features of the merger process itself including the number of cooperatives involved, the year in which the process took place, the sector which the organisations involved were in and the type of merger proposed.

The third part looked at the reasons why the merger failed. A Likert scale of 1 to 5 was chosen for measurement purposes, ranging from 1 (low influence of the factor analysed on the failure of the merger) to 5 (high influence of the factor analysed on the failure of the merger). Although a Likert-type scale has limitations, such as the absence of a continuous relationship between values and the distances there may be between them, it was chosen because it enables the multidimensionality of the variables to be appropriately addressed [78].

The fourth and last part of the questionnaire summarised the three main reasons which, in the respondent's opinion, led to the failure of the merger in order of importance. It also included other information which may have influenced the outcome of the merger, such as whether the cooperatives involved were engaged in shared business activities prior to the merger. Finally, they were asked whether they considered that their cooperative would have been better positioned if the merger had taken place or whether they had subsequently thought about taking part in other integration processes.

3. Results and Discussion

3.1. Descriptive Analysis of the Results

An analysis of the data extracted from the survey showed that the majority were first-tier cooperatives (95%) and only 5% were second-tier cooperatives (Table 3).

When classified by the core business of the organisations, most of the cooperatives which were involved in failed mergers in the sample were in the olive oil and table olive (39%), fruit and vegetable (19%) and wine (14%) industries (Table 3).

Table 3. Type of cooperatives in the sample.

Type	
First tier	95%
Federated cooperative	5%
Main activity	
Olive oil and olives	39%
Wine	14%
Fruit and vegetables	19%
Sheep and goats	7%
Supplies	5%
Other or a combination of the above	16%

The second part of the survey included questions about the merger process itself (Table 4). In 75% of the cases, the merger processes were conducted by two cooperatives, while only 25% involved more than two cooperatives.

As might have been expected, in 93% of the cases, the cooperatives were in the same production industry, and only the remaining 7% were engaged in distinct operations (Table 4).

The proximity between organisations was another of the factors analysed in the survey. In this case, 43% of the cooperatives were in the same municipality and 39% in neighbouring or very nearby municipalities (Table 4).

In terms of the type of merger proposed, 55% of the cooperatives opted for merger by acquisition, while 45% opted for merger by setting up a new company.

Table 4. Merger process information.

Number of Cooperatives Involved in the Merger Process	% of the Sample
Two	75%
More than two	25%
Were the cooperatives in the same sector?	
Yes	93%
No	7%
Had there been any previous relationship between the organisations prior to the merger process?	
Yes	68%
No	32%
The cooperatives involved in the merger were located in the same municipality	
In the same municipality	43%
In neighbouring or nearby municipalities	39%
In different and not so close municipalities, albeit in the same province	11%
In different provinces, albeit in the same autonomous region	5%
In different autonomous regions	2%
Type of integration envisaged	
Merger by setting up a new company	45%
Merger by acquisition	55%
Stage of the process in which the merger was halted	
Initial analysis or assessment	27%
Negotiation (not approved by boards of directors)	25%
Approval by the General Meeting	48%

Another relevant issue was the stage at which the process was halted. Forty-eight percent of the processes failed in the last stage, i.e., approval by the General Meeting. The rest of the mergers were not submitted to the General Meetings, 27% of them ended in the initial analysis or assessment stage and 25% in the negotiation stage, i.e., after the initial assessment, the boards of directors started talks about the merger but were unable to come to a satisfactory agreement.

The third section of the survey explored the reasons why the merger did not succeed. The highest scores were given to members' lack of commitment, localisms, communication failures and the lack of support and trust of the boards. The factors with the lowest scores were potential staff cost overruns after the merger, lack of previous business relationships and opportunism by the managing director (Table 5).

Table 5. Responses on the assessment of the reasons why the merger between cooperatives did not succeed, ranked from most to least perceived importance.

	Mean	Median	SD	% of Respondents Rating 4 or 5
Lack of members' commitment to and motivation for the merger	3.32	3	1.46	46%
Entirely due to localism	3.22	3	1.65	51%
Failure of communication and information management for members, managing directors and employees	3.21	3	1.39	44%
On the board of directors				
Lack of support for and confidence in the merger process	3.09	3	1.52	45%
Lack of commitment to and motivation for the merger	3	3	1.53	42%
Differences between corporate cultures (ways of interacting with members, working methods, etc.)	2.88	3	1.53	37%
The chair's lack of commitment to and motivation for the merger	2,84	2	1,63	37%
Members' fear of restructuring or relocation of assets	2.84	3	1.58	40%
Lack of leadership from chairs and directors to deal with the process	2.81	3	1.38	30%
Employees' lack of commitment to and motivation for the merger	2.76	3	1.36	28%
Lack of confidence in the viability of the common business project of the resulting cooperative	2.7	2	1.61	37%
Members putting personal interests before the cooperative's	2.67	2	1.43	30%
Disagreement on the equity valuation	2.6	2	1.68	35%
Lack of chairs' and directors' previous experience in dealing with the process	2.55	3	1.39	25%
The board putting personal interests before the cooperative's	2.52	2.5	1.42	34%
Lack of managerial leadership in dealing with the process	2.51	3	1.35	26%
Disagreement on the layout of the new workforce organisation chart	2.45	2	1.45	24%
Lack of training of chairs and directors to cope with the process	2.44	2	1.37	21%
The chair putting personal interests before the cooperative's	2.41	2	1.54	32%
Dissatisfaction with the makeup of the new board or its election system	2,41	2	1.37	23%
Staff or members' fear of workforce restructuring	2.40	2	1.4	24%
Managing director's lack of commitment to and motivation for the merger	2.38	2	1.36	19%
Member reluctance due to potential reduction in their payments after the merger	2.34	1	1.61	27%
Poor planning in process delivery	2.33	2	1.34	18%
Problems arising from slow decision-making in moving the process forward	2.33	2	1.39	21%
Employees putting personal interests before the cooperative's	2.28	2	1.14	12%
Underestimation of the costs of the merger process	2.25	1	1.59	30%
Lack of managing director's prior experience in dealing with the process	2.23	2	1.19	16%
Lack of support from government and/or financial institutions	2.16	2	1.45	21%
Member reluctance to make extra cash contributions to the merger so as to offset equity imbalances	2.09	1	1.43	20%
Dissatisfaction with the type of merger chosen	2.02	1	1.24	14%
Members' fear of losing services provided by the cooperative	2.02	1	1.32	21%
Lack of managerial training to cope with the process	2.02	2	1.16	10%
Lack of previous experience in mergers	2	2	1.44	24%
Managing director putting personal interests before the cooperative's	1.98	1.5	1.22	15%
Lack of pre-merger relations with the other cooperative	1.93	1	1.28	11%
Expected staff cost overruns due to the need to standardise salaries	1.84	1	1.15	11%

Consequently, four process management reasons were picked out with an average rating of more than 3, i.e., they were more to blame for the failure of the process. One was the members' lack of commitment to the merger with a rating of 3.32, and the other was what is called "localism" with an average score of 3.22. This is by no means surprising since many scholars point to this as one of the main causes of the failure of merger processes [18,62]. They all stress the importance of overcoming individualistic approaches and localism as a key issue in achieving greater business concentration in the cooperative industry.

Failures in communication and information management for members, employees and managing directors were the third problem highlighted, with 44% of respondents giving this factor a rating of 4 or 5. Again, these results match the findings of researchers such as BenDaniel and Rosenbloom [40] as well as Denisi and Shin in Stahl and Mendenhall [45], who advocate open, clear communication as a decisive factor in increasing a merger's chances of success.

Lack of support for and confidence in the merger process from boards of directors was the fourth most important reason with an average score of 3.09. As suggested by researchers such as Swanson [60] and Del Real Sánchez-Flor [56], this rejection may be driven by the fear of losing influence, prestige or authority in the cooperative resulting from the merger.

The reason with the lowest impact, with an average score of less than 2, was the lack of pre-merger relations, a response that is most likely due to the finding in Table 4 which revealed that most of the organisations had already conducted joint operations prior to the merger.

Finally, another two key factors in the failure of the merger process were the lack of commitment to and motivation for the merger, with an average rating of 3.32 in the case of members and 3 in the case of the boards of directors. As noted by Van Duren et al. [58], BenDaniel and Rosenbloom [40], Fulton et al. [59] and Vandeburg et al. [57], in human relations, confidence and commitment are crucial to the success of this type of process. The results suggest that members' lack of commitment to and motivation for the merger in 46% of cases, and on the part of the board of directors in another 42%, contributed to the failure of the processes. The rest of the social factors (personal interests, managing directors' or chairs' lack of capacity and lack of previous experience) were not seen as relevant.

The fourth and last block of the survey (Table 6) summarises the main reasons, in order of importance, cited by the respondents as the principal grounds for the failure of the merger so as to validate or add to the previous ones. The three most recurrent reasons are: (1) members' reluctance due to sheer localism (15%), (2) lack of support for and confidence in the merger process from boards of directors (12%) and (3) lack of confidence in the viability of the resulting cooperative's common business plan (11%). In this case, in addition to the localisms and lack of confidence in the merger process already identified in the previous section, the third most important factor was the lack of confidence in the financial viability of the plan.

Furthermore, it was also found that after the failed merger attempt, the cooperatives involved continued to keep up shared business activities in 69% of cases, where the most significant relationship was being part of the same second-tier cooperative (17%) or having members in common (12%) (Table 6).

The respondents' views as to whether they think that their cooperative would be in a better business situation if the merger had taken place are striking. The answer is overwhelmingly yes at 82.5%. Given that the main reasons pinpointed as accounting for the failure of the merger were non-financial (lack of confidence in the process, localism, communication and information management failures and lack of commitment to the merger by managing directors and board members), it is somewhat surprising that once the process had been called off and with the benefit of hindsight, they thought that their cooperative would have been better off if the merger had gone ahead. This suggests that the approach was almost certainly flawed in these processes. There was a failure to fully explore the real strategy and purpose of the merger, to explain and grasp the advantages and synergies it might bring in business terms with respect to cutting costs, expanding markets and so on, and fundamentally, it seems that communicating this strategy to members and directors failed, leading to their lack of motivation and commitment to the process. By contrast, other aspects which had nothing to do with the company's finances, such as localism, played a key role and contributed to the demise of the merger.

It is paradoxical that the very people involved in a merger who, during the process itself, show and convey little conviction and commitment to the process or prioritise aspects related to localism over business considerations are the very ones who after the process and over the years acknowledge their mistake and state that they should have made a greater effort to salvage the merger.

Finally, the question was also raised as to whether other merger processes with other cooperatives had been started up or considered after the failure. In this case, 53% of the cooperatives said yes and 47% said no, which adds to the paradox described above as it corroborates the need for integration due to economic circumstances, and the fact that the latter were outweighed by emotional reasons when the processes were undertaken.

Table 6. Summary and additional information on the process.

Please State in your View and in Order of Importance the Three Main Reasons for the Failure of the Merger out of the Following	% of Total Respondents
Reluctance of the members, entirely due to localism	15%
Lack of support and confidence in the merger process from the boards of directors	12%
Lack of confidence in the viability of the common business plan of the resulting cooperative	11%
After the failure of the merger, is there any kind of business activity between the cooperatives involved?	
Yes	82.5%
No	17.5%
Do you think that your cooperative would be in a better business position if the merger had been completed?	
Yes	82.5%
No	17.5%
Have you started or considered other merger processes with other cooperatives?	
Yes	53%
No	47%

3.2. Principal Component Analysis

Due to the large number of items, and drawing on the academic background in the literature, principal component analysis with Varimax rotation (KMO = 0.555; Barlett = 0.000; variance explained = 64.936) was performed in order to reduce the number of variables analysed (reasons why the merger did not succeed) by grouping them into factors with a more overarching significance. The number of factors was set at five on the basis of the scree plot. The results derived (Table 7) confirm the existence of five blocks or factors which pulled together qualitative information on the aspects impinging on the failure of the merger process.

The factors identified subsequently established whether they emerged in or had a different impact on the stage at which the merger process was called off or terminated.

The team ability block covers the assessment of the training, experience and leadership capacity of the managing director and the board of directors. The second component extracted—involvement of the board and the managing director—mainly includes aspects such as management and the board's commitment to and support for the project. The third factor contains issues related to communication failures and the involvement of members and employees: their commitment, fear of workforce restructuring and lack of confidence in the business plan. The fourth factor is made up of localisms and misgivings about the organisation chart and includes fear of the loss of services provided in the cooperative and of relocation. The last block reflects aspects related to the merger proposal, such as disagreement with the equity valuation and dissatisfaction with the type of merger together with disparities between corporate cultures (Table 5).

Table 7. Components derived from VARIMAX rotation in PCA.

Variables: Reasons Why the Merger Did Not Succeed	Identified Factors				
	F1 Team ability	F2 BD and managing director involvement	F3 Communication; non-involvement/opportunism of employees and members	F4 Localism, organisation chart and fear of change	F5 Merger plan/contract
Lack of capacity of the managing director in terms of training	**0.751**	−0.059	0.198	0.042	0.034
Lack of capacity of the managing director in terms of experience	**0.893**	0.001	0.208	0.181	−0.062
Lack of capacity of the managing director in terms of leadership	**0.745**	0.229	0.133	0.216	−0.339
Lack of capacity of the board of directors in terms of training	**0.715**	0.275	−0.241	0.227	0.183
Lack of capacity of the board of directors in terms of experience	**0.779**	0.241	−0.212	0.115	0.286
Lack of capacity of the board of directors in terms of leadership	**0.764**	0.389	−0.353	−0.060	0.040
Lack of support from the board of directors	0.192	**0.624**	−0.098	−0.335	−0.106
Lack of commitment and motivation of the board of directors	0.170	**0.809**	−0.064	−0.153	0.189
Lack of commitment and motivation of the chair	−0.046	**0.705**	−0.171	−0.135	0.334
Lack of commitment and motivation of the managing director	−0.150	**0.768**	−0.139	0.165	0.333
Chair putting personal interests first	0.173	**0.570**	0.043	−0.105	0.058
Managing director putting personal interests first	0.069	**0.797**	0.161	0.308	−0.067
Board of directors putting personal interests first	0.427	**0.701**	0.157	−0.111	−0.097
Failures in communication and information management	0.036	−0.250	**0.464**	0.380	−0.305
Lack of confidence in the business plan	0.046	−0.171	**−0.532**	0.234	0.323
Fear of workforce restructuring	−0.020	0.070	**0.570**	0.347	0.346
Members' lack of commitment	−0.037	−0.115	**0.889**	0.095	0.022
Employees' lack of commitment	−0.146	0.107	**0.822**	0.259	0.167
Members putting personal interests first	0.418	−0.249	**0.606**	−0.041	−0.230
Employees putting personal interests first	0.384	−0.120	**0.525**	0.297	0.242
Members' fear of having their payments diminished	−0.002	−0.331	0.020	**0.427**	−0.223
Members' fear of relocation	0.202	−0.384	0.169	**0.658**	−0.075
Members' fear of loss of services	0.214	−0.071	0.040	**0.655**	0.061
Due to localisms	−0.121	0.021	0.206	**0.724**	−0.281
Disagreement on the new workforce structure	0.103	0.193	0.446	**0.658**	0.278
Expected staff cost overruns to integrate workforces	0.344	0.036	−0.059	**0.772**	0.211
Divergences in corporate cultures	0.126	0.057	0.339	0.011	**0.575**
Disagreement with the type of merger chosen	−0.048	0.197	0.053	0.052	**0.538**
Disagreement on the valuation of cooperative equity	0.056	0.086	−0.133	−0.119	**0.802**

3.3. Discriminant Analysis

Discriminant analysis is a multivariate technique for explaining (or predicting) an individual's membership of an established category on the basis of observable characteristics or variables. In the case of this study, discriminant analysis helped to surmise at which point a merger process could become unsuccessful depending on the features of the cooperatives or the way in which the process was being approached by the participating undertakings.

The dependent variable in the analysis is the stage at which the failure of the merger is recorded, turned into a dichotomous variable (1: planning, which includes both the analysis and negotiation stages; 2: general meeting), while the following were included as independent variables: location, number of cooperatives taking part in the process, whether or not they were in the same sector, the existence of a previous business relationship, geographical proximity, type of merger, previous merger experience, whether they had already marketed their products together, and the five factors defined above: F1 (team), F2 (involvement of the BD and managing director), F3 (involvement of members and employees), F4 (localisms and fear of change) and F5 (plan proposal). After applying the step-by-step analysis, only previous merger experience, the existing business relationship and the F2 (involvement of the BD and managing director), F3 (involvement of members and employees) and F4 (localisms and fear of change) factors passed the test of equality of means, tolerance of variances and were significant in the classification.

A function was obtained that had a canonical correlation of 0.849 with a Wilks' lambda of 0.279 (0.000), yielding 100.00% correct classifications from the data (Table 8).

Table 8. Standardised coefficients of the derived discriminant function.

	Function 1
Existing business relationship	0.738
Previous merger experience	0.854
F2 (involvement of the BD and managing director)	0.736
F3 (involvement of members and employees)	−0.869
F4 (localisms and fear of change)	0.498

The centroids were 1.604 for the planning phase and −1.497 for the processes that were halted at the General Meeting.

Based on the coefficients and in view of the value of the centroids of the variables considered, it can be seen that experience in merger processes, knowledge of the organisations through previous business relations and factor 2 (involvement and commitment of the boards of directors and managing directors) coupled with factor 4 (aspects related to fear of change, localism and ones linked to workforce restructuring) were the factors that resulted in premature abandonment of merger processes. The first one, previous experience in merger processes, may be explained by overconfidence in one of the cooperatives involved which may have caused them to break off negotiations in the case of disagreement. The second, prior knowledge of the other organisations, may have had a constructive aspect due to greater propinquity; however, it also had a negative side in this study since if problems arose in the integration process, as the boards entrusted with the negotiation might have been reminded of previous adverse experiences, and this could have led them to scrap the process.

As for factors 2 and 4, it is striking that both included mostly non-financial components. In the case of the former, they were aspects related to the lesser involvement of managing directors and managers in the process and the fact that they put their personal interests before the interests of the cooperative itself which caused the process to be called off at an early stage. In the case of factor 4, it included members' and employees' fears of change, fear of relocation (due to potential transfer of some administration or production sites, loss of jobs or simple refusal to give up roots in the area, or localism) and fear of workforce restructuring (the new organisation chart).

When a merger process was not submitted to the General Meeting for approval, this means that members were denied the opportunity to examine, assess and discuss the benefits or drawbacks which this process could have had for the cooperative.

It should be stressed that the reasons why some processes were not even submitted for approval by the General Meeting, therefore depriving members of the option to make a choice, mostly concerned aspects which had nothing to do with the viability of the merger plan. Instead, they arose out of fears or mistrust of boards of directors and managing directors, putting the personal interests of all the groups in the cooperative before those of the merger itself, as well as sheer localism.

4. Conclusions

The key problem of agricultural cooperatives in Spain lies in their size. Integration processes have been encouraged in the last few decades, but some of them have not come to fruition. It is evident that, given the core role that members play in these organisations, qualitative factors are particularly relevant when it comes to explaining what leads to the failure of a merger process in cooperatives. Therefore, in line with several aspects such as mutual trust, commitment and taking risks towards a common goal, which are determinant factors in agri-food networks to achieve shared goals, are essential to integration processes [81,82].

For the first time, this study empirically shows that there are numerous aspects in Spanish agricultural cooperatives which influence the failed outcome of cooperative merger processes. It concludes that there are four influential factors: localism (of considerable or major importance in 48% of the cooperatives), the lack of support for and confidence in the merger process on the part of boards of directors (of considerable or major importance in the failure of 45% of the cooperatives surveyed), communication and information management failures (singled out as relevant and very relevant in 44% of the processes) and the lack of commitment and motivation before the merger on the part of members (identified in 46%) and boards of directors (in 42%).

Hence, localisms or regionalisms play a key role and often take precedence over well-founded financial grounds. They diminish the business vision of the cooperative by confining it to a small geographical area and thereby hinder its development and undermine its competitive capacity, which goes to confirm the hypothesis formulated in the study objectives.

Another shortcoming identified in aborted mergers is information and communication management, just as previous papers by Davenport and Barrow [44] pointed out. In a merger, the conflicting interests, and sometimes excessive zeal, of the parties mean that information is either not properly circulated or sends incorrect or incomplete messages, and this can generate opposition to the process on no other bases than ignorance, fear of embarking on new ventures and even the interests of groups which in no case are in a majority.

The third and last factor picked out as a cause of the failure of these processes is the lack of support and motivation for the merger from boards of directors and from members. This lack of support may stem from numerous sources yet should only really surface following painstaking scrutiny of the merger's consequences for the cooperative and its members. The results point to a number of factors which may influence this scant commitment. In 47% of the processes, respondents acknowledged that the lack of confidence in the viability of the joint business plan of the resulting cooperative was either fairly or very important, and in 40%, there were fears over the relocation of the cooperative's facilities. Likewise, this lack of support from members may be the upshot of the aforementioned poor information and communication management.

Principal component analysis was used to determine the extent to which the factors were involved in failed merger processes. Five factors were identified: team ability, board and management involvement, member commitment, social aspects such as localism and fear of loss of services by members, and the merger plan proposal. These blocks are consistent with proposals in previous studies such as the one by BenDaniel and Rosenbloom [40].

This work has analysed which variables may predict how likely a merger project will be aborted in its early stages. The issues which appear to be crucial in processes which are called off at an early stage, and therefore are not submitted for approval by members, are previous experience in merger processes and mutual knowledge through existing business relationships. The previous relationship and shared experiences (assumed to be off-putting) may mean the process is called off in the event of uncertainty, mistrust or disagreement without submitting it for discussion at the General Meeting. Thus, projects between cooperatives that already have business links and an integration team to guide the process would seem to be sensible recommendations.

The results suggest that the aspects triggering abandonment of merger projects in stages prior to approval by the General Meeting, and which deny members the chance to examine, evaluate and choose, are for the most part far removed from business strategy and more bound up with the personal interests of groups of people in the cooperative, which are called localisms, fear of change or simply rejection of new organisation charts. A cooperative's members and board of directors should either back a merger process or object to it on the basis of thorough economic, financial and social analysis which identifies how the cooperative's economic and financial position will be better or worse (cutting average costs, greater profitability, enhanced stability, achieving critical mass in production and/or marketing, etc.). It is these kinds of factors which ultimately shape and determine a member's return from the cooperative that should have the greatest bearing on whether the member is for or against the merger. However, and as noted above, a large proportion of processes are not even submitted for approval and, for the most part, on grounds unrelated to the foregoing points.

Nonetheless, one final aspect confirms these issues, which is that in 82.5% of the aborted processes and after a period of reflection, chairs and managing directors think that their cooperative would have been better off had the merger been successfully concluded. Several alternatives can be derived from this: (i) either the merger was not properly examined and this led to its rejection when actually it was a profitable plan; or (ii) the analysis was correct but the results of this analysis were not properly put across to members; or (iii) it is also possible that, having duly examined the merger and properly conveyed the findings to members, other aspects such as localisms or the fear of relocation of assets had a greater impact on their decision.

This paper contributes to increase existing knowledge and evidence about integration processes in cooperatives. It also provides valuable information about qualitative aspects that are relevant to the success of a merger, reinforcing the idea that cooperatives, like any other business, need to be managed by professionals. In this sense, training, education and guidance from representative organisations or government would help to ensure integration.

5. Study Limitations

The limitations of this study include, firstly, the fact that the survey respondents were 9% cooperative chairs, 37% managing directors and the remaining 54% other managers (not members of the board of directors). This spread unquestionably entails a bias towards managers and the results should therefore be interpreted in this light.

Another aspect to take into account is the potential endogeneity and thus the effect that unobserved or non-included factors (e.g., size) might have on the outcome of the merger process.

Furthermore, some of the cooperatives involved in the processes studied are no longer going concerns as they have been wound up, so it was impossible to gather their feedback about these processes. Another group of cooperatives declined to collaborate with this study. In some of these cases, the surveys were answered by the heads of the federations on behalf of these cooperatives as they were involved in the project at the time of the process.

Supplementary Materials: The following are available online at https://www.mdpi.com/article/10.3390/agriculture11111173/s1.

Author Contributions: Conceptualization, E.M.-M. and A.M.-G.; methodology, E.M.-M.; software, N.L.-C.; validation, E.M.-M., N.L.-C. and A.M.-G.; formal analysis, A.M.-G.; investigation, J.F.J.-I.; resources, J.F.J.-I.; data curation, N.L.-C.; writing—original draft preparation, J.F.J.-I.; writing—review and editing, E.M.-M. and N.L.-C.; visualization, J.F.J.-I.; supervision, J.F.J.-I.; project administration, E.M.-M.; All authors have read and agreed to the published version of the manuscript.

Funding: This research received no external funding.

Informed Consent Statement: Informed consent was obtained from all subjects involved in the study.

Acknowledgments: We would like to thank Cooperativas Agroalimentarias de España and its offices in Spain's autonomous regions for their wholehearted cooperation with this paper; without them, it would have been impossible to conduct this first empirical study as they identified the population of aborted mergers which enabled us to carry it out. We would also like to thank the cooperatives surveyed for their willingness to collaborate. We would like to thank the reviewers for their comments and suggestions, as they have definitely helped to improve the manuscript.

Conflicts of Interest: The authors declare no conflict of interest.

References

1. Dos-Santos, R.R.; Guarnieri, P. Social gains for artisanal agroindustrial producers induced by cooperation and collaboration in agri-food supply chain. *Soc. Responsib. J.* **2021**, *17*, 1131–1149. [CrossRef]
2. Ammirato, S.; Felicetti, A.M.; Ferrara, M.; Raso, C.; Violi, A. Collaborative Organization Models for Sustainable Development in the Agri-Food Sector. *Sustainability* **2021**, *13*, 2301. [CrossRef]
3. Renting, H.; Marsden, T.K.; Banks, J. Understanding alternative food networks: Exploring the role of short food supply chains in rural development. *Environ. Plan.* **2003**, *35*, 393–411. [CrossRef]
4. Allen, P.; FitzSimmons, M.; Goodman, M.; Warner, K. Shifting plates in the agrifood landscape: The tectonics of alternative agrifood initiatives in California. *J. Rural. Stud.* **2003**, *19*, 61–75. [CrossRef]
5. Marsden, T.; Banks, J.; Bristow, G. Food supply chain approaches: Exploring their role in rural development. *Sociol. Rural.* **2000**, *40*, 424–438. [CrossRef]
6. Melia-Marti, E.; Martinez-Garcia, A.M. Characterization and analysis of cooperative mergers and their results. *Ann. Public Coop. Econ.* **2015**, *86*, 479–504. [CrossRef]
7. Junni, P.; Sarala, R.; Tarba, S.; Weber, Y. The Role of Strategic Agility in Acquisitions. *Br. J. Manag.* **2015**, *26*, 596–616. [CrossRef]
8. Stahl, G.K.; Mendenhall, M.E.; Pablo, A.L.; Javidan, M. Sociocultural Integration in Mergers and Acquisitions. In *Mergers and Acquisitions: Managing Culture and Human Resources*; Stahl, G.K., Mendenhall, M.E., Eds.; Stanford Business Books: Stanford, CA, USA, 2005; pp. 3–17.
9. Papadakis, V.M. The role of broader context and the communication program in merger and acquisition implementation success. *Manag. Decis.* **2005**, *43*, 236–255. [CrossRef]
10. Epstein, M.J. The drivers of success in post-merger integration. *Organ. Dyn.* **2004**, *33*, 174–189. [CrossRef]
11. Romero, M.; Carmona, P.; Pozuelo, J. The prediction of the business failure of the Spanish cooperatives. Application of the Extreme Gradient Boosting Algorithm. *CIRIEC—España Rev. Econ. Pública Soc. Coop.* **2021**, *101*, 255–288. [CrossRef]
12. Grashuis, J.; Elliott, M. The role of capital capacity, spatial competition, and strategic orientation to mergers and acquisitions by U.S. farmer cooperatives. *J. Co-Oper. Organ. Manag.* **2018**, *6*, 78–85. [CrossRef]
13. Clark, K.; Ofek, E. Mergers as a Means of Restructuring Distressed Firms: An Empirical Investigation. *J. Financ. Quant. Anal.* **1994**, *29*, 541–565. [CrossRef]
14. Sharma, D.S.; Ho, J. The impact of acquisitions on operating performance: Some Australian evidence. *J. Bus. Financ. Account.* **2002**, *29*, 155–200. [CrossRef]
15. Kumar, R. Post-merger corporate performance: An Indian perspective. *Manag. Res. News* **2009**, *32*, 145–157. [CrossRef]
16. Server Izquierdo, R.; Meliá-Martí, E. Caracterización empresarial de los grupos y otras formas de integración cooperativa al amparo del nuevo marco legislativo. *REVESCO Rev. Estud. Coop.* **1999**, *69*, 199–215.
17. Arcas, N.; García, D.; Guzmán, I. Effect of size on performance of Spanish agricultural cooperatives. *Outlook Agric.* **2011**, *40*, 201–206. [CrossRef]
18. Giagnocavo, C.; Vargas-Vasserot, C. *Support for Farmers' Cooperatives. Country Report Spain*; Wageningen UR: Wageningen, The Netherlands, 2012.
19. Cos-Sánchez, P.; Escadíbul, B.; Colom, A. La diversificación geográfica de los destinos de exportación de las empresas y cooperativas agroalimentarias. Influencia de los factores externos para su selección. Ciriec-España. *Rev. Econ. Pública Soc. Coop.* **2021**, *102*, 161–195. [CrossRef]

20. Mozas-Moral, A.; Fernández-Uclés, D.; Medina-Viruel, M.J.; Bernal-Jurado, E. The role of the SDGs as enhancers of the performance of Spanish wine cooperatives. *Technol. Forecast. Soc. Change* **2021**, *173*, 121176. [CrossRef]
21. Montero García, A. *Cooperativismo Agrario de Segundo Grado*; IRYDA: Madrid, Spain, 1991.
22. Meliá-Martí, E.; Martínez-García, A. Aspectos sociales y culturales determinantes de los fracasos de fusiones de cooperativas. In Proceedings of the XVI Congreso de Investigadores en Economía Social y Cooperativa CIRIEC, Valencia, Spain, 19–21 October 2016.
23. Ribes Giner, G. Análisis de las variables organizativas que influyen en los procesos de fusión y adquisición de empresas. El caso de la industria auxiliar el automóvil. *Acad. Manag. Rev.* **2009**, *11*, 145–163.
24. Rao-Nicholson, R.; Salaber, J.; Cao, T.H. Long-term performance of mergers and acquisitions in ASEAN countries. *Res. Int. Bus. Financ.* **2016**, *36*, 373–387. [CrossRef]
25. Zhang, W.; Wang, K.; Li, L.; Chen, Y.; Wang, X. The impact of firms' mergers and acquisitions on their performance in emerging economies. *Technol. Forecast. Soc. Change* **2018**, *135*, 208–216. [CrossRef]
26. Agrawal, A.; Jaffe, J.F. The post-merger performance puzzle. In *Advances in Mergers and Acquisitions*; Emerald Group Publishing Limited: Bingley, UK, 2000.
27. Brouthers, K.; van Hastenburg, P.; van den Ven, J. If most mergers fail why are they so popular? *Long Range Plan.* **1998**, *31*, 347–353. [CrossRef]
28. Ismail, A.; Davidson, I. The Determinants of Target Returns in European Bank Mergers. *Serv. Ind. J.* **2007**, *27*, 617–634. [CrossRef]
29. Morck, R.; Schleifer, A.; Vishney, R. Do managerial objectives drive bad acquisitions? *J. Financ.* **1990**, *45*, 31–48. [CrossRef]
30. Ketelhöhn, N.; Martín, N. Determinantes de éxito en fusiones y adquisiciones. *INCAE Bus. Rev.* **2009**, *1*, 16–23.
31. Ogden, J.P.; Jen, F.C.; O'Connor, P.F. *Advanced Corporate Finance: Policies and Strategies*; Pearson Education Inc.: New Jersey, NJ, USA, 2003.
32. Goergen, M.; Renneboog, L. Shareholder Wealth Effects of European Domestic and Cross-border Takeover Bids. *Eur. Financ. Manag.* **2004**, *10*, 9–45. [CrossRef]
33. Harrison, D.T. Do Mergers Really Reduce Costs? Evidence from Hospitals. *Econ. Inq.* **2011**, *49*, 1054–1060. [CrossRef]
34. Healy, P.M.; Palepu, K.G. Effectiveness of accounting-based dividend covenants. *J. Account. Econ.* **1990**, *12*, 97–123. [CrossRef]
35. Child, J.; Faulkner, D.; Pitkethly, R. *The Management of International Acquisitions*; Oxford University Press: Oxford, UK, 2001.
36. Budden, G.; Slater, N.; Cooper, T. *Delivering the Merger Promise*; Telecommunications International: Geneva, Switzerland, 2002; Volume 36, No. 10; pp. 48–50.
37. Sarala, R.M.; Vaara, E.; Junni, P. Beyond merger syndrome and cultural differences: New avenues for research on the "human side" of global mergers and acquisitions (M&As). *J. World Bus.* **2019**, *54*, 307–321.
38. Bari, M.W.; Abrar, M.; Bashir, M.; Baig, S.A.; Fanchen, M. Soft issues during cross-border mergers and acquisitions and industry performance, China–Pakistan economic corridor based view. *SAGE Open* **2019**, *9*, 2158244019845180. [CrossRef]
39. Ratis, J.; Harikkala-Laihinen, R.; Hassett, M.; Nummela, N. *Socio-Cultural Integration in Mergers and Acquisitions*; PalgravePivot: London, UK, 2018.
40. BenDaniel, D.J.; RosenBloom, A.H. *International M&A, Joint Ventures, & Beyond*; John Wiley & Sons Inc.: New York, NY, USA, 1998.
41. Cho, I.; Park, H.; Dahlgaard-Park, S.M. The impacts of organisational justice and psychological resilience on employee commitment to change in an M&A context. *Total. Qual. Manag. Bus. Excell.* **2017**, *28*, 989–1002.
42. Gomes, E.; Mellahi, K.; Sahadev, S.; Harvey, A. Perceptions of justice and organisational commitment in international mergers and acquisitions. *Int. Mark. Rev.* **2017**, *34*, 582–605. [CrossRef]
43. Meynerts-Stiller, K.; Rohloff, C. Profile of an Integration Manager. In *Post-Merger Management*; Emerald Publishing Limited: Bingley, UK, 2019.
44. Davenport, J.; Barrow, S. *Employee Communication during Mergers and Acquisitions*; Routledge: Abingdon, UK, 2017.
45. Denisi, A.S.; Shin, S.J. Psychological Communication Interventions in Mergers and Acquisitions. In *Mergers and Acquisitions: Managing Culture and Human Resources*; Stahl, G.K., Mendenhall, M.E., Eds.; Stanford Business Books: Stanford, CA, USA, 2005; pp. 228–250.
46. Rodríguez-Sánchez, J.L.; Ortiz-de-Urbina-Criado, M.; Mora-Valentín, E.M. Thinking about people in mergers and acquisitions processes. *Int. J. Manpow.* **2019**, *40*, 643–657. [CrossRef]
47. Schuler, R.; Jackson, S. HR Issues and activities in mergers and acquisitions. *Eur. Manag. J.* **2001**, *19*, 239–253. [CrossRef]
48. Cuypers, I.R.; Cuypers, Y.; Martin, X. When the target may know better: Effects of experience and information asymmetries on value from mergers and acquisitions. *Strateg. Manag. J.* **2017**, *38*, 609–625. [CrossRef]
49. Hopkins, H.D. International Acquisitions: Strategic Considerations. *Int. Res. J. Financ. Econ.* **2008**, *15*, 261–270.
50. Greenberg, D.N.; Lane, H.W.; Bahde, K. Organizational Learning in Cross-Border Mergers and Acquisitions. In *Mergers and Acquisitions: Managing Culture and Human Resources*; Stahl, G.K., Mendenhall, M.E., Eds.; Stanford Business Books: Stanford, CA, USA, 2005; pp. 53–77.
51. Teerikangas, S.; Véry, P.; Pisano, V. Integration managers' value-capturing roles and acquisition performance. *Hum. Resour. Manag.* **2011**, *50*, 651–683. [CrossRef]
52. Oh, J.H.; Johnston, W.J. How post-merger integration duration affects merger outcomes. *J. Bus. Ind. Mark.* **2020**, *36*, 807–820. [CrossRef]

53. Reynolds, B.J.; Wadsworth, J.J.; Donald, A.F. *Cooperative Merger/Consolidation Negotiations. The Important Role of Facilitation*; Cooperative Information Report 52; United States Department of Agriculture, Rural Business/Cooperative Service: Washington, DC, USA, 1997.

54. Cartwright, S.; McCarthy, S. Developing a Framework for Cultural Due Diligence in Mergers and Acquisitions: Issues and Ideas. In *Mergers and Acquisitions: Managing Culture and Human Resources*; Stahl, G.K., Mendenhall, M.E., Eds.; Stanford Business Books: Stanford, CA, USA, 2005; pp. 253–268.

55. Buono, A.F.; Bowditch, J.L.; Lewis, J.W. When Cultures Collide: The Anatomy of a Merger. *Hum. Relat.* **1985**, *38*, 477–500. [CrossRef]

56. Del Real Sánchez-Flor, J.M. La dimensión como clave para la mejora de la competitividad. Cooperativas Agro-alimentarias Castilla La-Mancha. *Mediterráneo Económico* **2012**, *24*, 321–344.

57. Vandeburg, J.; Fulton, J.R.; Hine, S.; Mcnamara, K. Driving forces and success factors for mergers, acquisitions, joint ventures, and strategic alliances among local cooperatives. In Proceedings of the NCR 194 Annual Meeting, Las Vegas, NV, USA, 23–25 June 2000.

58. Van Duren, E.; Howard, W.; McKay, H. Forging Vertical Alliances. *Choices* **1995**, *10*, 31–33.

59. Fulton, J.R.; Popp, M.P.; Gray, C. Strategic Alliance and Joint Venture Agreements in Grain Marketing Cooperatives. *J. Coop.* **1996**, *11*, 1–14.

60. Swanson, B.L. *Merging Cooperatives: Planning, Negotiating, Implementing*; Research Report 43; Program Leader—Management and Operations Cooperative Management Division, Agricultural Cooperative Service: Washington, DC, USA, 1985.

61. Bijman, J. *Support for Farmers' Cooperatives. Country Report Spain*; Wageningen UR: Wageningen, The Netherlands, 2012.

62. Montero García, A. *El Cooperativismo Agroalimentario y Formas de Integración*; Ministerio de Agricultura, Pesca y Alimentación: Madrid, Spain, 1999.

63. Sánchez-Navarro, J.L.; Arcas-Lario, N.; Hernández-Espallardo, M. Antecedentes del oportunismo en las cooperativas agroalimentarias. *CIRIEC—España Rev. Econ. Pública Soc. Coop.* **2019**, *97*, 111–136. [CrossRef]

64. Watne, K.; Heide, B. Opportunism in Interfirm Relationships: Forms, Outcomes and Solutions. *J. Mark.* **2000**, *64*, 36–51. [CrossRef]

65. Maestrini, V.; Luzzini, D.; Caniato, F.; Ronchi, S. Effects of monitoring and in- centives on supplier performance: An agency theory perspective. *Int. J. Prod. Econ.* **2018**, *203*, 322–332. [CrossRef]

66. Wadsworth, J.; Chesnick, D. *Consolidation of Balance Sheet Components during Cooperative Mergers*; Research Report 139; USDA Rural Development: Washinton, DC, USA, 1995; p. 32.

67. García Sanz, D. *Concentración de Empresas Cooperativas. Aspectos Económico-Contables y Financieros*; CES Colección Documentación: Madrid, Spain, 2001.

68. Saisset, L.A.; Cheriet, F.; Couderc, J.P. Cognitive and partnership dimensions in merger processes in agricultural cooperatives: The case of winery cooperatives in Languedoc-Roussillon. *Int. J. Entrep. Small Bus.* **2017**, *32*, 181–207. [CrossRef]

69. Martínez Morillo-Velarde, L. La Internacionalización de las Cooperativas Agroalimentarias: Dificultades y perspectivas. Master's Thesis, Universidad de Córdoba, Córdoba, Spain, 2015.

70. Duft, K.D.; Zagelow, R.L. *Cooperative Director Liability Exposure: Issues and Resolutions*; Washington State University; Available online: http://www.agribusiness-mgmt.wsu.edu/ExtensionNewsletters/coop/CoopDirLiability.pdf (accessed on 15 October 2020).

71. Pffefer, J. *The Human Equation*; Harvard Business School Press: Boston, MA, USA, 1998.

72. Basterretxea, I.; Albizu, E. Management training as a source of perceived competitive advantage: The Mondragon Cooperative Group case. *Econ. Ind. Democr.* **2010**, *32*, 199–222. [CrossRef]

73. Morales, A.C. La dirección en la empresa de trabajo asociado: Una revisión de estudios empíricos. *CIRIEC—España Rev. Econ. Pública Soc. Coop.* **2004**, *48*, 99–122.

74. Münkner, H.H. Corporate governance in German co-operatives—What happenend to Co-op Dortmund? *Rev. Int. Co-Oper.* **2000**, *92–93*, 78–89.

75. Spear, R. Governance in Democratic Member Based Organisations. *Ann. Public Coop. Econ.* **2004**, *75*, 33–60. [CrossRef]

76. Calderón García, H.; Fayos Gardó, T.; Mir Piqueras, J. La internacionalización de las cooperativas agroalimentarias necesidad y problemática. *Mediterráneo Económico* **2012**, *24*, 61–76.

77. Cooperativas Agroalimentarias. Cooperativismo Agroalimentario en Cifras. 2020. Available online: http://www.agro-alimentarias.coop/cooperativismo_en_cifras (accessed on 10 July 2021).

78. Groves, R.M.; Fowler, F.J., Jr.; Couper, M.P.; Lepkowski, J.M.; Singer, E.; Tourangeau, R. *Survey Methodology*; John Wiley & Sons: Hoboken, NJ, USA, 2011; Volume 561.

79. Roberts, E.S. In defense of the survey method: An illustration from a study of user information satisfaction. *Account. Financ.* **1999**, *39*, 53–77. [CrossRef]

80. Dillman, D.A. *Mail and Internet Surveys: The Tailored Design Method*, 2nd ed.; John Wiley Co.: New York, NY, USA, 2000.

81. Camarinha-Matos, L.M.; Afsarmanesh, H. On reference models for collaborative networked organizations. *Int. J. Prod. Res.* **2008**, *46*, 2453–2469. [CrossRef]

82. Volpentesta, A.P.; Ammirato, S. Alternative agrifood networks in a regional area: A case study. *Int. J. Comput. Integr. Manuf.* **2013**, *26*, 55–66. [CrossRef]

 *agriculture*

Article

Trust in Collective Entrepreneurship in the Context of the Development of Rural Areas in Poland

Leszek Sieczko [1,*], Anna Justyna Parzonko [2] and Anna Sieczko [2]

[1] Department of Biometry, Institute of Agriculture, Warsaw University of Life Sciences-SGGW, 02-787 Warsaw, Poland

[2] Department of Tourism, Social Communication and Consulting, Institute of Economics and Finance, Warsaw University of Life Sciences-SGGW, 02-787 Warsaw, Poland; anna_parzonko@sggw.edu.pl (A.J.P.); anna_sieczko@sggw.edu.pl (A.S.)

[*] Correspondence: leszek_sieczko@sggw.edu.pl

Abstract: The aim of this research was to examine whether trust influences the functioning of various forms of collective entrepreneurship in rural areas. The study focused on organizations which are most common in rural Poland: agricultural producer organizations, rural women's circles, and local action groups. Hence, the survey sample included people engaged in these types of collective entrepreneurship. Data collection was based on a standardized questionnaire distributed online utilizing the computer-assisted web interviewing method. The statistical analysis of the empirical material obtained from 132 respondents involved Pearson and Spearman correlation and principal component analysis. The conducted research shows (1) the superior role of personal trust over institutional trust in the emergence and functioning of the studied forms of collective entrepreneurship in rural areas, (2) the greater importance of social than economic factors determining the functioning of rural collective entrepreneurship, (3) the positive impact of generalized trust on trust placed in the forms of entrepreneurship covered by the analysis, (4) the increase in trust over time of cooperation, and (5) the impact of trust on the functioning of collective entrepreneurship, in both the economic and the social dimensions, with a slight advantage of the latter. By focusing on trust, this article contributes to the literature on the role of trust in developing collective entrepreneurship in rural areas. The authors point out that this article only opens the space for a discussion on trust in the concept of the economics of trust.

Keywords: agricultural producer organizations; rural women's circles; local action groups

Citation: Sieczko, L.; Parzonko, A.J.; Sieczko, A. Trust in Collective Entrepreneurship in the Context of the Development of Rural Areas in Poland. *Agriculture* **2021**, *11*, 1151. https://doi.org/10.3390/agriculture11111151

Academic Editors: Adoración Mozas Moral and Domingo Fernandez Ucles

Received: 19 October 2021
Accepted: 11 November 2021
Published: 16 November 2021

1. Introduction

In Poland, rural areas have great social, environmental, and economic importance. This is mainly due to the fact that they cover the vast majority of the country's area and are inhabited by a high percentage of the population. The political transformation in Poland in 1989 revealed that, despite the fact that most of the farms were privately owned, farmers were not prepared for the changes that followed. One of the negative effects of the transformation was unemployment, practically unknown in Poland after World War II. It became obvious that unprofitable agricultural production had to be replaced by new off-farm jobs or additional on-farm activities. The structure of Polish agriculture is highly fragmented. Private sector farms dominate, including family farms, which account for slightly more than 99% of the total number of farms. As for the farm size, in 2019, more than half of the households were small farms with agricultural land not exceeding 5 ha (53.5%). The largest farms with an area of 50 ha of agricultural land accounted for only 2.4%. The average agricultural land per farm has remained at the level of approximately 10 ha for several years [1]. The agricultural policy implemented after the political transformation and after Poland's accession to the European Union has not led to an improvement in the structure of Polish agriculture. The productivity of the factors of production, especially labor, is unsatisfactory. It is visible mainly in the weak flow of land from farms that are not very productive or use land poorly to farms that are more effective in this respect. The agrarian structure of Polish agriculture remains flawed, especially when compared to other European Union countries. Currently, over 70% of economically active people living in rural areas do not derive any income from agriculture [2]. Micro-enterprises based on self-employment dominate in the employment structure. Thus, there is an increase in nonagricultural jobs in rural areas, but it is too low to meet the needs of the existing rural surplus of labor. For this reason, Polish rural areas and agriculture require continuous support for modernization and restructuring, as well as strengthening their competitive position in global markets.

The socioeconomic development of rural areas is a derivative of many factors, the importance of which cannot be overestimated. Some of them are exogenous, partly or totally independent of the activities of the local community, while others are endogenous. Research on the use of available resources (endogenous and exogenous) located in rural areas is necessary in order to stimulate positive attitudes toward entrepreneurial initiatives, including collective endeavors. In the literature, attitudes of this type are referred to as entrepreneurial attitudes [3–5].

The concept of entrepreneurship is often discussed in the literature. Entrepreneurship is a multidimensional phenomenon. All attempts to define the concept of "entrepreneurship" prove its interdisciplinarity. It has been a point of interest for researchers representing scientific disciples, such as psychology, sociology, economics, and management. Most often, entrepreneurship is associated with economic issues and a certain attitude involving looking for opportunities and achieving specific goals. In the literature on management and economics, entrepreneurship is most often treated as a process. In this case, the essence of entrepreneurship lies in initiating projects, as well as new forms of activity that would satisfy the needs and generate profits, and it would enable the reproduction and development of entrepreneurship. Entrepreneurship understood as a process involves activities undertaken in a specific space and time, with the use of specific resources by entities characterized by the features and skills desired in this process.

Entrepreneurship in the economic dimension involves the following [6,7]:

- creating new companies;
- increasing efficiency;
- creating new products, services, and techniques;
- revitalizing competition in the market.

Another understanding of entrepreneurship as a process assumes that it is the course of action involving setting up and running a business, including the following components: identifying opportunities and possibilities for operating on the market, developing a business plan and gathering the necessary resources, setting up an enterprise, and leading it through the subsequent phases of its development. Therefore, we can risk a statement that entrepreneurship is an organized activity, focused on the ability to generate and use innovative ideas in order to obtain measurable benefits, carried out under risk conditions [5,8–10].

When considering entrepreneurship in rural areas, one should not only focus on setting up and running a business [11]. Such an approach would considerably narrow the concept itself, as it also includes other activities contributing to the multidimensional development of rural areas, such as the activities of rural women's circles or local action groups. Then, entrepreneurship is associated with the attitude of a person toward the outside world and is expressed in a creative and active pursuit of new activities or expansion of the existing ones. Entrepreneurship can also be understood as a specific type of human activity, defined by behavior, the ability to use ideas, and opportunities that come unnoticed or are underestimated by others. Entrepreneurship, understood as a specific type of human endeavor, may take the form of an individual or collective activity. Collective entrepreneurship can be understood as organized, conscious, and voluntary cooperation of people aimed at achieving a common goal [12]. Its essence is the expression of the interests of the group and the representation of these interests in the external environment. It requires close working relationships between people, utilization of their talents and creative abilities, and mutual trust.

Trust can appear at different levels of social life. It may arise as a result of direct contacts and strong family and friendship ties in the traditional culture of the community, as well as in communities where secondary groups dominate, as a result of participation in the workplace environment, associations, and civic groups [13]. In the former case, the trust phenomenon derives from mutual relations: the more reciprocal the relations between individual members of the community or positive experiences in neighborly cooperation, the greater the credit of trust. In the latter case, trust is generalized and is based on a generally positive attitude toward others and public institutions. Trust in a situation of cooperation means that there is a widespread belief that others will participate in cooperation to a similar extent, that cooperation will be based on the principles of fair play, and that it is "profitable", i.e., brings benefits for all partners [14,15].

Collaborative networks are mutual relations that take place between individual members of a community, informal groups, and formal organizations. Through them, the community establishes strategic contacts, selects leaders, defines and agrees with local interests, identifies problems, and mobilizes material resources and residents to solve them [16]. People and organizations within such networks often gain two benefits: influence and power. They have greater opportunities to negotiate various types of transactions, because they have easier access to information, as well as greater control over its circulation and, thus better chances to use the emerging opportunities. Local leaders play a huge role in their use [14]. One of the programs facilitating the network cooperation in rural areas is the LEADER program, which contributes to the activation of rural communities by involving social and economic partners in the planning and implementation of local initiatives. It is a program implemented by local action groups (LAGs), which by linking representatives of three sectors, public, economic, and social, contribute to strengthening social capital and economic development of rural areas. Another manifestation of the cooperation network in rural areas, which has tradition dating back to the 19th century, is represented by rural women's circles. They represent the interests and work to improve the social and professional situation of rural women and their families, as well as support the comprehensive development of rural areas. Next, there are agricultural producer organizations, created to enable farmers to become partners for large buyers of agricultural products, as well as means of production. As an organization, farmers are able to meet the increasing requirements imposed by the market regarding the quality of agricultural and food products, but also share expenses on the purchase of equipment or organization of transport, negotiate contracts, trade together in the marketplace, or undertake marketing activities.

The abovementioned forms of collective entrepreneurship are part of the new direction of rural development by creating conditions for effective, efficient, and partnership-based implementation of territorially oriented development activities. The literature suggests that, in this area, it is important to support the so-called collective trust, so that the participants can share their experiences, be open to creative ideas, and find new directions for their enterprises through mutual help and support [17–20].

A manifestation of favorable changes in the perception of the impact of various factors on the economy, including the "soft" ones, is the development of a new trend in economic sciences—the economics of trust. Traditional (neoclassical) economics, despite being a social science, took into account only one type of human behavior, based on rational principle and driven by the maximization of utility [21,22]. The exclusion of social aspects from the analyses, which has been described in the literature as "desocialization" [23], was one of the levels of criticism of mainstream economics and the reason for a significant limitation of its ability to explain many phenomena. Evidence provided by behavioral economics [24], experimental economics [25], or game theory [26] shows that individuals not only assume the existence of noneconomic values, but also make choices guided by these values. One of such values is trust, which plays a crucial role in both social and economic life [27].

In business, trust is recognized as an important component of success [28,29]. The economics of trust is detectable at various levels of social and economic life, from saving money that would have to be spent on security, to improving the functioning of the political system. However, primarily, trust fosters business relationships. Employees who trust each other are more collaborative and more willing to share ideas and information, which facilitates innovation and ultimately increases productivity [30,31]. Lack of trust may stimulate hesitant attitudes toward teamwork or even antisocial attitudes when team members hold back good ideas or vital observations and do not share their experiences with each other [32]; they also have lower tolerance for organizational change [33]. Consequently, absence of trust leads to lower wages, profits, and employment, while its presence facilitates trade and encourages activities that add economic value [34]. In this sense, trust is like an interdependent network that connects all actors in an economy and determines how they interact to drive the economic growth [35,36].

Research in the field of the economics of trust is based on the thesis that trust is related to equality understood in two dimensions: economic equality and equal opportunities [37]. In countries with a high level of generalized trust, the level of economic equality and equal opportunities is high [38], in contrast to countries with a low level of generalized trust. It is also visible in the approach to entrepreneurship. Specifically, societies characterized by economic equality and equal opportunities, i.e., open, knowledge-based societies with a well-developed welfare state, have the best conditions for intrapreneurship and ultimately economic growth [39]. It is worth emphasizing that trust, both institutional and individual, tends to decline when socioeconomic conditions deteriorate [40,41].

Nature of the relationship between generalized trust and economic dynamics is still minimally explored and understood, and the research carried out so far is mainly positioned in a macroeconomic perspective [42].

The aim of the presented research was to find an answer to the question whether and to what extent trust influences the functioning of various forms of collective entrepreneurship in rural areas. We focused on those forms of collective entrepreneurship that are common in Poland—agricultural producer organizations, rural women's circles, and local action groups.

The issue of trust and its role in the creation and management of an enterprise have been widely discussed in the literature. However, the researchers mostly focused on its role in three areas of economic behavior: consumption of goods and services, relations between enterprises, and the superior–subordinate relationship. Unfortunately, the rural context has largely been neglected. As emphasized by Gillath, Ai, Branicky, Keshmiri, Davison, and Spaulding [43], the lack of trust (whatever the cause) can result in limited cooperation, efficiency, and productivity. Few researchers have focused on trust in collective entrepreneurship, which makes this study fit into the existing research gap.

Given the lack or a very limited scale of research focused on trust in collective entrepreneurship in rural areas, and having taken into account its role in business management and in economic and social development, the following research hypotheses were formulated:

Hypothesis 1 (H1). *People engaged in the studied forms of collective entrepreneurship in rural areas have greater personal than institutional trust.*

Hypothesis 2 (H2). *Social factors determine the functioning of the studied forms of collective entrepreneurship to a greater extent than economic factors.*

Hypothesis 3 (H3). *According to the concept of the economics of trust, the higher the level of generalized trust is, the higher the trust in various forms of cooperation will be.*

2. Research Methods

The empirical research made it possible to verify the formulated research hypotheses. The survey based on a standardized questionnaire was carried out online using the computer-assisted web interviewing method (CAWI), which, in the situation of the COVID-19 pandemic and restrictions on direct social contacts, was one of very few possibilities to collect empirical data. The CAWI method enables quantitative measurement through questionnaires provided via the Internet. The method makes it possible to reach large samples, while ensuring the respondents' anonymity [44]. The proprietary questionnaire was prepared on the ProfiTest platform between June and July 2021. The link to the survey was made available via social media on profiles related to agriculture and rural areas, as well as posted to organizations whose e-mail addresses were obtained from registers kept by the Agency for Restructuring and Modernization of Agriculture: Register of Pre-Recognized Producer Organizations, Recognized Producer Organizations, and their Associations, as well as Transnational Producer Organizations and their Associations on the Fruit and Vegetables Market, Register of Agricultural Producer Groups, National Register of Rural Women's Circles, and Local Action Groups Register 2014–2020 held by the National Network of Rural Areas. We also planned to include members of the voluntary fire brigades, which have a large representation in the rural areas. The link to the questionnaire was posted on thematic Internet forums and sent by e-mail to all the Voivodship Branches of the Volunteer Fire Brigade Associations of the Republic of Poland. Unfortunately, only a few incomplete questionnaires were received in response; therefore, they could not be included in further analysis.

The survey questionnaire consisted mainly of closed rating questions with an 11-point Likert scale. In total, correctly completed questionnaires were obtained from 132 respondents representing the types of organizations selected for the study.

As can be seen from the data presented in Table 1, both the size of the studied population and its structure indicate that the sample cannot be treated as representative of a larger population. However, the overall number of correctly completed questionnaires justifies further analysis of the results.

Table 1. Characteristics of the sample.

		Form of Collective Entrepreneurship		
		Agricultural Producer Organization (G1)	Local Action Group (G3)	Rural Women's Circle (G4)
Gender	Women	6	42	59
	Men	11	14	-
Education	Primary/lower secondary/vocational	1		4
	Secondary/post-secondary	8		24
	Higher	8	56	31
Average length of participation in the organization (years)		8.9	10.7	8.6

The questionnaire examining various aspects of trust related to the studied forms of entrepreneurship was designed in such a way that the respondents, when answering each question, had the same 11-degree span. In questions Q8, Q9, Q10, and Q13 (for code descriptions, see Table 2), the scale was defined from −5 through 0 to +5, so that the respondent could give negative and positive ratings. Neutral opinions could be indicated by choosing 0 or values close to it. Question Q11, which examined the importance of trust impact on various aspects of collective activities, involved a scale from 0 (not important) to 10 (very important). The initial dataset collected in this way was cleared of incomplete questionnaires. Finally, 132 data records were left for statistical analysis. Initially, the results were divided into three groups according to the type of organization represented by the respondent: agricultural producer organizations (G1), rural women's circles (G3), and local action groups (G4), for which basic statistics were calculated, including mean scores (Table 2).

Pearson and Spearman correlation analyses were conducted between the age of the respondents, the length of participation in an organization, and all answers from questions Q8 to Q13. A correlation matrix with dimensions of 54 × 54 was obtained (Appendix A, Tables A1–A4).

In the survey, we asked the respondents to assess their level of generalized trust. Depending on their answers, they fell into one of the three categories: (1) people who are generally trustful, (2) people who are cautious and distrustful in relations with others, and (3) people who cannot identify their attitude in this respect. As the declared trust was a very important factor in our research, further analyses were performed on the basis of a further division of respondents within the three existing groups (G1, G3, and G4) into three more groups depending on the declared trust (T1—trustful, T2—distrustful, and T3—unable to identify their attitude). Therefore, in further analyses, a division into nine groups was used, which were combinations of forms of collective entrepreneurship and the declared level of trust (G1T1, G1T2, G1T3, ... , G4T3).

For question Q8, in which the respondents assessed the degree of trust in a given organization at three stages of their engagement, an analysis of variance was performed, where the factor at nine levels was the division of respondents according to the form of entrepreneurship and the declared level of trust. A significance level of 0.05 was assumed.

For questions Q9 to Q13, due to a fairly large and detailed set of possible answers, principal component analysis (PCA) was performed. The main purpose of PCA was to facilitate the presentation of the relationship between the studied groups and the obtained results. In addition, a varimax rotation was used, which changes the position of the tested objects so that the individual components contain objects that are strongly correlated with each other and have little correlation with other components. The synthesis and the basis for the discussion were the biplots that contained plotted variants of responses correlated with the components in the two-dimensional space of the first two components (PC1 and PC2). The position of the tested nine groups in relation to the answers and components was also added to biplots. All analyses and graphical presentation of data were executed using IBM SPSS Statistics ver. 26.

Table 2. Mean values of ratings by three types of organizations (G1, G3, and G4).

Question/Answer Code	Organization Type Code		
	G1	**G3**	**G4**
Q8—Question 8 (scale −5 to 5) Please assess the degree of trust in the organization of which you are a member at three stages.			
Q8.1—before deciding to join	2.824	3.109	3.175
Q8.2—at the beginning of cooperation	3.471	3.655	3.684
Q8.3—currently	3.471	3.618	3.614
Q9—Question 9 (scale −5 to 5) Do you trust or not trust the following?			
Q9.1—immediate family—parents, children, spouse	4.824	4.691	4.702
Q9.2—your friends	3.059	3.455	3.456
Q9.3—extended family	2.882	3.182	3.175
Q9.4—people with whom you work on a daily basis	2.647	2.636	2.649
Q9.5—neighbors	2.353	2.309	2.333
Q9.6—the local parish priest	0.706	1.236	1.281
Q9.7—people who work voluntarily in your place of residence	1.471	2.455	2.474
Q9.8—local entrepreneurs	1.294	2.127	2.140
Q9.9—local teachers	2.176	2.127	2.140
Q9.10—people with whom you cooperate in the organization	3.176	3.345	3.368

Table 2. *Cont.*

Question/Answer Code	Organization Type Code		
	G1	G3	G4
Q10—Question 10 (scale −5 to 5) **Do you trust or not trust the following institutions?**			
Q10.1—agricultural advisory centers	2.412	2.782	2.860
Q10.2—private consulting/advisory firms	0.588	1.345	1.368
Q10.3—local government	1.118	1.236	1.298
Q10.4—agricultural chambers	1.353	1.927	1.965
Q10.5—Polish Agency for Enterprise Development (PARP)	0.765	1.545	1.596
Q10.6—non-governmental organizations supporting entrepreneurship, e.g., business incubators and technology parks	0.882	1.909	1.930
Q10.7—commercial banks	0.529	0.509	0.544
Q10.8—cooperative banks	1.706	1.836	1.842
Q10.9—scientific institutions/universities	1.706	2.491	2.491
Q11—Question 11 (scale 0 to 10) **How do you think trust influences the functioning of the collective activity?**			
Q11.1—improves financial results	8.059	7.727	7.719
Q11.2—influences the optimization of the use of resources available to the members of the organization	7.588	7.836	7.825
Q11.3—optimizes business processes	7.059	7.255	7.263
Q11.4—facilitates teamwork	8.294	8.673	8.649
Q11.5—enables knowledge transfer	7.706	8.000	8.000
Q11.6—enhances entrepreneurial behavior	7.059	7.655	7.667
Q11.7—reduces the risk of failure	7.765	7.818	7.807
Q11.8—strengthens the sense of social identity (belonging to the social environment)	8.412	8.436	8.421
Q11.9—enhances self-esteem and development opportunities	8.412	8.618	8.596
Q11.10—promotes integration	9.294	8.836	8.807
Q13—Question 13 (scale −5 to 5) **Please assess the impact of the proposed factors on the development of collective entrepreneurship.**			
Q13.1—personality of the leader	3.765	4.236	4.246
Q13.2—ability to manage a team by the leader	4.353	4.400	4.421
Q13.3—close interpersonal relationships	3.941	4.091	4.088
Q13.4—creativity/resourcefulness of the group	4.294	4.255	4.281
Q13.5—focus on achieving a common goal	4.412	4.236	4.263
Q13.6—individual risk appetite	1.824	2.400	2.456
Q13.7—using talents in the team	4.471	4.455	4.474
Q13.8—qualifications of group members	4.294	3.873	3.912
Q13.9—family patterns	3.000	3.600	3.632
Q13.10—quick and flexible adaptation to changes	4.118	3.964	4.000
Q13.11—the climate of social trust	4.059	4.127	4.158
Q13.12—the environment	2.941	3.364	3.404
Q13.13—access to external financing sources	3.824	3.909	3.947
Q13.14—macroeconomic situation	2.824	3.273	3.333
Q13.15—legibility and knowledge of legal acts	3.588	3.800	3.842
Q13.16—fiscal/tax system	1.941	2.473	2.474
Q13.17—access to economic information	2.824	2.964	2.947
Q13.18—condition and development of local technical infrastructure	3.176	3.291	3.298
Q13.19—activities of business support institutions	3.000	3.127	3.140
Q13.20—activities of local authorities	3.588	3.473	3.421

3. Results and Discussion

3.1. Personal and Institutional Trust

The main question relating to trust is whether one should put trust in someone or something. According to Munns [45], "trust can be defined as a decision to become vulnerable to or dependent on another in return for the possibility of a shared positive outcome". Trust has many levels and dimensions. It is also dynamic and changes over time and in different environments. Blomqvist [46] noted that, in psychology, some researchers see trust as a personal trait, while others stress its social aspect. This means that each person has their own predisposition to trust. One can trust another person or a system (an institution). In economics, trust is seen as a mutual trust or "informal agreement" in which an individual or an organization trusts another person or organization that they will act as promised [47–49]. Trust, therefore, is the foundation for building relationships between business partners. However, the very existence of trust inherently entails risk and, consequently, risk is embedded in any entrepreneurial activity [50].

When considering the issue of trust in the context of entrepreneurship, two categories of trust—personal and institutional—need to be distinguished. Personal trust is based on unwritten rules and values that are shared by people in informal relationships. It derives from personal experience in contact with the other party and stimulates further actions according to the assumption of good will of all parties. Institutional trust concerns the relationship between an individual and organizations, as well as formal structures; moreover, apart from generally accepted operating standards, it also has its source in the rules applicable in a given industry or sector. Personal trust is particularly important at the beginning of an entrepreneur's activity on the market. Nevertheless, along with the development of the company, institutional trust also starts playing an important role. In order to perform successfully, an organization needs to have both types of trust [51]. Lack of trust in representatives of various institutions may be a significant obstacle in undertaking entrepreneurial activities.

The literature shows that the higher level of social trust is positively correlated with the level of entrepreneurship [52]. For example, the Japanese trust strangers less compared to the Americans. Managers in collectivist societies (such as Japan) show a lower overall level of trust than their counterparts in individualistic societies (such as the United States) but tend to have more trust in other managers in the same group. According to the research by Ding et al. [53], Turkey has the lowest level of social trust, and Sweden has the highest. Thus, what is the level of trust in the case of Poland? The study of interpersonal trust among the Polish population showed that, despite relatively limited trust in strangers, Poles generally trust people with whom they maintain regular contacts. The vast majority of respondents trust their immediate family (98%), their friends (95%), extended family (89%), people they work with on a daily basis (88%), and neighbors (80%) [54].

For the purpose of this analysis, we distinguished two types of trust, personal and institutional. A hypothesis was formulated that people operating within the studied forms of collective entrepreneurship in rural areas show more personal than institutional trust.

Personal trust, also known as interpersonal trust, is placed in the people with whom we have close relationships. The most trusted group is usually the family tied by close and intimate relationships, followed by friends and acquaintances, neighbors, coworkers, and business partners, i.e., people we know personally and with whom we interact directly. This type of trust is also placed in people fulfilling specific social and professional roles. In Polish literature, it is referred to as position-related trust [55].

Institutional trust is aimed at institutions and organizations, understood as sets of structural rules within which actions and interactions take place. This group includes schools, universities, the army, churches, courts, the police, banks, stock exchanges, the government, the parliament, and enterprises.

The scale from −5 to 5, designed to measure trust, allowed the respondents to indicate either the level of existing trust or the lack of it. All the mean scores obtained in questions Q9 and Q10 in the three types of organizations were positive (Table 2). The highest mean score (higher than 4.69) represented trust in family and the loved ones (Q9.1). Interestingly, the next highly trusted groups included friends (Q9.2) and people cooperating in a given organization (Q9.10); the mean score ranged from 3.059 to 3.456. This may indicate that members of rural organizations are also friends. People like the extended family, coworkers, and neighbors (Q9.3, Q9.4, Q9.5, and Q9.7 (for two groups)) received average trust scores of 2.30–3.18. The remaining groups of people from question Q9 and most of institutions from question Q10 also received positive mean scores, but at a level closer to the neutral rating. In question Q10, a relatively rating of 2.4–2.9 was given to agricultural advisory centers, i.e., the institutions that implement the mission of supporting the development of rural areas and their residents.

As the survey covered many dimensions, divided into three groups and three sub-groups (nine groups in total), principal component analysis was performed to reduce dimensionality. As a result, 19 variants of answers were reduced to first five principal components (Appendix A, Table 5). PC1 explained 25.21% of the general variance by including variants of trust assessment such as those included in questions Q9.1, Q9.4, Q9.8, Q9.9, Q10.3, and Q10.7. The second component explained 18.94% of the general variance, including variants of trust assessment such as those included in questions Q9.6, Q9.10, Q10.8, and Q10.9 (Figure 1).

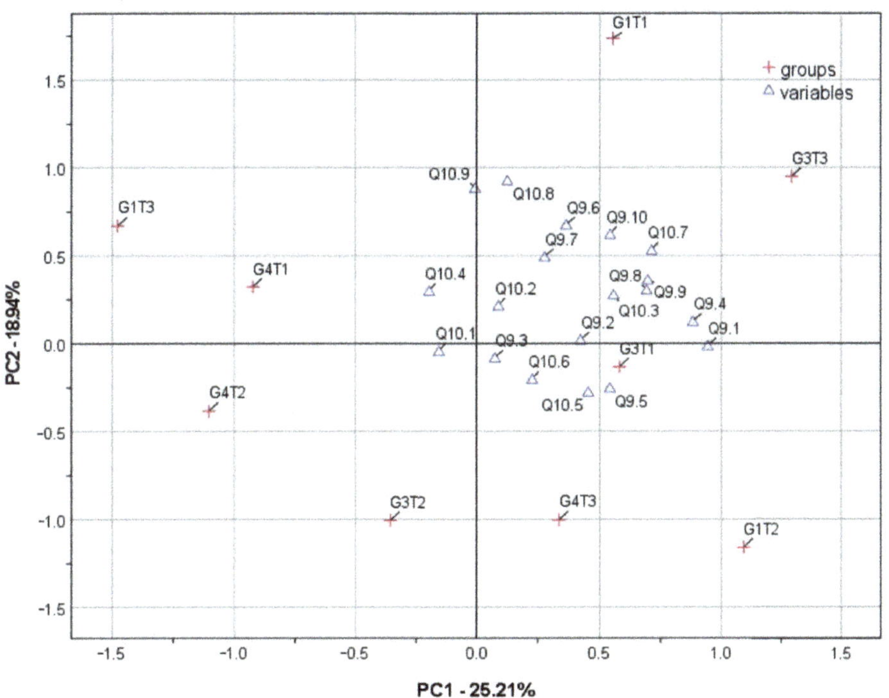

Figure 1. Biplot of relationships of variables (Q9.1, . . . , Q9.10, Q10.1, . . . , Q10.9) for groups (G1T1, . . . , G1T3, G3T1, . . . , G4T3) representing types of organization subdivided according to the declared level of trust in the space of the first two components PC1 and PC2.

The analysis of the results obtained from questions Q9 and Q10 allowed us to positively verify Hypothesis 1, assuming a greater importance of personal trust over institutional trust in the formation and functioning of the studied forms of collective entrepreneurship in rural areas. As our research is one of the few that concerns trust in collective entrepreneurship in rural areas, we can only refer our results to the available analyses carried out as part of the report "2018 Edelman Trust Barometer" [56]. This study showed that the trust in institutions in Poland in 2018 was below global average (the index for Poland was 39/100 compared to the average global index of 48/100).

We believe that, in the case of the studied forms of entrepreneurship, these two types of trust are correlated and, therefore, should not be analyzed separately.

3.2. Social and Economic Determinants of Collective Entrepreneurship Development

An attempt was made to evaluate the factors determining the development of collective entrepreneurship. The designed rating scale was from −5 (inhibiting development) through 0 (having no influence) to 5 (stimulating development). Twenty determinants were presented to be evaluated. All determinants obtained average positive scores, which suggests that, according to the respondents, they stimulate the development of collective entrepreneurship in rural areas. The highest mean scores (within the range from 4.00 to 4.65 depending on a group) were achieved by factors related to the leader—personality and managerial skills (Q13.1, Q13.2). Other highly stimulating determinants included working toward a common goal (Q13.5), as well as talents and creativity present in the team (Q13.4, Q13.7), which, in individual groups, were assessed at an average level of 2.94 to 4.42. The remaining determinants were also assessed positively but depending on the studied group at a lower level (Table 2).

The applied PCA reduced the studied space from 20 dimensions to the first four principal components. PC1 accounted for 37.73% of the overall variance, concentrating on determinants such as Q13.4, Q13.5, Q13.6, Q13.7, Q13.8, Q13.10, Q13.12, Q13.13, and Q13.20 (Appendix A, Table 6). This group contained high- and medium-rated determinants. The applied varimax rotation revealed that they are mutually correlated. PC2 accounted for 25.62% of the total variance and the determinants most strongly correlated with it were variables Q13.1, Q13.2, Q13.3, Q13.11, Q13.14, Q13.18, and Q13.19. The first three determinants also had some of the highest ratings. Therefore, this component can be interpreted as gathering the most important features of collective entrepreneurship development (Figure 2). PC3 accounted for 17.76% of the overall variance. Determinants related to family patterns (Q13.9), fiscal systems (Q13.16), and the availability of economic information (Q13.17) correlated with it most strongly. Interestingly, PC4 was strongly correlated with only one feature related to the legibility and knowledge of legal acts, and it explained 10.57% of the overall variance.

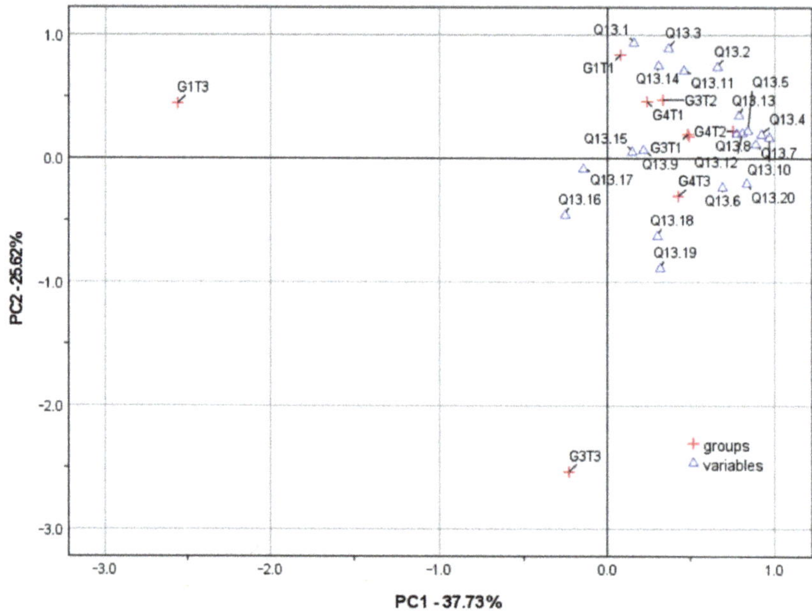

Figure 2. Biplot of relationships of variables (Q13.1, . . . , Q13.20) for groups (G1T1, . . . , G1T3, G3T1, . . . , G4T3) representing types of organization subdivided according to the declared level of trust in the space of the first two components PC1 and PC2.

The reduction of 20 initial variables to four principal components enabled explanation in a simplified manner of 91.69% of the total variance.

The PCA allowed for the positive verification of Hypothesis 2, assuming that social factors determine the functioning of the studied forms of collective entrepreneurship to a greater extent than economic factors.

A significant correlation (at the level of 0.35) was revealed between how the respondents assessed the impact of family relationships, neighborly relations, friendship, and the performance of social roles on the level of trust and the assessment of the importance of social factors influencing the development of collective entrepreneurship (Appendix A, Table 7).

A similar comparative analysis of the impact of social and economic factors on the development of cooperatives was carried out by Pan [57]. This study showed that the influence of social factors on this collective form of entrepreneurship is stronger than of the economic ones. According to Brodziński, the influence of social and economic factors on the development of entrepreneurship among agricultural producers is the same, although the author pointed to creativity, optimism and commitment as the primary drive of entrepreneurship [58]. Furthermore, studies of other researchers indicated a greater impact of these factors on the development of entrepreneurship in rural areas [59,60]. According to Björklund, it is easier for entrepreneurs to overcome barriers to innovation if they have specific cognitive abilities. These abilities include sufficient knowledge, access to information, and decision flexibility. Positive attitudes facilitate the implementation of innovations, while negative attitudes make it difficult. A positive attitude toward work and others can enhance individual performance and creativity, foster new relationships, and expand the use of intellectual and social resources [61]. Colombo and Perujo-Villanueva, on the other hand, emphasized that relations between agricultural neighbors are the main determinant of cooperation, not only between small entities [62]. In the research by Krzyżanowska [63] conducted on a group of 132 leaders of agricultural producer groups, trust was assessed as one of the most important interpersonal skills essential in leadership processes. It was rated 3.98 on a Likert scale from 1 to 5, where 1 meant unimportant and 5 meant very important. The importance of trust in entrepreneurial teams was also confirmed by the results of the research conducted by Falkowski et al. [64] among members of agricultural producer organizations in Poland. In this case, 70% of the surveyed farmers most often indicated lack of trust as a factor hindering the creation of producer organizations. Only in the case of producer organizations based on family ties was this aspect not so important. This is probably a manifestation of the mentality barrier resulting from the negative experiences of farmers in the times of centrally planned economy and forced collectivization of agricultural sector, which now still influences the preference for private (family) farming. In rural areas in Poland, the generalized trust index is still relatively low compared to larger cities (on a seven-point scale from -3 to $+3$, the mean value was -0.88 for rural areas, 0.22 for larger cities, and -0.90 for farmers) [54].

3.3. Trust in Collective Entrepreneurship

It was assumed in the study (H3) that, according to the concept of the economics of trust, a higher level of generalized trust would result in a greater trust in various forms of cooperation; furthermore, as participants gain common experiences, the level of trust in group cooperation increases. The analysis performed with the use of descriptive statistics, i.e., the statistics of the distribution of the feature value (location and dispersion) confirmed this hypothesis (H3). The data presented in Figure 3 show that those people who believed that most people could be trusted (T1) (and, therefore, showed a higher level of generalized trust than the remaining respondents) were more likely to trust joint ventures. It is interesting, however, that, regardless of the level of generalized trust, after starting cooperation under the researched forms of entrepreneurship, the trust in the organization visibly increases. The highest increase was recorded in the G3T3 group. The exceptions were respondents representing rural women's circles, who did not define their level of trust by indicating the answer "I do not know" (G4T3). In this group, a slight decrease in trust in the circle was noticeable during its operation.

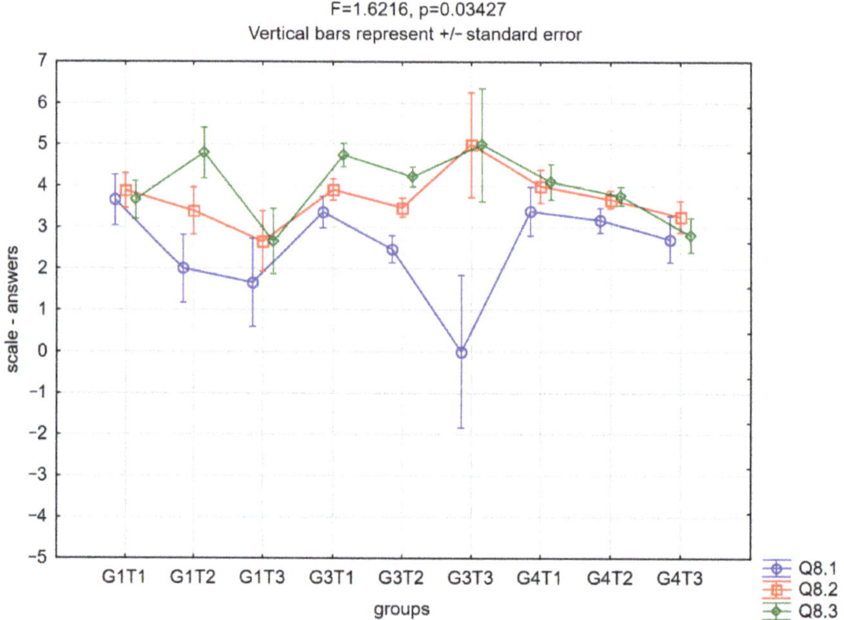

Figure 3. Average trust assessment values (scale from −5 to 5) at three stages of participation by organization type. ANOVA at a significance level of 0.05.

The confirmation of Hypothesis 3 is extremely important from the point of view of the development of collective entrepreneurship in rural areas. Since trust affects the quality of cooperation, it lowers transaction costs and has a positive impact on its durability. In addition, trust can reduce intergroup conflicts and strengthen collaboration between different social groups [53]. This was confirmed by the results of studies by Adro and Franco [65]. The study looked at local and national collaborative networks based on the example of Casa Agrícola dos Arais (CAA). The authors described how a rural farming network operates in a specific sector of Protected Denomination of Origin (PDO) and what its success factors are. Economic benefits, self-confidence, trust, mutual respect, and resilience were the words that the respondents used to identify the factors behind the network's success and the ones that helped them to endure more difficult periods. Despite the fact that the contracts are not formalized, all partners (CAA, milk suppliers, and Tio Careca) had great confidence in the network. On the Likert scale, where 1 meant no trust and 5 meant high trust, all rated the level of trust at 4 or 5 points. Older or less educated producers trusted the given word more than a written contract, which they said they did not need.

The aim of this study was to find an answer to the question whether and to what extent trust affects the functioning of collective entrepreneurship in rural areas. Taking into account the conducted analyses, it can be concluded that trust plays a crucial role, and that the level of trust increases with the length of cooperation.

In Q10, we sought an answer to the question of how, in the respondents' opinion, trust affects the functioning of selected forms of collective activities. Ten aspects that may be affected by trust were selected. The impact was measured on an 11-point scale from 0 (no impact at all) to 10 (very high impact). This scale made it possible to flexibly assess the degree of this impact. All variants of the answers were highly rated by the respondents. For individual groups, the mean score was from 7.28 to 8.98 (Table 2). The range of assessment, expressed in the mean values of the answers, was about 1.5, which, on an 11-point scale, can be considered small. PCA was performed, where 10 dimensions were reduced to the first two components which accounted for 85.62% of the overall variance. PC1 explained 44.89% of the overall variance and was most strongly correlated with the following response variants: Q11.5, Q11.6, Q11.7, Q11.9, and Q11.10. PC2 explained 40.73% of the overall variance and was most strongly correlated with the following responses: Q11.1, Q11.2, Q11.3, Q11.4, and Q11.8 (Figure 4).

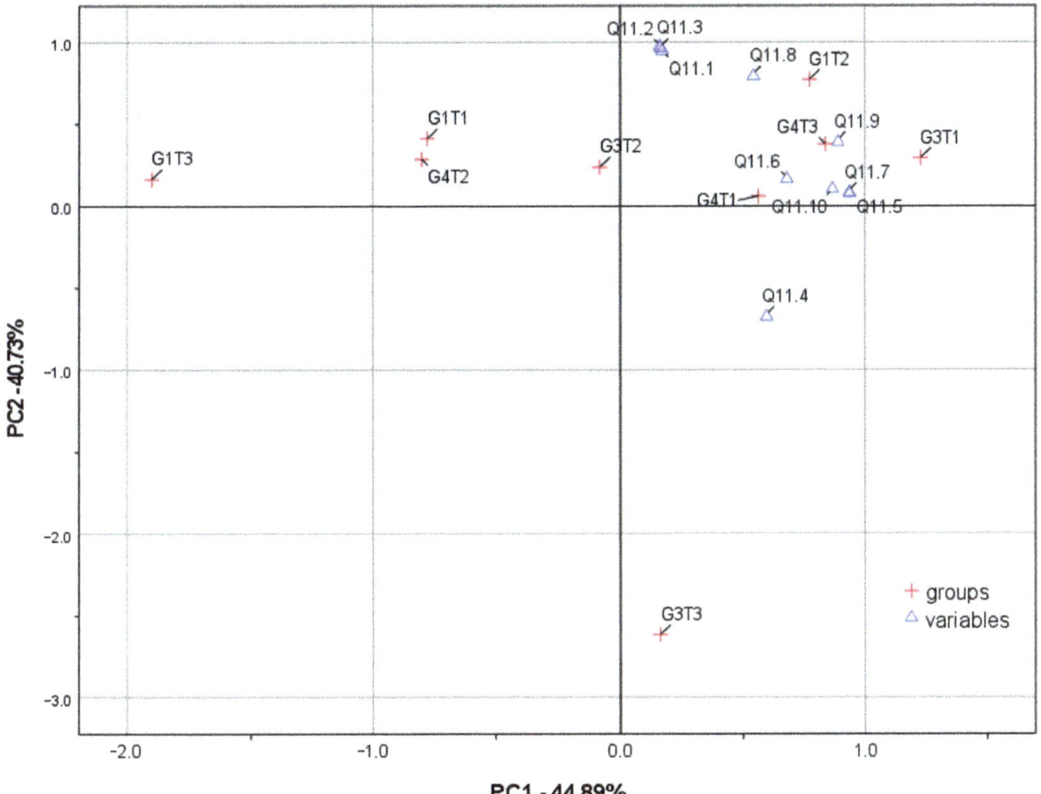

Figure 4. Biplot of relationships of variables (Q11.1, … , Q13.10) for groups (G1T1, … , G1T3, G3T1, … , G4T3) representing types of organization subdivided according to the declared level of trust in the space of the first two components PC1 and PC2.

The conducted analysis showed that, in the opinion of the respondents, the impact of trust on the functioning of the selected forms of collective entrepreneurship is noticeable in both the economic and the social dimensions. In our study, the respondents rated the impact of trust higher in social areas, such as strengthening the sense of social identity, self-esteem, and development opportunities, as well as a positive impact on teamwork or social integration. The exceptions were people representing the G1T3 group (members of agricultural producer organizations who did not define their level of trust), for whom these aspects were not as important as the economic ones, e.g., improving financial results or optimizing the use of resources. A completely different opinion was expressed by the respondents qualified to the G3T3 group (women from rural women's circles who did not define their level of generalized trust), for whom these economic aspects were the least important.

4. Conclusions

The development of entrepreneurship in rural areas has been, for many reasons, a more difficult process than in urbanized areas [7,66]. This is caused by many factors including location, dispersion, small scale of activity, shallowness of local markets, the level of infrastructure development, and the level of education of rural residents.

Due to the fact that the sample selected for this research was not representative, it does not allow for generalizations. However, as our findings correspond with the results of other authors, as shown in the paper, they provide sufficient grounds for verification of the research hypotheses.

The literature implies that entrepreneurial activity is related to both the most basic trust we place in people with whom we maintain social relations and other types of trust, including trust in institutions. Trust is a value that facilitates cooperation between entrepreneurs and their partners [67–69]. This was confirmed by our research. The conducted PCA showed that the respondents are more likely to trust their family, friends, people they work with, local entrepreneurs, and teachers. In the conducted analysis, institutional trust was rated much lower. The exceptions were agricultural advisory centers, which obtained higher trust mean scores (2.4–2.9) compared to other institutions. This result can be explained by the fact that their activity is directly dedicated to rural residents, and the respondents are familiar with it. Moreover, the PCA analysis revealed that the respondents were also willing to trust institutions such as local authorities, commercial and cooperative banks, and scientific centers. Hypothesis 1 was, thus, confirmed.

The socioeconomic development of rural areas is, inter alia, determined by the effective cooperation between entities operating locally. There is no doubt that the main factor limiting cooperation is insufficient knowledge and, above all, the uncertainty resulting from limited trust. This study shows that trust is a necessary condition for effective cooperation. This was also confirmed by the studies of other researchers [70]. Apart from the influence of trust on the development of collective entrepreneurship, this study shows that other factors, of both a social and an economic nature, also play an important role. The respondents highly rated the impact of social factors including personality traits and skills such as the leader's ability to manage a team, creativity of the members, and orientation toward a common goal. Moreover, conditions such as the climate of trust play an important role. In the opinion of the respondents who actively participate in the analyzed forms of collective entrepreneurship, equally important were economic factors such as the availability of external sources of financing, the macroeconomic situation of the country, the condition of infrastructure, and access to business support institutions. The abovementioned social factors, highly rated by respondents, basically constitute the social capital, which in rural areas is identified mainly with local organizations such as associations, civic groups or producer organizations, rural women's circles, and local action groups covered by this research. Summing up, Hypothesis 2 was positively verified, as the prevalence of social factors over economic factors was demonstrated.

There seems to be a consensus among researchers that a higher level of trust facilitates forming multi-actor networks of entrepreneurs and their successful performance [71,72]. Trust has a considerable impact on cooperation between entrepreneurs in small networks where the relationships are more personal [64]. Our research also shows that people with a higher level of generalized trust are more likely to trust joint ventures, especially before joining a collective endeavor. This confirms Hypothesis 3. However, a new relationship emerged. The local action group members who did not indicate their level of generalized trust declared that their trust increased with the length of participation. Unfortunately, in the remaining groups, i.e., agricultural producers and rural women's circles, such trust growth was not statistically confirmed. The most surprising results were obtained in the group of people who indicated a low level of generalized trust. In this case, in all the studied groups, the level of trust increased along with the length of participation in the organization.

We are aware that our study is one of the first undertaken in this field. The positively verified hypotheses formulated in the research do not, however, justify the formulation of general conclusions. Firstly, to make this possible, further in-depth research would be required in this area, with particular emphasis on the dominant character of the activities typical for different forms of entrepreneurship. Agricultural producer organizations pursue mainly economic goals, while rural women's circles and local action groups focus rather on the social aspects of local community life. Secondly, there was an overrepresentation of women in the sample. This was due to the following reasons: the survey was carried out using the CAWI method, which subjected the process of sampling to a large extent to fate and, despite targeting a diverse population including producer groups, local action groups, rural women's circles, and volunteer fire brigades, responses were obtained mostly from rural women's circles and local action groups represented mainly by women. While our empirical research contributes to the existing literature on trust in collective entrepreneurship in rural areas, it does show some limitations. We believe that the conducted analysis will inspire future research on the role of trust in the development of collective entrepreneurship in rural areas. It seems important to consider the following research areas:

(1) the role of trust in the economics of trust paradigm in the perspective of rural development,
(2) the conditions for the formation and functioning of collective entrepreneurship in rural areas, which is an impulse stimulating the local community to undertake entrepreneurial activity,
(3) defining the factors strengthening the social trust of rural residents.

We hope that the presented research areas will serve as inspiration for future studies. By focusing on trust, this paper contributes to the literature examining the role of trust in the development of collective entrepreneurship in rural areas. Trust is a complex, difficult-to-study topic, both theoretically and methodologically. The limitation of this type of research is the credibility of the respondents in terms of competence and perception. To counteract that, in-depth, qualitative research using the interview technique is required. We believe that our study, which is one of the first in this field, should be seen as a pilot study.

Author Contributions: Conceptualization, L.S., A.J.P. and A.S.; methodology, L.S., A.J.P. and A.S.; formal analysis, L.S., A.J.P. and A.S.; investigation, L.S., A.J.P. and A.S.; writing—original draft preparation, L.S., A.J.P. and A.S.; visualization, L.S. All authors have read and agreed to the published version of the manuscript.

Funding: This research received no external funding.

Institutional Review Board Statement: Not applicable.

Data Availability Statement: Not applicable.

Conflicts of Interest: The authors declare no conflict of interest.

Appendix A

Table A1. Pearson's correlation matrix, * significant at the alpha 0.05 level, ** significant at the alpha 0.01 level.

	Age	Years Partyc	Q8.1	Q8.2	Q8.3	Q9.1	Q9.2	Q9.3	Q9.4	Q9.5	Q9.6	Q9.7	Q9.8	Q9.9	Q9.10
age	1	0.301 **	0.058	0.108	0.113	0.036	−0.006	0.013	0.027	0.067	0.067	0.024	0.105	0.114	0.110
years partyc	0.301 **	1	−0.062	−0.120	0.138	0.137	0.115	−0.017	−0.050	0.036	0.038	−0.021	0.032	0.071	0.090
Q8.1	0.058	−0.062	1	0.540 **	0.079	0.026	0.279 **	0.200 *	0.008	0.192 *	−0.007	0.233 **	0.315 **	0.273 **	0.138
Q8.2	0.108	−0.120	0.540 **	1	0.208 *	0.015	0.166	0.212 *	0.100	0.220 *	0.059	0.240 **	0.353 **	0.261 **	0.224 **
Q8.3	0.113	0.138	0.079	0.208 *	1	0.005	−0.005	−0.012	0.127	0.141	−0.024	0.158	0.220 *	0.195 *	0.419 **
Q9.1	0.036	0.137	0.026	0.015	0.005	1	0.416 **	0.331 **	0.185 *	0.111	−0.035	0.138	0.145	0.111	0.068
Q9.2	−0.006	0.115	0.279 **	0.166	−0.005	0.416 **	1	0.670 **	0.345 **	0.440 **	0.206 *	0.373 **	0.456 **	0.392 **	0.305 **
Q9.3	0.013	−0.017	0.200 *	0.212 *	−0.012	0.331 **	0.670 **	1	0.401 **	0.445 **	0.277 **	0.486 **	0.498 **	0.477 **	0.236 **
Q9.4	0.027	−0.050	0.008	0.100	0.127	0.185 *	0.345 **	0.401 **	1	0.442 **	0.312 **	0.251 **	0.298 **	0.300 **	0.345 **
Q9.5	0.067	0.036	0.192 *	0.220 *	0.141	0.111	0.440 **	0.445 **	0.442 **	1	0.503 **	0.390 **	0.506 **	0.435 **	0.372 **
Q9.6	0.067	0.038	−0.007	0.059	−0.024	−0.035	0.206 *	0.277 **	0.312 **	0.503 **	1	0.375 **	0.455 **	0.501 **	0.228 **
Q9.7	0.024	−0.021	0.233 **	0.240 **	0.158	0.138	0.373 **	0.486 **	0.251 **	0.390 **	0.375 **	1	0.738 **	0.597 **	0.423 **
Q9.8	0.105	0.032	0.315 **	0.353 **	0.220 *	0.145	0.456 **	0.498 **	0.298 **	0.506 **	0.455 **	0.738 **	1	0.722 **	0.467 **
Q9.9	0.114	0.071	0.273 **	0.261 **	0.195 *	0.111	0.392 **	0.477 **	0.300 **	0.435 **	0.501 **	0.597 **	0.722 **	1	0.370 **
Q9.10	0.110	0.090	0.138	0.224 **	0.419 **	0.068	0.305 **	0.236 **	0.345 **	0.372 **	0.228 **	0.423 **	0.467 **	0.370 **	1
Q10.1	−0.005	0.072	0.155	0.143	−0.067	0.006	0.178 *	0.093	0.046	0.189 *	0.125	0.099	0.200 *	0.145	0.243 **
Q10.2	0.035	−0.049	0.203 *	0.174 *	0.022	0.145	0.235 **	0.322 **	0.217 *	0.283 **	0.175 *	0.275 **	0.439 **	0.362 **	0.228 **
Q10.3	0.049	0.045	0.157	0.261 **	0.154	0.177 *	0.248 **	0.205 *	0.210 *	0.292 **	0.203 *	0.372 **	0.439 **	0.327 **	0.304 **
Q10.4	−0.001	0.071	0.160	0.076	−0.063	0.133	0.234 **	0.264 **	0.088	0.273 **	0.194 *	0.268 **	0.365 **	0.284 **	0.295 **
Q10.5	−0.062	0.099	0.220 *	0.272 **	0.111	0.077	0.308 **	0.242 **	0.201 *	0.402 **	0.231 **	0.287 **	0.544 **	0.318 **	0.228 **
Q10.6	−0.054	0.108	0.071	0.176 *	0.131	0.109	0.270 **	0.237 **	0.295 **	0.290 **	0.175 *	0.339 **	0.416 **	0.278 **	0.312 **
Q10.7	0.042	0.048	0.010	0.084	0.101	0.238 **	0.334 **	0.302 **	0.266 **	0.254 **	0.161	0.294 **	0.428 **	0.403 **	0.355 **
Q10.8	0.125	0.170	−0.033	0.083	0.031	0.179 *	0.238 **	0.265 **	0.112	0.242 **	0.254 **	0.299 **	0.360 **	0.368 **	0.256 **
Q10.9	0.029	0.067	0.082	0.168	0.126	0.027	0.263 **	0.210 *	0.257 **	0.243 **	0.230 **	0.313 **	0.331 **	0.294 **	0.339 **
Q11.1	−0.156	−0.281 **	0.142	0.010	−0.026	0.078	0.288 **	0.309 **	0.067	0.138	0.187 *	0.153	0.122	0.156	0.113
Q11.2	−0.147	−0.211 *	0.066	−0.034	0.063	−0.019	0.225 **	0.250 **	0.051	0.155	0.181 *	0.223 *	0.177 *	0.162	0.270 **
Q11.3	−0.128	−0.328 **	0.024	−0.055	0.103	−0.110	0.093	0.174 *	0.023	0.060	0.065	0.106	0.115	0.104	0.146
Q11.4	−0.193 *	−0.254 **	−0.053	−0.050	−0.002	0.050	0.253 **	0.224 **	0.137	0.211 *	0.184 *	0.222 *	0.143	0.137	0.241 **
Q11.5	−0.190 *	−0.161	−0.039	−0.009	0.025	0.111	0.257 **	0.258 **	0.154	0.135	0.081	0.237 **	0.195 *	0.239 **	0.236 **
Q11.6	−0.127	−0.205 *	0.028	0.013	0.028	0.104	0.317 **	0.343 **	0.201 *	0.188 *	0.157	0.318 **	0.269 **	0.281 **	0.222 *
Q11.7	−0.059	−0.234 **	0.159	0.166	0.102	0.093	0.273 **	0.270 **	0.148	0.143	0.112	0.217 *	0.126	0.157	0.246 **

Table A1. *Cont.*

	Age	Years Partyc	Q8.1	Q8.2	Q8.3	Q9.1	Q9.2	Q9.3	Q9.4	Q9.5	Q9.6	Q9.7	Q9.8	Q9.9	Q9.10
Q11.8	0.024	−0.106	0.095	0.050	0.105	−0.026	0.222 *	0.136	0.053	0.195 *	0.123	0.208 *	0.191 *	0.163	0.227 **
Q11.9	0.016	−0.069	0.161	0.096	0.099	0.059	0.287 **	0.233 **	0.086	0.180 *	0.105	0.241 **	0.229 **	0.239 **	0.270 **
Q11.10	0.000	−0.004	0.073	0.149	0.108	0.022	0.086	0.019	−0.066	0.070	0.028	0.095	0.125	0.150	0.072
Q13.1	0.134	−0.146	0.092	0.040	0.022	0.085	0.233 **	0.118	0.052	0.204 *	0.187 *	0.260 **	0.226 **	0.179 *	0.140
Q13.2	0.140	−0.042	0.154	0.164	0.075	0.319 **	0.260 **	0.180 *	0.062	0.173 *	0.157	0.250 **	0.267 **	0.265 **	0.124
Q13.3	0.152	0.073	0.117	0.079	0.160	0.074	0.089	0.039	0.114	0.290 **	0.212 *	0.126	0.163	0.161	0.159
Q13.4	0.098	−0.032	0.064	0.154	0.089	0.042	0.064	0.046	0.142	0.173 *	0.107	0.078	0.164	0.105	0.016
Q13.5	0.136	−0.108	0.022	0.223 *	0.083	0.081	0.042	0.142	0.135	0.067	0.075	0.139	0.216 *	0.193 *	0.049
Q13.6	−0.036	−0.093	0.187 *	0.337 **	0.256 **	0.017	0.048	0.122	0.226 **	0.113	0.040	0.191 *	0.221 *	0.183 *	0.182 *
Q13.7	0.102	0.002	0.086	0.144	0.180 *	0.048	−0.012	0.049	−0.068	0.032	−0.018	0.116	0.141	0.161	−0.011
Q13.8	0.158	−0.057	0.063	0.098	0.082	0.065	0.063	0.041	−0.080	0.069	0.111	0.113	0.097	0.182 *	0.098
Q13.9	0.141	0.094	0.224 **	0.245 **	0.121	0.197 *	0.255 **	0.331 **	0.010	0.312 **	0.231 **	0.312 **	0.359 **	0.328 **	0.203 *
Q13.10	0.094	−0.113	0.249 **	0.279 **	0.102	0.158	0.220 *	0.253 **	0.105	0.143	0.077	0.266 **	0.313 **	0.283 **	0.141
Q13.11	0.172 *	−0.035	0.190 *	0.209 *	0.107	0.015	0.136	0.079	−0.110	0.193 *	0.099	0.235 **	0.284 **	0.208 *	0.179 *
Q13.12	−0.010	0.085	0.221 *	0.196 *	0.087	0.187 *	0.192 *	0.109	0.106	0.189 *	0.112	0.194 *	0.275 **	0.273 **	0.054
Q13.13	0.155	−0.042	0.170	0.130	0.023	0.115	0.192 *	0.155	0.087	0.073	0.048	0.222 *	0.257 **	0.285 **	0.061
Q13.14	0.171	0.168	0.249 **	0.151	−0.018	0.140	0.148	0.071	0.052	0.020	−0.055	0.133	0.206 *	0.194 *	0.028
Q13.15	0.149	0.008	0.126	0.192 *	−0.007	0.147	0.214 *	0.248 **	0.042	0.225 **	0.104	0.236 **	0.303 **	0.359 **	0.054
Q13.16	0.087	−0.025	0.076	0.184 *	0.078	0.040	0.147	0.211 *	0.139	0.260 **	0.077	0.229 **	0.359 **	0.353 **	0.249 **
Q13.17	0.079	−0.044	0.050	0.167	0.029	0.129	0.208 *	0.206 *	0.149	0.205 *	0.058	0.204 *	0.288 **	0.275 **	0.136
Q13.18	0.055	−0.051	0.136	0.149	0.048	0.176 *	0.276 **	0.215 *	0.011	0.116	0.048	0.252 **	0.315 **	0.330 **	0.164
Q13.19	0.025	−0.100	0.048	0.150	−0.007	0.249 **	0.355 **	0.348 **	0.181 *	0.168	0.106	0.319 **	0.376 **	0.381 **	0.172 *
Q13.20	0.010	−0.037	0.016	−0.006	0.043	0.033	0.105	0.129	−0.093	0.040	0.055	0.160	0.219 *	0.196 *	−0.025

Table A2. Pearson's correlation matrix, * significant at the alpha 0.05 level, ** significant at the alpha 0.01 level.

	Q10.1	Q10.2	Q10.3	Q10.4	Q10.5	Q10.6	Q10.7	Q10.8	Q10.9	Q11.1	Q11.2	Q11.3	Q11.4	Q11.5	Q11.6
age	−0.005	0.035	0.049	−0.001	−0.062	−0.054	0.042	0.125	0.029	−0.156	−0.147	−0.128	−0.193 *	−0.190 *	−0.127
years partyc	0.072	−0.049	0.045	0.071	0.099	0.108	0.048	0.170	0.067	−0.281 **	−0.211 *	−0.328 **	−0.254 **	−0.161	−0.205 *
Q8.1	0.155	0.203 *	0.157	0.160	0.220 *	0.071	0.010	−0.033	0.082	0.142	0.066	0.024	−0.053	−0.039	0.028
Q8.2	0.143	0.174 *	0.261 **	0.076	0.272 **	0.176 *	0.084	0.083	0.168	0.010	−0.034	−0.055	−0.050	−0.009	0.013
Q8.3	−0.067	0.022	0.154	−0.063	0.111	0.131	0.101	0.031	0.126	−0.026	0.063	0.103	−0.002	0.025	0.028
Q9.1	0.006	0.145	0.177 *	0.133	0.077	0.109	0.238 **	0.179 *	0.027	0.078	−0.019	−0.110	0.050	0.111	0.104
Q9.2	0.178 *	0.235 **	0.248 **	0.234 **	0.308 **	0.270 **	0.334 **	0.238 **	0.263 **	0.288 **	0.225 **	0.093	0.253 **	0.257 **	0.317 **

Table A2. *Cont.*

	Q10.1	Q10.2	Q10.3	Q10.4	Q10.5	Q10.6	Q10.7	Q10.8	Q10.9	Q11.1	Q11.2	Q11.3	Q11.4	Q11.5	Q11.6
Q9.3	0.093	0.322 **	0.205 *	0.264 **	0.242 **	0.237 **	0.302 **	0.265 **	0.210 *	0.309 **	0.250 **	0.174 *	0.224 **	0.258 **	0.343 **
Q9.4	0.046	0.217 *	0.210 *	0.088	0.201 *	0.295 **	0.266 **	0.112	0.257 **	0.067	0.051	0.023	0.137	0.154	0.201 *
Q9.5	0.189 *	0.283 **	0.292 **	0.273 **	0.402 **	0.290 **	0.254 **	0.242 **	0.243 **	0.138	0.155	0.060	0.211 *	0.135	0.188 *
Q9.6	0.125	0.175 *	0.203 *	0.194 *	0.231 **	0.175 *	0.161	0.254 **	0.230 **	0.187 *	0.181 *	0.065	0.184 *	0.081	0.157
Q9.7	0.099	0.275 **	0.372 **	0.268 **	0.287 **	0.339 **	0.294 **	0.299 **	0.313 **	0.153	0.223 *	0.106	0.222 *	0.237 **	0.318 **
Q9.8	0.200 *	0.439 **	0.439 **	0.365 **	0.544 **	0.416 **	0.428 **	0.360 **	0.331 **	0.122	0.177 *	0.115	0.143	0.195 *	0.269 **
Q9.9	0.145	0.362 **	0.327 **	0.284 **	0.318 **	0.278 **	0.403 **	0.368 **	0.294 **	0.156	0.162	0.104	0.137	0.239 **	0.281 **
Q9.10	0.243 **	0.228 **	0.304 **	0.295 **	0.228 **	0.312 **	0.355 **	0.256 **	0.339 **	0.113	0.270 **	0.146	0.241 **	0.236 **	0.222 *
Q10.1	1	0.507 **	0.318 **	0.471 **	0.385 **	0.341 **	0.292 **	0.367 **	0.323 **	0.301 **	0.065	0.055	0.007	−0.003	0.026
Q10.2	0.507 **	1	0.310 **	0.424 **	0.455 **	0.312 **	0.462 **	0.394 **	0.269 **	0.224 **	0.044	0.100	−0.009	−0.060	0.086
Q10.3	0.318 **	0.310 **	1	0.464 **	0.434 **	0.421 **	0.476 **	0.448 **	0.404 **	0.093	0.115	0.057	0.135	0.088	0.025
Q10.4	0.471 **	0.424 **	0.464 **	1	0.514 **	0.504 **	0.427 **	0.421 **	0.376 **	0.244 **	0.128	0.020	0.107	0.100	0.056
Q10.5	0.385 **	0.455 **	0.434 **	0.514 **	1	0.669 **	0.426 **	0.418 **	0.459 **	0.163	0.099	0.104	0.058	0.141	0.118
Q10.6	0.341 **	0.312 **	0.421 **	0.504 **	0.669 **	1	0.532 **	0.386 **	0.506 **	0.159	0.153	0.105	0.111	0.195 *	0.113
Q10.7	0.292 **	0.462 **	0.476 **	0.427 **	0.426 **	0.532 **	1	0.713 **	0.490 **	0.179 *	0.230 **	0.147	0.160	0.201 *	0.249 **
Q10.8	0.367 **	0.394 **	0.448 **	0.421 **	0.418 **	0.386 **	0.713 **	1	0.610 **	0.109	0.156	0.035	0.143	0.092	0.150
Q10.9	0.323 **	0.269 **	0.404 **	0.376 **	0.459 **	0.506 **	0.490 **	0.610 **	1	0.170	0.309 **	0.161	0.298 **	0.274 **	0.255 **
Q11.1	0.301 **	0.224 **	0.093	0.244 **	0.163	0.159	0.179 *	0.109	0.170	1	0.685 **	0.702 **	0.483 **	0.493 **	0.550 **
Q11.2	0.065	0.044	0.115	0.128	0.099	0.153	0.230 **	0.156	0.309 **	0.685 **	1	0.793 **	0.715 **	0.691 **	0.707 **
Q11.3	0.055	0.100	0.057	0.020	0.104	0.105	0.147	0.035	0.161	0.702 **	0.793 **	1	0.645 **	0.641 **	0.657 **
Q11.4	0.007	−0.009	0.135	0.107	0.058	0.111	0.160	0.143	0.298 **	0.483 **	0.715 **	0.645 **	1	0.796 **	0.731 **
Q11.5	−0.003	−0.060	0.088	0.100	0.141	0.195 *	0.201 *	0.092	0.274 **	0.493 **	0.691 **	0.641 **	0.796 **	1	0.829 **
Q11.6	0.026	0.086	0.025	0.056	0.118	0.113	0.249 **	0.150	0.255 **	0.550 **	0.707 **	0.657 **	0.731 **	0.829 **	1
Q11.7	0.033	0.025	0.078	0.091	0.070	0.160	0.128	0.117	0.237 **	0.601 **	0.661 **	0.579 **	0.644 **	0.629 **	0.655 **
Q11.8	0.112	−0.003	0.100	0.022	0.058	0.080	0.137	0.105	0.189 *	0.436 **	0.602 **	0.502 **	0.704 **	0.609 **	0.586 **
Q11.9	0.045	−0.061	0.115	0.098	0.068	0.139	0.146	0.119	0.274 **	0.407 **	0.574 **	0.443 **	0.679 **	0.708 **	0.642 **
Q11.10	−0.015	−0.029	0.180 *	0.099	0.125	0.193 *	0.174 *	0.136	0.147	0.296 **	0.424 **	0.396 **	0.551 **	0.546 **	0.428 **
Q13.1	0.074	0.131	0.140	0.167	0.005	0.091	0.164	0.108	0.196 *	0.334 **	0.433 **	0.293 **	0.470 **	0.294 **	0.303 **
Q13.2	0.050	0.071	0.132	0.202 *	0.076	0.101	0.204 *	0.120	0.159	0.307 **	0.307 **	0.203 *	0.345 **	0.337 **	0.292 **
Q13.3	0.139	0.205 *	0.113	0.352 **	0.138	0.251 **	0.224 **	0.165	0.195 *	0.343 **	0.232 **	0.144	0.226 **	0.191 *	0.153
Q13.4	0.059	0.095	0.116	0.130	0.171 *	0.171 *	0.088	0.011	0.079	0.206 *	0.200 *	0.232 **	0.245 **	0.212 *	0.151
Q13.5	−0.037	0.036	0.051	0.036	0.093	0.120	0.093	−0.005	0.014	0.238 **	0.205 *	0.243 **	0.224 **	0.265 **	0.262 **
Q13.6	−0.061	0.142	0.176 *	0.191 *	0.145	0.179 *	0.158	0.019	0.216 *	0.300 **	0.210 *	0.287 **	0.107	0.231 **	0.252 **
Q13.7	−0.057	−0.030	0.017	0.140	−0.018	0.018	0.002	−0.046	−0.057	0.209 *	0.261 **	0.220 *	0.225 **	0.257 **	0.185 *
Q13.8	−0.035	0.125	0.155	0.056	−0.138	−0.109	−0.002	−0.064	−0.133	0.176 *	0.151	0.196 *	0.213 *	0.121	0.037
Q13.9	0.107	0.254 **	0.217 *	0.369 **	0.172 *	0.209 *	−0.003	0.079	0.015	0.156	0.095	0.047	0.144	0.114	0.044
Q13.10	0.046	0.140	0.090	0.141	0.111	0.055	0.061	−0.059	0.128	0.356 **	0.288 **	0.240 **	0.288 **	0.343 **	0.328 **

Table A2. *Cont.*

	Q10.1	Q10.2	Q10.3	Q10.4	Q10.5	Q10.6	Q10.7	Q10.8	Q10.9	Q11.1	Q11.2	Q11.3	Q11.4	Q11.5	Q11.6
Q13.11	0.109	0.150	0.209 *	0.247 **	0.135	0.105	0.177 *	0.138	0.105	0.279 **	0.307 **	0.189 *	0.304 **	0.225 **	0.268 **
Q13.12	0.129	0.194 *	0.300 **	0.404 **	0.299 **	0.413 **	0.307 **	0.165	0.120	0.199 *	0.060	0.009	0.060	0.197 *	0.092
Q13.13	0.225 **	0.279 **	0.217 *	0.210 *	0.110	0.158	0.225 **	0.183 *	0.159	0.189 *	0.122	0.058	0.151	0.162	0.156
Q13.14	0.188 *	0.142	0.229 **	0.253 **	0.179 *	0.274 **	0.230 **	0.211 *	0.245 **	0.111	0.103	0.070	0.069	0.170	0.130
Q13.15	0.052	0.218 *	0.314 **	0.217 *	0.103	0.172 *	0.395 **	0.369 **	0.228 **	0.145	0.174 *	0.084	0.201 *	0.291 **	0.265 **
Q13.16	0.196 *	0.282 **	0.356 **	0.309 **	0.213 *	0.273 **	0.422 **	0.357 **	0.231 **	0.201 *	0.165	0.169	0.136	0.227 **	0.241 **
Q13.17	0.118	0.277 **	0.230 **	0.165	0.214 *	0.169	0.337 **	0.260 **	0.184 *	0.299 **	0.232 **	0.283 **	0.216 *	0.384 **	0.399 **
Q13.18	0.156	0.238 **	0.276 **	0.197 *	0.138	0.172 *	0.317 **	0.192 *	0.131	0.303 **	0.229 **	0.231 **	0.220 *	0.301 **	0.326 **
Q13.19	0.167	0.339 **	0.284 **	0.253 **	0.242 **	0.318 **	0.313 **	0.209 *	0.181 *	0.399 **	0.297 **	0.298 **	0.314 **	0.411 **	0.425 **
Q13.20	0.117	0.154	0.332 **	0.241 **	0.221 *	0.131	0.243 **	0.173 *	0.041	0.266 **	0.216 *	0.226 **	0.240 **	0.221 *	0.193 *

Table A3. Pearson's correlation matrix, * significant at the alpha 0.05 level, ** significant at the alpha 0.01 level.

	Q11.7	Q11.8	Q11.9	Q11.10	Q13.1	Q13.2	Q13.3	Q13.4	Q13.5	Q13.6
age	−0.059	0.024	0.016	0.000	0.134	0.140	0.152	0.098	0.136	−0.036
years partyc	−0.234 **	−0.106	−0.069	−0.004	−0.146	−0.042	0.073	−0.032	−0.108	−0.093
Q8.1	0.159	0.095	0.161	0.073	0.092	0.154	0.117	0.064	0.022	0.187 *
Q8.2	0.166	0.050	0.096	0.149	0.040	0.164	0.079	0.154	0.223 *	0.337 **
Q8.3	0.102	0.105	0.099	0.108	0.022	0.075	0.160	0.089	0.083	0.256 **
Q9.1	0.093	−0.026	0.059	0.022	0.085	0.319 **	0.074	0.042	0.081	0.017
Q9.2	0.273 **	0.222 *	0.287 **	0.086	0.233 **	0.260 **	0.089	0.064	0.042	0.048
Q9.3	0.270 **	0.136	0.233 **	0.019	0.118	0.180 *	0.039	0.046	0.142	0.122
Q9.4	0.148	0.053	0.086	−0.066	0.052	0.062	0.114	0.142	0.135	0.226 **
Q9.5	0.143	0.195 *	0.180 *	0.070	0.204 *	0.173 *	0.290 **	0.173 *	0.067	0.113
Q9.6	0.112	0.123	0.105	0.028	0.187 *	0.157	0.212 *	0.107	0.075	0.040
Q9.7	0.217 *	0.208 *	0.241 **	0.095	0.260 **	0.250 **	0.126	0.078	0.139	0.191 *
Q9.8	0.126	0.191 *	0.229 **	0.125	0.226 **	0.267 **	0.163	0.164	0.216 *	0.221 *
Q9.9	0.157	0.163	0.239 **	0.150	0.179 *	0.265 **	0.161	0.105	0.193 *	0.183 *
Q9.10	0.246 **	0.227 **	0.270 **	0.072	0.140	0.124	0.159	0.016	0.049	0.182 *
Q10.1	0.033	0.112	0.045	−0.015	0.074	0.050	0.139	0.059	−0.037	−0.061
Q10.2	0.025	−0.003	−0.061	−0.029	0.131	0.071	0.205 *	0.095	0.036	0.142
Q10.3	0.078	0.100	0.115	0.180 *	0.140	0.132	0.113	0.116	0.051	0.176 *
Q10.4	0.091	0.022	0.098	0.099	0.167	0.202 *	0.352 **	0.130	0.036	0.191 *
Q10.5	0.070	0.058	0.068	0.125	0.005	0.076	0.138	0.171 *	0.093	0.145
Q10.6	0.160	0.080	0.139	0.193 *	0.091	0.101	0.251 **	0.171 *	0.120	0.179 *

Table A3. *Cont.*

	Q11.7	Q11.8	Q11.9	Q11.10	Q13.1	Q13.2	Q13.3	Q13.4	Q13.5	Q13.6
Q10.7	0.128	0.137	0.146	0.174 *	0.164	0.204 *	0.224 **	0.088	0.093	0.158
Q10.8	0.117	0.105	0.119	0.136	0.108	0.120	0.165	0.011	−0.005	0.019
Q10.9	0.237 **	0.189 *	0.274 **	0.147	0.196 *	0.159	0.195 *	0.079	0.014	0.216 *
Q11.1	0.601 **	0.436 **	0.407 **	0.296 **	0.334 **	0.307 **	0.343 **	0.206 *	0.238 **	0.300 **
Q11.2	0.661 **	0.602 **	0.574 **	0.424 **	0.433 **	0.307 **	0.232 **	0.200 *	0.205 *	0.210 *
Q11.3	0.579 **	0.502 **	0.443 **	0.396 **	0.293 **	0.203 *	0.144	0.232 **	0.243 **	0.287 **
Q11.4	0.644 **	0.704 **	0.679 **	0.551 **	0.470 **	0.345 **	0.226 **	0.245 **	0.224 **	0.107
Q11.5	0.629 **	0.609 **	0.708 **	0.546 **	0.294 **	0.337 **	0.191 *	0.212 *	0.265 **	0.231 **
Q11.6	0.655 **	0.586 **	0.642 **	0.428 **	0.303 **	0.292 **	0.153	0.151	0.262 **	0.252 **
Q11.7	1	0.666 **	0.626 **	0.549 **	0.326 **	0.307 **	0.206 *	0.233 **	0.350 **	0.250 **
Q11.8	0.666 **	1	0.842 **	0.637 **	0.427 **	0.348 **	0.246 **	0.319 **	0.288 **	0.071
Q11.9	0.626 **	0.842 **	1	0.638 **	0.351 **	0.452 **	0.280 **	0.305 **	0.297 **	0.197 *
Q11.10	0.549 **	0.637 **	0.638 **	1	0.311 **	0.363 **	0.363 **	0.466 **	0.502 **	0.191 *
Q13.1	0.326 **	0.427 **	0.351 **	0.311 **	1	0.704 **	0.483 **	0.457 **	0.427 **	0.040
Q13.2	0.307 **	0.348 **	0.452 **	0.363 **	0.704 **	1	0.567 **	0.623 **	0.591 **	0.193 *
Q13.3	0.206 *	0.246 **	0.280 **	0.363 **	0.483 **	0.567 **	1	0.580 **	0.421 **	0.300 **
Q13.4	0.233 **	0.319 **	0.305 **	0.466 **	0.457 **	0.623 **	0.580 **	1	0.766 **	0.322 **
Q13.5	0.350 **	0.288 **	0.297 **	0.502 **	0.427 **	0.591 **	0.421 **	0.766 **	1	0.369 **
Q13.6	0.250 **	0.071	0.197 *	0.191 *	0.040	0.193 *	0.300 **	0.322 **	0.369 **	1
Q13.7	0.228 **	0.298 **	0.384 **	0.494 **	0.358 **	0.536 **	0.435 **	0.598 **	0.517 **	0.347 **
Q13.8	0.151	0.231 **	0.210 *	0.303 **	0.348 **	0.457 **	0.362 **	0.491 **	0.373 **	0.172 *
Q13.9	0.142	0.124	0.231 **	0.199 *	0.267 **	0.304 **	0.346 **	0.315 **	0.266 **	0.353 **
Q13.10	0.327 **	0.376 **	0.489 **	0.353 **	0.390 **	0.609 **	0.361 **	0.530 **	0.548 **	0.518 **
Q13.11	0.329 **	0.444 **	0.397 **	0.440 **	0.554 **	0.437 **	0.482 **	0.411 **	0.423 **	0.147
Q13.12	0.174 *	0.133	0.292 **	0.406 **	0.106	0.256 **	0.416 **	0.314 **	0.300 **	0.428 **
Q13.13	0.182 *	0.307 **	0.307 **	0.256 **	0.341 **	0.380 **	0.372 **	0.452 **	0.380 **	0.059
Q13.14	0.099	0.188 *	0.306 **	0.258 **	0.144	0.303 **	0.242 **	0.365 **	0.324 **	0.339 **
Q13.15	0.283 **	0.281 **	0.365 **	0.419 **	0.349 **	0.425 **	0.428 **	0.346 **	0.340 **	0.259 **
Q13.16	0.220 *	0.242 **	0.289 **	0.225 **	0.211 *	0.264 **	0.293 **	0.193 *	0.207 *	0.326 **
Q13.17	0.309 **	0.314 **	0.361 **	0.278 **	0.282 **	0.327 **	0.302 **	0.243 **	0.261 **	0.254 **
Q13.18	0.370 **	0.347 **	0.358 **	0.312 **	0.300 **	0.368 **	0.296 **	0.223 *	0.273 **	0.185 *
Q13.19	0.405 **	0.324 **	0.354 **	0.327 **	0.338 **	0.366 **	0.309 **	0.297 **	0.339 **	0.257 **
Q13.20	0.203 *	0.305 **	0.222 *	0.322 **	0.302 **	0.263 **	0.270 **	0.220 *	0.197 *	0.043

Table A4. Pearson's correlation matrix, * significant at the alpha 0.05 level, ** significant at the alpha 0.01 level.

	Q13.7	Q13.8	Q13.9	Q13.10	Q13.11	Q13.12	Q13.13	Q13.14	Q13.15	Q13.16	Q13.17	Q13.18	Q13.19	Q13.20
age	0.102	0.158	0.141	0.094	0.172 *	−0.010	0.155	0.171	0.149	0.087	0.079	0.055	0.025	0.010
years partyc	0.002	−0.057	0.094	−0.113	−0.035	0.085	−0.042	0.168	0.008	−0.025	−0.044	−0.051	−0.100	−0.037
Q8.1	0.086	0.063	0.224 **	0.249 **	0.190 *	0.221 *	0.170	0.249 **	0.126	0.076	0.050	0.136	0.048	0.016
Q8.2	0.144	0.098	0.245 **	0.279 **	0.209 *	0.196 *	0.130	0.151	0.192 *	0.184 *	0.167	0.149	0.150	−0.006
Q8.3	0.180 *	0.082	0.121	0.102	0.107	0.087	0.023	−0.018	−0.007	0.078	0.029	0.048	−0.007	0.043
Q9.1	0.048	0.065	0.197 *	0.158	0.015	0.187 *	0.115	0.140	0.147	0.040	0.129	0.176 *	0.249 **	0.033
Q9.2	−0.012	0.063	0.255 **	0.220 *	0.136	0.192 *	0.192 *	0.148	0.214 *	0.147	0.208 *	0.276 **	0.355 **	0.105
Q9.3	0.049	0.041	0.331 **	0.253 **	0.079	0.109	0.155	0.071	0.248 **	0.211 *	0.206 *	0.215 *	0.348 **	0.129
Q9.4	−0.068	−0.080	0.010	0.105	−0.110	0.106	0.087	0.052	0.042	0.139	0.149	0.011	0.181 *	−0.093
Q9.5	0.032	0.069	0.312 **	0.143	0.193 *	0.189 *	0.073	0.020	0.225 **	0.260 **	0.205 *	0.116	0.168	0.040
Q9.6	−0.018	0.111	0.231 **	0.077	0.099	0.112	0.048	−0.055	0.104	0.077	0.058	0.048	0.106	0.055
Q9.7	0.116	0.113	0.312 **	0.266 **	0.235 **	0.194 *	0.222 *	0.133	0.236 **	0.229 **	0.204 *	0.252 **	0.319 **	0.160
Q9.8	0.141	0.097	0.359 **	0.313 **	0.284 **	0.275 **	0.257 **	0.206 *	0.303 **	0.359 **	0.288 **	0.315 **	0.376 **	0.219 *
Q9.9	0.161	0.182 *	0.328 **	0.283 **	0.208 *	0.273 **	0.285 **	0.194 *	0.359 **	0.353 **	0.275 **	0.330 **	0.381 **	0.196 *
Q9.10	−0.011	0.098	0.203 *	0.141	0.179 *	0.054	0.061	0.028	0.054	0.249 **	0.136	0.164	0.172 *	−0.025
Q10.1	−0.057	−0.035	0.107	0.046	0.109	0.129	0.225 **	0.188 *	0.052	0.196 *	0.118	0.156	0.167	0.117
Q10.2	−0.030	0.125	0.254 **	0.140	0.150	0.194 *	0.279 **	0.142	0.218 *	0.282 **	0.277 **	0.238 **	0.339 **	0.154
Q10.3	0.017	0.155	0.217 *	0.090	0.209 *	0.300 **	0.217 *	0.229 **	0.314 **	0.356 **	0.230 **	0.276 **	0.284 **	0.332 **
Q10.4	0.140	0.056	0.369 **	0.141	0.247 **	0.404 **	0.210 *	0.253 **	0.217 *	0.309 **	0.165	0.197 *	0.253 **	0.241 **
Q10.5	−0.018	−0.138	0.172 *	0.111	0.135	0.299 **	0.110	0.179 *	0.103	0.213 *	0.214 *	0.138	0.242 **	0.221 *
Q10.6	0.018	−0.109	0.209 *	0.055	0.105	0.413 **	0.158	0.274 **	0.172 *	0.273 **	0.169	0.172 *	0.318 **	0.131
Q10.7	0.002	−0.002	−0.003	0.061	0.177 *	0.307 **	0.225 **	0.230 **	0.395 **	0.422 **	0.337 **	0.317 **	0.313 **	0.243 **
Q10.8	−0.046	−0.064	0.079	−0.059	0.138	0.165	0.183 *	0.211 *	0.369 **	0.357 **	0.260 **	0.192 *	0.209 *	0.173 *
Q10.9	−0.057	−0.133	0.015	0.128	0.105	0.120	0.159	0.245 **	0.228 **	0.231 **	0.184 *	0.131	0.181 *	0.041
Q11.1	0.209 *	0.176 *	0.156	0.356 **	0.279 **	0.199 *	0.189 *	0.111	0.145	0.201 *	0.299 **	0.303 **	0.399 **	0.266 **
Q11.2	0.261 **	0.151	0.095	0.288 **	0.307 **	0.060	0.122	0.103	0.174 *	0.165	0.232 **	0.229 **	0.297 **	0.216 *
Q11.3	0.220 *	0.196 *	0.047	0.240 **	0.189 *	0.009	0.058	0.070	0.084	0.169	0.283 **	0.231 **	0.298 **	0.226 **
Q11.4	0.225 **	0.213 *	0.144	0.288 **	0.304 **	0.060	0.151	0.069	0.201 *	0.136	0.216 *	0.220 *	0.314 **	0.240 **
Q11.5	0.257 **	0.121	0.114	0.343 **	0.225 **	0.197 *	0.162	0.170	0.291 **	0.227 **	0.384 **	0.301 **	0.411 **	0.221 *
Q11.6	0.185 *	0.037	0.044	0.328 **	0.268 **	0.092	0.156	0.130	0.265 **	0.241 **	0.399 **	0.326 **	0.425 **	0.193 *
Q11.7	0.228 **	0.151	0.142	0.327 **	0.329 **	0.174 *	0.182 *	0.099	0.283 **	0.220 *	0.309 **	0.370 **	0.405 **	0.203 *
Q11.8	0.298 **	0.231 **	0.124	0.376 **	0.444 **	0.133	0.307 **	0.188 *	0.281 **	0.242 **	0.314 **	0.347 **	0.324 **	0.305 **
Q11.9	0.384 **	0.210 *	0.231 **	0.489 **	0.397 **	0.292 **	0.307 **	0.306 **	0.365 **	0.289 **	0.361 **	0.358 **	0.354 **	0.222 *
Q11.10	0.494 **	0.303 **	0.199 *	0.353 **	0.440 **	0.406 **	0.256 **	0.258 **	0.419 **	0.225 **	0.278 **	0.312 **	0.327 **	0.322 **

Table A4. *Cont.*

	Q13.7	Q13.8	Q13.9	Q13.10	Q13.11	Q13.12	Q13.13	Q13.14	Q13.15	Q13.16	Q13.17	Q13.18	Q13.19	Q13.20
Q13.1	0.358 **	0.348 **	0.267 **	0.390 **	0.554 **	0.106	0.341 **	0.144	0.349 **	0.211 *	0.282 **	0.300 **	0.338 **	0.302 **
Q13.2	0.536 **	0.457 **	0.304 **	0.609 **	0.437 **	0.256 **	0.380 **	0.303 **	0.425 **	0.264 **	0.327 **	0.368 **	0.366 **	0.263 **
Q13.3	0.435 **	0.362 **	0.346 **	0.361 **	0.482 **	0.416 **	0.372 **	0.242 **	0.428 **	0.293 **	0.302 **	0.296 **	0.309 **	0.270 **
Q13.4	0.598 **	0.491 **	0.315 **	0.530 **	0.411 **	0.314 **	0.452 **	0.365 **	0.346 **	0.193 *	0.243 **	0.223 *	0.297 **	0.220 *
Q13.5	0.517 **	0.373 **	0.266 **	0.548 **	0.423 **	0.300 **	0.380 **	0.324 **	0.340 **	0.207 *	0.261 **	0.273 **	0.339 **	0.197 *
Q13.6	0.347 **	0.172 *	0.353 **	0.518 **	0.147	0.428 **	0.059	0.339 **	0.259 **	0.326 **	0.254 **	0.185 *	0.257 **	0.043
Q13.7	1	0.519 **	0.428 **	0.565 **	0.370 **	0.367 **	0.321 **	0.339 **	0.385 **	0.194 *	0.201 *	0.253 **	0.254 **	0.282 **
Q13.8	0.519 **	1	0.357 **	0.394 **	0.360 **	0.191 *	0.361 **	0.178 *	0.279 **	0.161	0.155	0.232 **	0.229 **	0.257 **
Q13.9	0.428 **	0.357 **	1	0.440 **	0.362 **	0.436 **	0.292 **	0.347 **	0.286 **	0.227 **	0.139	0.179 *	0.304 **	0.147
Q13.10	0.565 **	0.394 **	0.440 **	1	0.510 **	0.362 **	0.391 **	0.302 **	0.356 **	0.286 **	0.369 **	0.375 **	0.440 **	0.288 **
Q13.11	0.370 **	0.360 **	0.362 **	0.510 **	1	0.427 **	0.527 **	0.269 **	0.467 **	0.354 **	0.366 **	0.412 **	0.460 **	0.431 **
Q13.12	0.367 **	0.191 *	0.436 **	0.362 **	0.427 **	1	0.368 **	0.461 **	0.450 **	0.293 **	0.230 **	0.320 **	0.378 **	0.166
Q13.13	0.321 **	0.361 **	0.292 **	0.391 **	0.527 **	0.368 **	1	0.581 **	0.538 **	0.316 **	0.388 **	0.509 **	0.457 **	0.376 **
Q13.14	0.339 **	0.178 *	0.347 **	0.302 **	0.269 **	0.461 **	0.581 **	1	0.518 **	0.378 **	0.278 **	0.368 **	0.336 **	0.139
Q13.15	0.385 **	0.279 **	0.286 **	0.356 **	0.467 **	0.450 **	0.538 **	0.518 **	1	0.704 **	0.678 **	0.706 **	0.607 **	0.490 **
Q13.16	0.194 *	0.161	0.227 **	0.286 **	0.354 **	0.293 **	0.316 **	0.378 **	0.704 **	1	0.781 **	0.730 **	0.649 **	0.581 **
Q13.17	0.201 *	0.155	0.139	0.369 **	0.366 **	0.230 **	0.388 **	0.278 **	0.678 **	0.781 **	1	0.807 **	0.767 **	0.648 **
Q13.18	0.253 **	0.232 **	0.179 *	0.375 **	0.412 **	0.320 **	0.509 **	0.368 **	0.706 **	0.730 **	0.807 **	1	0.833 **	0.690 **
Q13.19	0.254 **	0.229 **	0.304 **	0.440 **	0.460 **	0.378 **	0.457 **	0.336 **	0.607 **	0.649 **	0.767 **	0.833 **	1	0.621 **
Q13.20	0.282 **	0.257 **	0.147	0.288 **	0.431 **	0.166	0.376 **	0.139	0.490 **	0.581 **	0.648 **	0.690 **	0.621 **	1

Table 5. Eigenvalues and proportion of the total variance in 9 groups of collective entrepreneurship, as explained by the first three principal components for the original 19 traits and the correlation coefficients between these traits and the first three PCs on for questions Q9 and Q10.

	Component				
	1	**2**	**3**	**4**	**5**
Q9.1	0.945	−0.018	0.146	0.217	0.115
Q9.2	0.423	0.015	0.715	0.202	0.462
Q9.3	0.075	−0.086	0.901	0.213	0.259
Q9.4	0.884	0.119	−0.079	0.283	−0.227
Q9.5	0.542	−0.257	0.630	0.066	0.204
Q9.6	0.365	0.670	−0.528	0.246	−0.022
Q9.7	0.278	0.489	0.009	0.774	0.021
Q9.8	0.693	0.298	−0.015	0.641	0.113
Q9.9	0.698	0.357	−0.041	0.471	0.164
Q9.10	0.543	0.616	0.212	0.299	0.347
Q10.1	−0.154	−0.049	0.299	−0.113	0.917
Q10.2	0.090	0.209	0.188	−0.026	0.873
Q10.3	0.557	0.271	0.633	0.050	−0.421
Q10.4	−0.194	0.293	0.862	0.049	0.180
Q10.5	0.454	−0.281	0.208	0.695	−0.122
Q10.6	0.228	−0.208	0.246	0.893	−0.134
Q10.7	0.714	0.526	0.197	0.178	−0.318
Q10.8	0.126	0.922	−0.002	−0.327	0.090
Q10.9	−0.006	0.879	0.106	0.072	0.031
Total Variance Explained—Rotation Sums of Squared Loadings					
Total	4.789	3.598	3.495	3.042	2.504
% of Variance	25.207	18.936	18.397	16.009	13.178
Cumulative %	25.207	44.143	62.540	78.548	91.727

Extraction Method: Principal Component Analysis. Rotation Method: Varimax with Kaiser Normalization.

Table 6. Eigenvalues and proportion of the total variance in 9 groups of collective entrepreneurship, as explained by the first three principal components for the original 20 traits and the correlation coefficients between these traits and the first three PCs on for questions Q13.

	Component			
	1	**2**	**3**	**4**
Q13.1	0.157	0.935	0.111	−0.131
Q13.2	0.657	0.740	0.012	−0.034
Q13.3	0.366	0.891	0.165	0.061
Q13.4	0.922	0.194	0.303	−0.043
Q13.5	0.840	0.224	0.455	−0.021
Q13.6	0.687	−0.235	−0.231	0.286
Q13.7	0.888	0.114	0.284	0.218
Q13.8	0.771	0.203	0.529	0.233
Q13.9	0.215	0.066	0.856	0.047
Q13.10	0.967	0.170	0.069	−0.155
Q13.11	0.458	0.711	0.507	0.045
Q13.12	0.807	0.206	0.410	0.162
Q13.13	0.786	0.349	0.378	−0.064
Q13.14	0.306	0.751	0.272	0.473
Q13.15	0.147	0.053	0.076	0.980
Q13.16	−0.255	−0.457	−0.630	0.553
Q13.17	−0.145	−0.084	−0.862	0.008
Q13.18	0.301	−0.629	−0.443	0.462
Q13.19	0.317	−0.889	0.079	0.049
Q13.20	0.833	−0.204	−0.269	0.378

Table 6. *Cont.*

	Component			
	1	**2**	**3**	**4**
	Total Variance Explained—Rotation Sums of Squared Loadings			
Total	7.546	5.124	3.552	2.115
% of Variance	37.732	25.621	17.761	10.574
Cumulative %	37.732	63.353	81.114	91.688

Extraction Method: Principal Component Analysis. Rotation Method: Varimax with Kaiser Normalization.

Table 7. Eigenvalues and proportion of the total variance in 9 groups of collective entrepreneurship, as explained by the first three principal components for the original 10 traits and the correlation coefficients between these traits and the first three PCs on for questions Q11.

	Component	
	1	**2**
Q11.1	0.171	0.949
Q11.2	0.160	0.976
Q11.3	0.164	0.963
Q11.4	0.597	−0.674
Q11.5	0.938	0.082
Q11.6	0.679	0.166
Q11.7	0.935	0.081
Q11.8	0.544	0.795
Q11.9	0.889	0.392
Q11.10	0.866	0.106
	Total Variance Explained—Rotation Sums of Squared Loadings	
Total	4.489	4.073
% of Variance	44.889	40.733
Cumulative %	44.889	85.622

Extraction Method: Principal Component Analysis. Rotation Method: Varimax with Kaiser Normalization.

References

1. Statistics Poland. *Agriculture in 2019*; Statistics Poland: Warsaw, Poland, 2020; p. 15. Available online: https://stat.gov.pl/obszary-tematyczne/rolnictwo-lesnictwo/rolnictwo/rolnictwo-w-2019-roku,3,16.html (accessed on 8 October 2021).
2. Nurzyńska, I.; Drygas, M. Rural Development Policy in Poland (and in the EU)—A Needless Expense or a Must? *Wieś Rol.* **2018**, *179*, 169–187. [CrossRef]
3. Boldureanu, G.; Ionescu, A.M.; Bercu, A.-M.; Bedrule-Grigoruță, M.V.; Boldureanu, D. Entrepreneurship Education through Successful Entrepreneurial Models in Higher Education Institutions. *Sustainability* **2020**, *12*, 1267. [CrossRef]
4. Huang, B.; Tani, M.; Zhu, Y. Does higher education make you more entrepreneurial? Causal evidence from China. *JBR* **2021**, *135*, 543–558. [CrossRef]
5. Nowakowska-Grunt, J.; Parzonko, A.J.; Kiełbasa, B. *Determinants of Managing Networks of Organizations in Rural Areas*; Publishing Office of Faculty of Management Częstochowa University of Technology: Częstochowa, Poland, 2016; pp. 51–54.
6. Korsgaard, S.; Müller, S.; Wittorff Tanvig, H. Rural entrepreneurship or entrepreneurship in the rural—Between place and space. *IJEBR* **2015**, *21*, 5–26. [CrossRef]
7. Galvão, A.R.; Mascarenhas, C.; Marques, C.S.E.; Braga, V.; Ferreira, M. Mentoring entrepreneurship in a rural territory—A qualitative exploration of an entrepreneurship program for rural areas. *J. Rural. Stud.* **2020**, *78*, 314–324. [CrossRef]
8. Minniti, M.; Lévesque, M. Entrepreneurial types and economic growth. *J. Bus. Ventur.* **2010**, *25*, 305–314. [CrossRef]
9. Ardichvili, A.; Cardozo, R.; Sourav, R. A Theory of Entrepreneurial Opportunity Identification and Development. *J. Bus. Ventur.* **2003**, *18*, 105–123. [CrossRef]
10. Shane, S. *A General Theory of Entrepreneurship: The Individual-Opportunity Nexus*; Edward Elgar: Northampton, MA, USA, 2003.
11. Krzyżanowska, K.; Parzonko, A.J.; Sieczko, A. *Przedsiębiorczość Zespołowa na Obszarach Wiejskich. Stan i Perspektywy Rozwoju*; WULS Press: Warszawa, Poland, 2020; pp. 16–19. Available online: https://www.ieif.sggw.pl/wp-content/uploads/2020/10/Przedsiebiorczosc_Krzyzanowska_Parzonko_Sieczko.pdf (accessed on 8 October 2021).
12. Parzonko, A.J.; Sieczko, A. Agricultural producer groups as manifestation of team entrepreneurship in Poland. In *Rural Development and Entrepreneurship: Production and Co-Operation in Agriculture, Proceedings of the International Scientific Conference,*

    Vladivostok, Russia, 2–4 October 2018; Auzina, A., Ed.; Latvia University of Life Sciences and Technologies: Jelgava, Latvia, 2018; pp. 221–228.

13. Sztompka, P. *Trust: A Sociological Theory*; Cambridge University Press: Warsaw, Poland, 2000.
14. Lewenstein, B. Nowe paradygmaty rozwoju układów lokalnych. In *Oblicza Lokalności. Różnorodność Miejsc i Czasu*; Kurczewska, J., Ed.; Wydawnictwo IFiS PAN: Warszawa, Poland, 2006; p. 229.
15. Ayupova, S.; Bents, D.; Kozlova, E. Trust as a Factor of Interaction Efficiency. In *Smart Technologies and Innovations in Design for Control of Technological Processes and Objects: Economy and Production. Smart Innovation, Systems and Technologies (139)*; Solovev, D., Ed.; Springer: Cham, Switzerland, 2019. [CrossRef]
16. Camarinha-Matos, L.M.; Afsarmanesh, H. Collaborative Networks. In *Knowledge Enterprise: Intelligent Strategies in Product Design, Manufacturing, and Management. IFIP International Federation for Information Processing (207)*; Wang, K., Kovacs, G.L., Wozny, M., Fang, M., Eds.; Springer: Boston, MA, USA, 2006. [CrossRef]
17. Levin, D.Z.; Cross, R. The Strength of Weak Ties You Can Trust: The Mediating Role of Trust in Effective Knowledge. *Transfer. Manag. Sci.* **2004**, *50*, 1477–1490. [CrossRef]
18. Taylor, B.M.; Van Grieken, M. Local institutions and farmer participation in agri-environmental schemes. *J. Rural. Stud.* **2015**, *37*, 10–19. [CrossRef]
19. Quinn, B.; McKitterick, L.; Tregear, A.; McAdam, R. Trust in the programme: An exploration of trust dynamics within rural group-based support programmes. *J. Rural. Stud.* **2021**. [CrossRef]
20. Welter, F. "All you need is trust?" A critical review of the trust and entrepreneurship literature. *Int. Small Bus. J.* **2012**, *30*, 193–212. [CrossRef]
21. Persky, J.J. Retrospectives The Ethology of Homo Economicus. *J. Econ. Perspect.* **1995**, *9*, 221–231. [CrossRef]
22. Bruni, L.; Guala, F. Vilfredo Pareto and the Epistemological Foundations of Choice Theory. *HOPE* **2001**, *33*, 21–49. Available online: http://muse.jhu.edu/journals/hpe/summary/v033/33.1bruni.html (accessed on 8 October 2021). [CrossRef]
23. Milonakis, D.; Fine, B. *From Political Economy to Economics. Method, the Social and the Historical in the Evolution of Economic Theory*; Routledge: London, UK; New York, NY, USA, 2009; Available online: http://pauladaunt.com/books/From_Political_Economy_to_Freakonomics.pdf (accessed on 8 October 2021).
24. Thaler, R.H. Behavioral Economics: Past, Present, and Future. *AER* **2016**, *106*, 1577–1600. [CrossRef]
25. Smith, V. Microeconomic systems as an experimental science. *AER* **1982**, *72*, 923–955. Available online: https://www.jstor.org/stable/1812014 (accessed on 8 October 2021).
26. Samuelson, L. Game Theory in Economics and Beyond. *JEP* **2016**, *30*, 107–130. [CrossRef]
27. Mikłowicz, D. Zaufanie jako wartość społeczna. *Studia Ekon.* **2016**, *259*, 80–88.
28. Besser, T.; Miller, N. The structural, social and strategic factors associated with successful business networks. *Enterpren. Reg. Dev.* **2011**, *23*, 113–133. [CrossRef]
29. Newbery, R.; Sauer, J.; Gorton, M.; Phillipson, J.; Atterton, J. Determinants of the performance of business associations in rural settlements in the United Kingdom: An analysis of members' satisfaction and willingness-to-pay for association survival. *Environ. Plann.* **2013**, *45*, 967–985. [CrossRef]
30. Rauch, J.E. *Productivity Gains from Geographic Concentration of Human Capital: Evidence from the Cities*; Working Paper No. 3905; National Bureau of Economic Research: Cambridge, MA, USA, 1991; Available online: https://www.nber.org/system/files/working_papers/w3905/w3905.pdf (accessed on 8 October 2021).
31. Chalker, M.; Loosemore, M. Trust and productivity in Australian construction projects: A subcontractor perspective. *ECAM* **2016**, *23*, 192–210. [CrossRef]
32. Kayes, A.; Kayes, D.C. Nurturing Trust. In *The Learning Advantage*; Palgrave Macmillan: London, UK, 2011; pp. 154–170. [CrossRef]
33. Sørensen, O.H.; Hasle, P.; Pejtersen, J.H. Trust relations in management of change. *SJM* **2011**, *27*, 405–417. [CrossRef]
34. Dincer, O.; Uslaner, E.M. Trust and growth. *Public Choice* **2010**, *142*, 59–67. [CrossRef]
35. Deelmann, T.; Loos, P. Trust economy: Aspects of reputation and trust building for SMEs in e-business in e-business. *AMCIS 2002 Proc.* **2002**, *2002*, 302. Available online: http://aisel.aisnet.org/amcis2002/302 (accessed on 8 October 2021).
36. Khalifa, S. Trust, Landscape, And Economic Development. *JED* **2016**, *41*, 19–32. Available online: http://www.jed.or.kr/full-text/41-1/2.pdf (accessed on 8 October 2021).
37. Bjørnskov, C. Determinants of generalized trust: A cross-country comparison. *Public Choice* **2007**, *130*, 1–21. [CrossRef]
38. Algan, Y.; Cahuc, P. Inherited Trust and Growth. *AER* **2010**, *100*, 2060–2092. [CrossRef]
39. Elert, N.; Stam, E.; Stenkula, M. *Intrapreneurship and Trust*; Working Papers 19-08. U.S.E.; Research Institute: Utrecht, The Netherlands, 2019; pp. 1–27. Available online: https://www.uu.nl/sites/default/files/rebo_use_wp_2019_1908.pdf (accessed on 8 October 2021).
40. Dotti Sani, G.M.; Magistro, B. Increasingly unequal? The economic crisis, social inequalities and trust in the European Parliament in 20 European countries. *EJPR* **2016**, *55*, 246–264. [CrossRef]
41. Lipps, J.; Schraff, D. Regional inequality and institutional trust in Europe. *EJPR* **2021**, *60*, 892–913. [CrossRef]
42. Burciu, A.; Kicsi, R.; Bostan, I. Social Trust and Dynamics of Capitalist Economies in the Context of Clashing Managerial Factors with Risks and Severe Turbulence: A Conceptual Inquiry. *Sustainability* **2020**, *12*, 8794. [CrossRef]

43. Gillath, O.; Ai, T.; Branicky, M.S.; Keshmiri, S.; Davison, R.B.; Spaulding, R. Attachment and trust in artificial intelligence. *Comput. Hum. Behav.* **2021**, *115*, 106607. [CrossRef]
44. Parzonko, A.; Balińska, A.; Sieczko, A. Pro-Environmental Behaviors of Generation Z in the Context of the Concept of Homo Socio-Oeconomicus. *Energies* **2021**, *14*, 1597. [CrossRef]
45. Munns, A. Potential influence of trust on the successful completion of a project. *IJPM* **1995**, *13*, 19–24. [CrossRef]
46. Blomqvist, K. The many faces of trust. *SJM* **1997**, *13*, 271–286. [CrossRef]
47. Hotte, N.; Kozak, R.; Wyatt, S. How institutions shape trust during collective action: A case study of forest governance on Haida Gwaii. *For. Policy Econ.* **2019**, *107*, 101921. [CrossRef]
48. Miller, G.J.; Whitford, A.B. Trust and incentives in principal-agent negotiations: The "insurance/incentive trade-off". *JTP* **2002**, *14*, 231–267. [CrossRef]
49. Cerić, A.; Vukomanović, M.; Ivić, I.; Kolarić, S. Trust in megaprojects: A comprehensive literature review of research trends. *IJPM* **2020**, *39*, 325–338. [CrossRef]
50. Paliszkiewicz, J. *Zaufanie w Zarządzaniu*; WN PWN: Warsaw, Poland, 2013.
51. Krawczyk-Bryłka, B. Zaufanie do siebie jako jeden z aspektów zaufania w aktywności przedsiębiorczej. In *Przedsiębiorczość: Jednostka, Organizacja, Kontekst*; Postuła, A., Majczyk, J., Darecki, M., Eds.; Scientific Publishing House FMUW: Warsaw, Poland, 2015; pp. 90–102. [CrossRef]
52. Kwon, S.-W.; Arenius, P. Nations of entrepreneurs: A social capital perspective. *JBV* **2010**, *25*, 315–330. [CrossRef]
53. Ding, Z.; Au, K.; Chiang, F. Social trust and angel investors' decisions: A multilevel analysis across nations. *JBV* **2015**, *30*, 307–321. [CrossRef]
54. Centrum Badania Opinii Społecznej. *Zaufanie Społeczne. Komunikat z Badań nr 43/2020*; CBOS: Warszawa, Poland, 2020; Available online: https://cbos.pl/SPISKOM.POL/2020/K_043_20.PDF (accessed on 8 October 2021).
55. Moczydłowska, J.M. Zaufanie jako determinanta relacji przedsiębiorcy—Przedstawiciele administracji samorządowej. *Zarządzanie Finanse* **2012**, *10*, 75–85. Available online: http://jmf.wzr.pl/pim/2012_4_2_6.pdf (accessed on 8 October 2021).
56. Trust Barometer. Global Report. Edelman. 2018. Available online: https://www.edelman.com/sites/g/files/aatuss191/files/2018-10/2018_Edelman_Trust_Barometer_Global_Report_FEB.pdf (accessed on 21 December 2020).
57. Pan, L. Willingness of Farmers Joining Professional Cooperatives. Based on the Questionnaire Survey of Nanjing. *Asian Agric. Res.* **2011**, *3*, 1–5. [CrossRef]
58. Brodziński, Z. Entrepreneurship of Agricultural Producers—Determinants of Development. *Ann. PAAAE* **2019**, *3*. [CrossRef]
59. Phillipson, J.; Gorton, M.; Laschewski, L. Local Business Co-operation and the Dilemmas of Collective Action: Rural Micro-business Networks in the North of England. *Sociol. Rural.* **2006**, *46*, 40–60. [CrossRef]
60. Sgroi, F.; Trapani, A.M.D.; Testa, R.; Tudisca, S. The Rural Tourism as Development Opportunity or Farms. The Case of Direct Sales in Sicily. *AJABS* **2014**, *9*, 407–419. [CrossRef]
61. Björklund, J.C. Barriers to Sustainable Business Model Innovation in Swedish Agriculture. *JEMI* **2018**, *14*, 65–90. [CrossRef]
62. Colombo, S.; Perujo-Villanueva, M. Analysis of the spatial relationship between small olive farms to increase their competitiveness through cooperation. *Land Use Policy* **2017**, *63*, 226–235. [CrossRef]
63. Krzyżanowska, K. *Ekonomiczno—Społeczne Uwarunkowania Innowacji w Zespołowym Działaniu w Rolnictwie*; WULS Press: Warsaw, Poland, 2016; Available online: https://www.ieif.sggw.pl/wp-content/uploads/2021/05/Ekonomiczno-spoleczne_KrzyzanowskaK.pdf (accessed on 8 October 2021).
64. Fałkowski, J.; Chlebicka, A.; Łopaciuk-Gonczaryk, B. Social relationships and governing collaborative actions in rural areas: Some evidence from agricultural producer groups in Poland. *J. Rural. Stud.* **2017**, *49*, 104–116. [CrossRef]
65. do Adro, F.; Franco, M. Rural and agri-entrepreneurial networks: A qualitative case study. *Land Use Policy* **2020**, *99*, 105117. [CrossRef]
66. Fortunato, M.W.P. Supporting rural entrepreneurship: A review of conceptual developments from research to practice. *Community Dev.* **2014**, *45*, 387–408. [CrossRef]
67. Skandrani, H.; Triki, A.; Baratli, B. Trust in supply chains, meanings, determinants and demonstrations: A qualitative study in an emerging market context. *Qual. Mark. Res.* **2011**, *14*, 391–409. [CrossRef]
68. Fawcett, S.E.; Jones, S.L.; Fawcett, A.M. Supply chain trust: The catalyst for collaborative innovation. *Bus. Horiz.* **2012**, *55*, 163–178. [CrossRef]
69. Day, M.; Fawcett, S.E.; Fawcett, A.M.; Magnan, G.M. Trust and relational embeddedness: Exploring a paradox of trust pattern development in key supplier relationships. *Ind. Mark. Manag.* **2013**, *42*, 152–165. [CrossRef]
70. Szabó, G.G. The importance and role of trust in agricultural marketing co-operatives. *Stud. Agric. Econ.* **2010**, *112*, 5–22. [CrossRef]
71. Pulfer, I.; Möhring, A.; Dobricki, M.; Lips, M. Success factors for farming collectives. In Proceedings of the 12th Congress of the European Association of Agricultural Economists—EAAE 2008, Ghent, Belgium, 26–29 August 2008. [CrossRef]
72. Davies, P.; Mason-Jones, R. Communities of interest as a lens to explore the advantage of collaborative behaviour for developing economies: An example of the Welsh organic food sector. *IJEIM* **2017**, *18*, 5–13. [CrossRef]

agriculture

Article

Profile of the Small-Scale Farms Willing to Cooperate—Evidence from Lithuania

Jolanta Droždz [1,*], Vlada Vitunskienė [2] and Lina Novickytė [3]

[1] Faculty of Economics and Business Administration, Vilnius University, 10222 Vilnius, Lithuania
[2] Faculty of Bioeconomy Development, Vytautas Magnus University, 44248 Kaunas, Lithuania; vlada.vitunskiene@vdu.lt
[3] Government Strategic Analysis Center, 01108 Vilnius, Lithuania; lina.novickyte@gmail.com
* Correspondence: jolanta.drozdz@cr.vu.lt

Abstract: Cooperatives cover a large part of the agricultural sectors and have substantial market shares in agri-food supply chains in the EU Western countries. They account for approximately half of agricultural trade in the EU. By contrast, in the EU Western countries, where farmer cooperatives are widespread and successful, agricultural cooperation in Lithuania has developed intermittently in the last century. We still have very limited knowledge of why the country's agricultural producers (especially smallholder farmers) are reluctant to cooperate in Lithuania. The aim of this study is to assess the level of the willingness to cooperate among smallholder farmers in Lithuania and to draw up the profiles of small-scale farms that participate in and intend to join cooperatives and, conversely, that do not participate in cooperatives and do not intend to do so. To achieve this goal, a representative survey of small-scale farms was conducted. Results of surveys carried out in 2019 in Lithuania on a group of 1002 small-scale farms showed that only 8% of the surveyed farms participate in producer groups or cooperatives, while another 8% intend to participate. Small-scale farms in Lithuania have weak market integration, with no bargaining power on input and output markets. The vast majority of small-scale farms are reluctant to participate in cooperative activities in Lithuania. Therefore, this study aimed to determine the profile of a small farm that tends to cooperate. The main social characteristics of farm managers and economic factors of farms willing to cooperate have been identified.

Keywords: small-scale farm; cooperation; contractual integration; willingness to cooperate; farm profile; Lithuanian case

Citation: Droždz, J.; Vitunskienė, V.; Novickytė, L. Profile of the Small-Scale Farms Willing to Cooperate—Evidence from Lithuania. *Agriculture* 2021, 11, 1071. https://doi.org/10.3390/agriculture11111071

Academic Editors: Adoración Mozas Moral and Domingo Fernandez Ucles

Received: 4 October 2021
Accepted: 29 October 2021
Published: 30 October 2021

Publisher's Note: MDPI stays neutral with regard to jurisdictional claims in published maps and institutional affiliations.

1. Introduction

There are more than forty thousand agricultural cooperatives in Europe with nine million farmer members [1]. Cooperatives cover a large part of the agricultural sectors and have substantial market shares in agri-food supply chains in the European Union (EU) western countries. They account for approximately half of agricultural trade in the EU and over half in some member states such as Austria, Denmark, Finland, France, Ireland, the Netherlands, and Sweden. Moreover, the market shares of cooperatives differ considerably with respect to sectors [2,3]. According to Ollila [4], not only ideological or sociological but also economic reasons justify the existence of cooperatives. The existence of cooperative organizations in today's business environment, particularly in agriculture, signals their continued ability to provide value to their members [5] by increasing farms' (especially small-scale farms) competitiveness on the national and international markets [6,7]. Agricultural cooperatives have provided a model for overcoming the disadvantages of small-scale farming for more than 150 years [8]. It should be added that the process of farm cooperation in modern Lithuania was decisively influenced by the historical path of agricultural development and the experience of the agricultural community.

By contrast, in the EU Western countries where farmers' cooperatives are widespread and successful, agricultural cooperation in Lithuania has developed intermittently in the last century. The cooperation of farmers developed in the second and third decades of the 20th century, based on the classical principles of cooperation, was completely destroyed by the collectivization of farms into pseudo-cooperative kolkhoz at the beginning of the Soviet era. Moreover, farmers' households were allowed to use only up to 60 acres of land. The collective farms that operated during the Soviet regime (1945–1989) still maintain a non-beneficial image of anti-cooperation, deterring farmers from joining cooperatives. After the restructuring of independent Lithuanian agriculture in the early 1990s, cooperation between newly established family farms and corporate companies in agriculture has not been expanding and is still very weak. Cooperation among farmers is not widespread in Lithuania—only about 12% of the country's farmers are involved in cooperatives [9]. The research of Borychowski et al. [10] included Lithuania and showed that the production scale was the key determinant of the resilience of small-scale farms in the countries. Moreover, the main way to achieve the higher benefits of increasing the production should be combined with strengthening the market integration of agricultural producers, so the cooperation issue becomes even more relevant. This study extends an established direction of the resilient development of the small-scale farms in Lithuania.

EU Western researchers are looking for the reasons why agricultural producers in the 21st century are choosing a cooperative structure based on classical principles of cooperation [11], and we still have very limited knowledge of why the country's agricultural producers (especially smallholder farmers) are reluctant to cooperate in Lithuania. The main aim of this study is to assess the level of the willingness to cooperate among smallholder farmers in Lithuania and to draw up the profiles of small-scale farms that participate in and intend to join cooperatives and, conversely, that do not participate in cooperatives and do not intend to do so.

To achieve this aim, a representative survey of small-scale farms was conducted. We employed the results of surveys carried out in 2019 in Lithuania on a group of 1002 small-scale farms. The study results contribute to a better understanding of the cooperative or non-cooperative behaviour of small-scale farms.

2. Literature Overview

2.1. Reluctance to Cooperate

An overview of the research conducted on farm co-operation in Lithuania revealed that the main reasons for the attitude of Lithuanian farmers towards co-operation have not changed for a long time. The farmer and expert survey data show [12] that the main reasons for farmers' reluctance to cooperate and change are their individuality, lack of trust in collective (cooperative) actions and new ideas, internal competition, and inability to find a joint agreement on different issues.

Both farmers and managers of agricultural companies do not consider cooperation to be a matter of necessity; for them, according to experts, daily work on the farm is of greater value than changes in the area of cooperation. In addition, they feel that there is a lack of time for collaborative actions due to the high workload. Other authors have also identified distrust between people, unwillingness to change their habits, and a lack of time for cooperative activities as the main barriers of the development of farm cooperation processes in Lithuania [13–15]. Tuna and Karantininis [15] conducted a social network analysis and found that there are low levels of social capital (structural, which refers to the presence of a network of access to people and resources, as well as cognitive aspects, such as norms, values, trust, attitudes, and beliefs) in agricultural cooperatives in post-socialist countries. A lack of time and money for cooperative activities, a lack of leaders, and bureaucratic shortcomings have been emphasized as the main obstacles for cooperation in Russia [7].

In some studies, a low awareness of the benefits of contractual integration among farmers was observed along with a weak willingness to cooperate because of the low bargaining power of farmers [16,17]. Kahneman and Tversky's Prospect Theory [18] shows that individuals tend to avoid potential losses rather than seek potential benefits. Agricultural cooperatives in post-socialistic countries often fail to justify their purpose [15]. Czyzewski et al. [19] confirm that human capital plays a significant role in contractual governance and requires special attention. The positive correlations between the length of education and willingness to join agricultural cooperatives have been observed in several studies [20–23]. Education contributes to the quality of human capital. Martey et al. [24] argue that, through educational processes, farmers gain the ability to cooperate and to participate in social activities, while others conclude that better educated and self-confident farmers appreciate contractual integration more than others [20–22].

Other research claims that it can be difficult to organize work and communication processes in cooperative enterprises and it is time consuming to establish new collaborations [25]. On the other hand, an expansion in production is expected to increase production costs unless it is achieved solely through an increase in productivity via costless improved management practices [26]. In addition, exchanges of raw agricultural products are governed by stable contractual relations between farms and buyers when talking about modern markets [27], which also hinders the merging of farms into cooperatives in Lithuania. The data also show that farmers tend to compete with each other and cannot understand the benefits of cooperation [28]. The study from Miceikienė et al. [29] shows that difficult access to financial funds also limits the growth potential of agricultural cooperatives. Agricultural cooperatives face the problem of financing due to higher operational risk.

The low involvement of small-scale family farms in the activities of formal cooperatives is influenced by informal mutual assistance when farmers share machinery and experience in agricultural production and help each other with a high workload during the season [25]. Informal cooperation is of continual importance for small-scale farms [25,30]. Such informal cooperation, similar to mutual assistance and based on trust and the constant fostering of personal relationships, often occurs in Lithuania between neighbours or relatives. Informal cooperation between farmers was identified as one of the main reasons for non-cooperation of farmers in Lithuania [14]. This can be thought of as being particularly common for small-scale farmers.

In summary, three groups of factors can be distinguished, due to which farmers in Lithuania are reluctant to cooperate:

- Psychological factors include a bad association with the Soviet-era "kolchozes", producers' distrust of each other, a low level of economic awareness among farmers, a lack of leaders, a lack of successful stories, and exaggeration of negative experiences being a cooperative member.
- Economic factors and legal issues include a lack of financial funds to start a successful economic activity, the employment of specialists, a lack of financial support, the lack of a system for regulating the equitable distribution of value added throughout the value chain, and changing law.
- Organisational factors include a lack of awareness of the benefits cooperatives bring, a lack of professional consulting and coaching facilities available to cooperative members, and a diverse level of knowledge and skills of existing consultants.

Nevertheless, small-scale farms have little opportunity to compete in the traditional market through individual activities, so one of the options is local food systems, and the way they are created is through farmers' cooperation [25,31]. As long as the level of farmers' cooperation is low, the market share is low as well [32], so market integration through cooperation is essential for small-scale farms in order to improve their market position.

2.2. Incentives for Cooperation

The establishment and development of different forms of business cooperation, particularly in small-scale agriculture, must depend on the initiative and willingness of the farmer to actively participate in and join the different forms of business cooperation [6]. Many authors have raised the problem of contractual integration of farms and their drivers.

Recent examples are Abate [33], Kispál-Vitai et al. [11], Ciliberti et al. [34], Ncube [35], Souza et al. [36], and Candemir et al. [3]. Many positive effects have been found on farms participating in cooperation activities, which should be an incentive for other farms to get involved in the process as well. The efficiency of farm activities increases after they join a cooperative and the financial situation of all farms improves [37]. Cooperatives can help producers in several ways, two of which are specific to their activities in the market: countervailing power and competitive criterion, i.e., market price regulator [11]. In the case of horticulture, producer groups and organisations play a significant role in the modernisation processes [38]. Cooperatives or other producer organizations give small-scale farms the opportunity to get involved in modern agricultural value chains, especially as traditional markets are dominated by large farms [39]. Producer organizations help to reduce barriers to market entry for small producers [40]. The experience of Western European and Scandinavian countries proves that small and medium-sized farms operating through a cooperative increase their bargaining power in the market, become more competitive, and reduce production and logistics costs.

It is therefore important to emphasize, that cooperation not only contributes to the reduction of production costs, but also helps to organize certain markets [36]. Cooperatives can improve smallholder farmers' access to both input and output markets and strengthen their competitive position in different ways, both on an internal and international level [6,13]. For instance, cooperatives enable farmers to bargain collectively with both sellers of inputs and buyers of farm products, can decrease transaction costs and improve transaction efficiency, and can support the information flow between farmers and the market and thus help farmers to meet the specific requirements of high value-added food markets [41]. Ortega et al. [42] proved this positive effect for coffee producers in Rwanda, where cooperative membership was linked to greater access to inputs and an increase in income. Cooperatives can reduce market risk for their members and joint liability groups to enable access to microfinance when there is limited collateral [43], can use a collective quality label or create their own brands and create a product differentiation, and can help farmers to cope with market imperfections [3]. Cooperatives are as response to the weak bargaining power of individual farms on a market [41,44].

Cooperation promotes the development of common infrastructure (machinery, logistics, and transport), integrated food production and processing methods, and common agricultural practices [25]. Joint activities help farmers to improve production processes and logistics management and reduce food loss and waste along the entire supply chain [45]. Cooperatives and other types of organization in agriculture provide increased access to information, but also access to credits, equipment, and other types of subsidies and support [15]. Agarwal and Dorin's [46] study on group farming in France identified reasons why farmers in some regions are more likely to cooperate than in others: cooperation is more prevalent in regions with low economic inequality and with a predominance of small or medium-sized farms; more labour-intensive farms engaged in agricultural activities, i.e., livestock farmers are more likely to cooperate than cereal farmers; and it is influenced by demographic factors, such as the agricultural education of farmers. It can be added that this is influenced by the historical conditions of agricultural development in the regions or countries. Members of the agricultural cooperatives to a large extent assess their groups in social terms rather than only on economic ones [47]. Cooperation activities not only have a positive impact on the welfare of the members of cooperatives, but in general, joint activities also increase living standards in rural areas, preserve and influence the development of rural lifestyles, and prevent some rural territories from becoming extinct [7]. Thus, the social aspect of the analysed problem is also important. Based on the analysed literature, both social and economic factors were selected for further analysis.

3. Materials and Methods

In the preparation stage for this study, we defined small-scale farms in Lithuania and performed a regression analysis to determine whether there is a relationship between the physical and economic size of the farm, which could help to identify typical small-scale farms in Lithuania. The definition of a small-scale farm makes it easier to understand which farms are covered in the study. Small-scale farms are classified as farms with less than EUR 25,000 of standard output (SO). Moreover, the criterion of the physical farm size under hectares of the UAA was applied in this analysis, and its cut-off threshold determination was based on the analysis of the relationship between the economic and physical size of farms (up to 20 ha) [48]. The further research process was carried out in the three stages presented in Table 1.

Table 1. Research stages, main tasks, results, and data sources.

Research Stages	Task	Result	Data Source
Preparation stage	Define the small-scale farm size in Lithuania	Definition of the small-scale farm of the total Lithuanian farms	Farm Structures Survey data 2016 (Statistics Lithuania, 2018 and EUROSTAT, 2019)
Stage I	Explore the main characteristics of the small-scale farms in Lithuania according the official statistics	Small-scale farms' performance in Lithuania	Farm Structures Survey data 2016 (Statistics Lithuania, 2018 and EUROSTAT, 2019)
Stage II	Explore the main characteristics and market integration level of the small-scale farms in Lithuania according to the primary data Investigate the situation of the Agricultural Cooperatives and Cooperative Companies in Lithuania	Dataset on various characteristics and market integration level of small-scale farms in Lithuania Dataset of Agricultural Cooperatives and Cooperative Companies acting in Lithuania (number, size, types. and sectors of activity)	Representative survey of small-scale farms in Lithuania, 2019 (N = 1002) Cooperative survey data, 2019 (N = 102)
Stage III	Draw up a profile of a smallholder farmer, small-scale farm willing to cooperate and compare to those reluctant to cooperate	Economic characteristics of farms and social profile of smallholder farmer investigated	Representative survey of small-scale farms in Lithuania, 2019 (N = 1002)

In the first stage, the main characteristics of the small-scale farms in Lithuania were set according to the 2016 Farm Structures Survey data (Statistics Lithuania, 2018 and EU-ROSTAT, 2019). This part of the analysis allowed us to investigate the main characteristics of the small-scale farms in Lithuania: the number of small-scale farms, percentage of the total farms, utilized agricultural area covered by the small-scale farms, level of subsistence, economic size, and employment level in the small-scale farms in Lithuania.

In the second stage of the study, a survey of small-scale farms was conducted in 2019 (sample of 1002 small farms from all Lithuanian Counties as determined by the stratified selection process, based on the above definition). The random sample is representative of a 95% confidence level, 0.5 fraction, and 3% maximum error. In order to determine the spread of operating cooperatives and their structural features and the coverage of farmers' involvement in them, the Lithuanian Agricultural Cooperatives and Cooperative Companies' 2019 survey conducted by the Chamber of Agriculture of the Republic of Lithuania was used as additional information survey data (N = 102). The study was supplemented by a survey of cooperatives, conducted in 2019 (sample of 102 cooperatives from Lithuania). The sample covers 58% of the cooperatives operating in Lithuania in 2019. Additional data help to determine which cooperatives are operating in Lithuania, their size (by members), and the activities they are engaged in.

We used the data, which allowed us to set the level of willingness to cooperate among the small-scale farmers in Lithuania, to set up the market integration level (both on input and output markets) position, the level of bargaining power, the level of vertical and horizontal integration, the management of production and price risks, and the willingness to participate in cooperative actions and contractual integration in the broader sense. All data collected for the study are under the international FAMFAR Project "The role of the small farms of the sustainable development of agri-food sector in the countries of Central and Eastern Europe", financed by the Polish Agency for Academic Exchange (NAWA).

In the third stage of the study, chosen variables from the survey data of small-scale farms (N = 1002) were used to define the characteristics and profile of farm managers and farms who are willing or are reluctant to cooperate. The descriptive research method is an important part of the analysis of primary data and provides a basis for comparing variables with derived statistical tests. In many instances, description can also point toward causal understanding and to the mechanisms behind causal relationships [49]. Thus, this study (and the chosen survey methodology) is a first step in isolating a sample of cooperative small-scale farms from the overall sample of the whole survey and compiling a profile of a cooperative farm based on the methods of descriptive statistics. This type of descriptive research can be especially informative when we do not yet have enough understanding of a phenomenon.

The question in the survey was: What are your plans for participating in the cooperative and/or producer group? Possible answers were as follows: (a) I am participating and plan to continue my membership; (b) I am participating but plan to terminate my membership; (c) I do not participate, but I plan to get involved in the activities of the cooperative and/or producer group; (d) I am not present and do not plan to participate; (e) I have not heard anything about it; (f) I have no opinion. Three groups of farms were created: (1) farms willing to cooperate (123 farms, covers "a" and "c" answers); (2) farms reluctant to cooperate (576 farms, covers "b" and "d" answers), and (3) farms having no opinion (263 farms, covers "e" and "f" answers). The study sought to identify the main differences between these groups. Both social and economic components of the profile were examined to ensure the fullest possible picture. Social factors covered in the study pertained to the farm managers and included age, gender, level of education, socio-economic group, and participation in social and/or cultural events. The farm profile was defined based on economic variables such as total farm area (ha), market value of the farm (in euros), total agricultural production value (in euros), income structure (as a percentage of income from agriculture, or work, self-employment, pension, social transfers, remittances, or other sources in total farm income), and the level of direct support in agriculture income (%). All these variables were selected based on the literature review and can explain the essential features of farms' behaviour. Descriptive statistics of the dataset are presented in Tables A1 and A2.

4. Results

4.1. Small-Scale Farms' Performance in Lithuania

According to the Farm Structures Survey in 2016, there were 150,320 farms in Lithuania, excluding farms with less than 1 ha of UAA and from agricultural activity generated revenue of less than EUR 1520 per year (Statistics Lithuania, 2018). Most of them were small in physical or economic nature. In terms of physical size, half of all farms had less than 5 hectares of utilized agricultural area (UAA), while a further one-third farmed on an area 10–20 hectares in size. At the other end of the physical size scale, only 7.2% of farms had more than 50 hectares.

Along with their small physical size, most farms in Lithuania are small in economic terms as well. Table 2 presents data (in absolute and relative terms) for farms with less than EUR 25,000 of SO per year, which for this analysis will be considered as a cut-off threshold for economically small farms in this article. In 2016, there was nearly 103.5 thousand farms in Lithuania with a standard output less than EUR 8000, while a further 22.6 thousand farms had a standard output within the range from EUR 15,000 to EUR 24,999 per year. Together, very small and medium-small farms accounted for more than four-fifths (84%) of all farms in Lithuania, whereas their share of standard output was slightly more than a quarter (26%). Together they cover about a third of Lithuania's UAA.

Table 2. The main statistical characteristics of economically small-scale farms * in Lithuania, 2016.

Economic Size Based on Standard Output (SO)		Number of Farms	% of Total Farms	% of Farms Where Household Consumes >50% of the Final Production **	UAA in 1000 ha	% of Total UAA	SO in 1000 €	% of Total SO	Number of AWU	% of Total AWU
Very small farms (<€8000)	<€2000	46,300	30.8	58.8	164.4	5.6	41,303	1.9	28,840	19.4
	€2000–<€4000	30,890	20.5	60.4	142.1	4.9	89,571	4.0	23,170	15.6
	€4000–<€8000	26,330	17.5	50.9	221.4	7.6	148,804	6.7	21,810	14.7
	Total	103,520	68.8	57.3	527.9	18.1	279,678	12.6	73,820	49.7
Medium-small farms (€8000–<€25,000)	€8000–<€15,000	16,390	10.9	24.0	257.6	8.8	177,334	8.0	17,260	11.6
	€15,000–<€25,000	6250	4.2	3.4	175.5	6.0	118,598	5.3	7480	5.0
	Total	22,640	15.1	18.3	433.1	14.8	295,932	13.3	24,740	16.7
Total number of small-scale farms (<€25,000)		126,160	83.9	50.3	961.0	32.9	575,610	25.9	98,560	66.4

* Includes farms that produce agricultural products (crop or livestock), i.e., with SO higher than zero. According to the Farm Structures Survey in 2016, there were over 10,000 farms with zero SO in Lithuania. ** % of all farms in each specified economic size class. Source: own calculations based on EUROSTAT data.

As indicated in Table 2, half of the small-scale farms are subsistence-oriented, meaning their households consume more than half of the final farm production. The highest proportion of subsistence and semi-subsistence farms is in the group of very small farms in economic terms and in physical size.

In terms of the relationship between physical and economic size, the linear regression analysis based on the sample from the whole population of family farms (N = 1298) data from the Lithuanian FADN [48] shows that the physical farm size in UAA hectares has a positive relationship with the economic farm size in euros of SO ($r^2 = 0.844$, $p < 0.000$). Meanwhile, in the sample of economically small-scale farms (N = 461), a strong dependence of farm standard output on physical farm size expressed in UAA was not found ($r^2 = 0.303$, $p < 0.000$). This indicates that small-scale farms of the same size in physical terms can be extremely different by size in economic terms for various reasons, in particular the type of farming.

4.2. Extent of the Involvement of Small Family Farms in Cooperation in Lithuania

According to the physical size of the farm, the farms participating in the survey were distributed as follows: 14.3% were "three-hectare" farms (up to 3 ha of land), so named according to the Seimas of the Republic of Lithuania resolution of 1990 "On the expansion of homestead plots of rural residents to 2–3 hectares"; 39.5% of farms were a physical size of 3 to 10 ha; and the remainder (46.2%) were 10 to 20 ha.

Based on the data of the Lithuanian Chamber of Agriculture, 323 cooperative units engaged in agricultural activities were registered at the beginning of 2019. However, the investigation revealed that 147 cooperatives were suspended, terminated, or in liquidation and only 176 of them remained operating. Most agricultural cooperatives are relatively small. For example, a survey of 102 cooperatives (58% of the cooperatives operating at the time) showed that two-thirds of cooperatives have no more than ten members each, and over one hundred members have only around 8% of cooperatives (Figure 1a). Half of the cooperatives are engaged in crop and livestock production (Figure 1b). "Various agricultural activities" means that members of the cooperatives are mixed farming farms (i.e., non-specialized crop and non-specialized livestock farms). "Other activities" means that agricultural cooperatives perform various other functions such as providing other services to farms (e.g., tillage, harvesting, purchase of fertilizers).

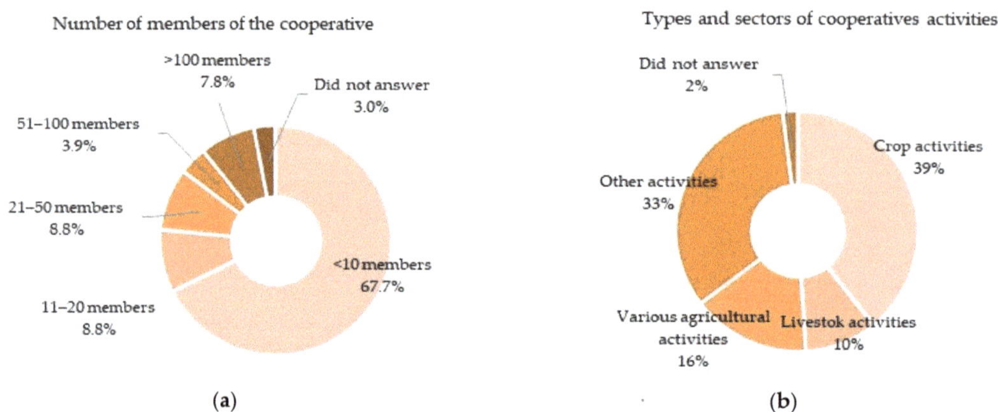

(**a**)   (**b**)

Figure 1. Structure of Lithuanian agricultural cooperatives by the number of members (**a**) and types and sectors of activity (**b**) (N = 102 Cooperative survey data, 2019).

Particularly after having joined the EU in 2004, Lithuania promoted agricultural service cooperatives among small-scale farmers. However, cooperative development was not as successful as anticipated. Their market share remained relatively low. The Lithuanian Rural Development Programs provide benefits to agricultural cooperatives, so the cooperative development depends on state support.

The vast majority of farmers or almost half of the farms surveyed sell food and agricultural products on the market without any agreement, and only 4% sell products within a producer group or a cooperative group (Figure 2a) while, another 10% base sales on long-term contracts. However, when asked to indicate which sales channel allows for a higher price, almost one-fifth indicated that these were sales through producer organizations or cooperatives (Figure 2b).

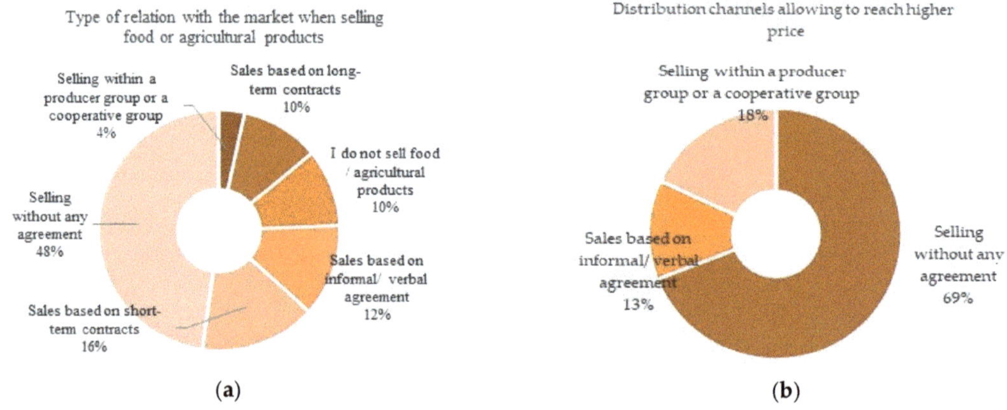

(**a**)   (**b**)

Figure 2. Type of relations with the market of Lithuanian small farms when selling food or agricultural products (**a**) or choosing the distribution channel to help reach higher selling prices (**b**) (N = 1002).

More than 45% of the small-scale farms surveyed obtain the necessary raw materials and other inputs from their own farm, and only 2% have regular agreements with input suppliers (Figure 3a); an even smaller proportion of smallholder farmers in Lithuania set the terms of the contract (Figure 3b). The position of small-scale farms in the input market is very weak, without making a significant contribution to forming terms of contracts or negotiating more favourable input prices.

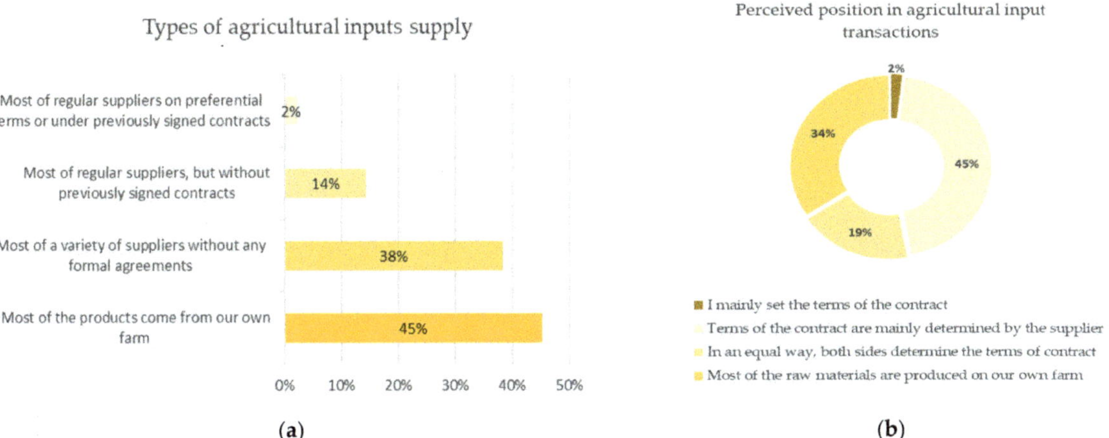

Figure 3. Types of agricultural inputs supply (**a**) and positioning in agricultural input transactions (**b**) of the small-scale farms in Lithuania (N = 1002).

Taking into consideration the production process and price risk management the vast majority of small-scale farms in Lithuania use only the compulsory insurance, which is usually mandatory when receiving investment support for a farm. Participation in producer groups or cooperatives is also not considered as a possible risk management tool for price fluctuations and is used by only a small part (only 2%) of Lithuanian smallholder farmers (Figure 4a). As many as 57% of small family farms do not participate in cooperatives and do not intend to become members, and 22% have no opinion on this issue at all (Figure 4b).

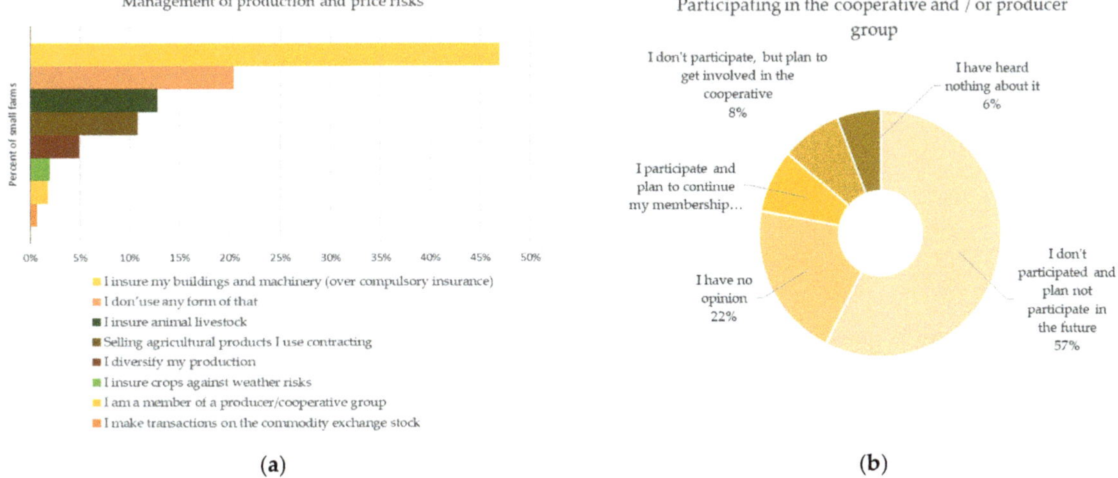

Figure 4. Management of production and price risks (**a**) and willingness to participate in cooperative actions (**b**) in Lithuanian small-scale farms (N = 1002).

The research shows that very small "three-hectare" farms are not involved in the activities of any formal cooperatives. Only about a tenth of them plan to do so. Almost one-sixth of 3–10 ha working farms and almost one-fifth of 10–20 ha working farms participate or plan to participate in a cooperative.

4.3. The Profile of Small-Scale Farms Willing to Cooperate in Lithuania

The profile of the small-scale farm managers in Lithuania was assessed from a social point of view: the age structure, gender, education, socio-economical group, and participation in social and/or cultural events were assessed. Meanwhile, the profile of small-scale farms was assessed in economic terms by farm size and type, income structure, share of support in farm income, farm capital assets, and self-sufficiency in capital resources.

The distribution of small-scale farms in Lithuania by the age of farm managers corresponds to the normal distribution. The share of farm managers under the age of 50 (51% of farm managers) and over 50 (49%) is roughly equal among the small-scale farms surveyed. The age of the farm manager seems to be an important factor in fostering the cooperation processes in Lithuania. As presented in Figure 5, the average age of those who are willing to cooperate or have no opinion (not decided yet) is lower than the age of those farm managers who are reluctant to cooperate, reaching on average about 46 years and 49 years, respectively.

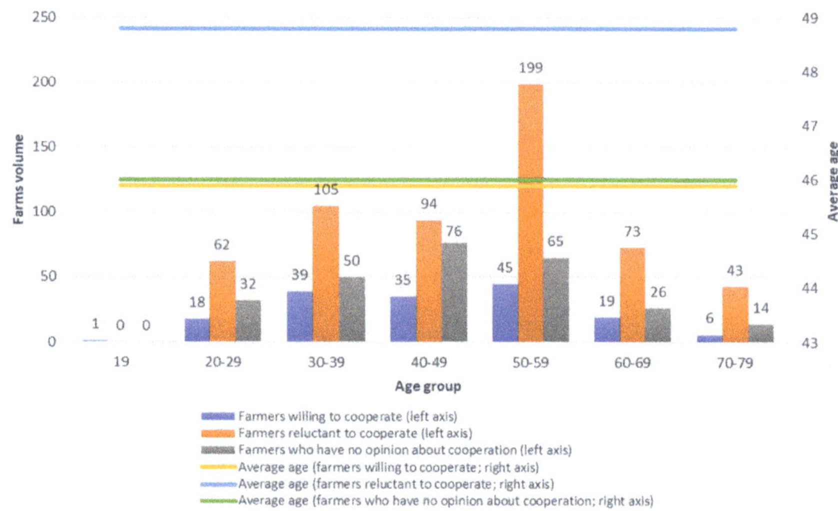

Figure 5. Distribution of smallholder farmers according to their age (N = 1002).

No significant differences are observed when analysing the gender distribution of farm managers. Both women and men as farm managers are almost equally distributed into the three groups examined (Figure 6a). A small difference is observed in the group that tends to cooperate, in which there is a higher number of women among farm managers than men. However, it would be difficult to answer from the current analysis whether this is a significant or random difference, so it is suggested that the gender aspect be analysed in detail in further studies to determine the extent to which this influences the final decisions of the farm manager in contractual integration.

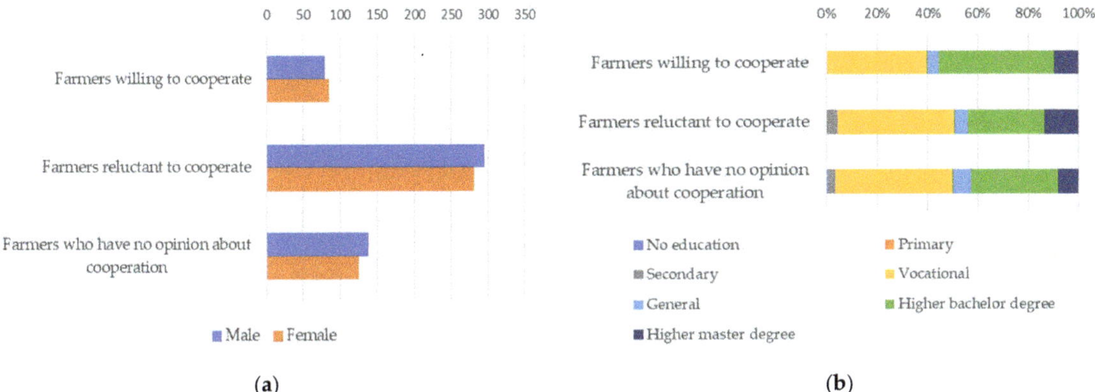

Figure 6. Distribution of smallholder farmers according to their gender (**a**) and the level of education (**b**) (N = 1002).

Assessing the education of smallholder farmers in Lithuania, it was found that there are no farm managers who have only primary education. The majority of small-scale farm managers in Lithuania have a professional or higher education (Figure 6b). A bachelor's degree education predominates in the group where the tendency to cooperate is assessed. A master's level education suggests that a farm manager will choose a profession outside of agriculture. Agricultural education does not have a significant impact on willingness to cooperate. Generally, farmers with higher education are more active in cooperative activities, especially those with a bachelor's degree.

The socio-economic group of smallholder farmers in Lithuania varies among the three analysed groups. The groups of farmers reluctant to cooperate and farmers who have no opinion about cooperation are rather similar, with the predominance of farm manager employment and a higher share in total income from other agricultural activities (Figure 7a). Farm managers who are willing to cooperate are self-employed in agriculture and earn a higher income from agricultural activities even when they have an additional income from hired work.

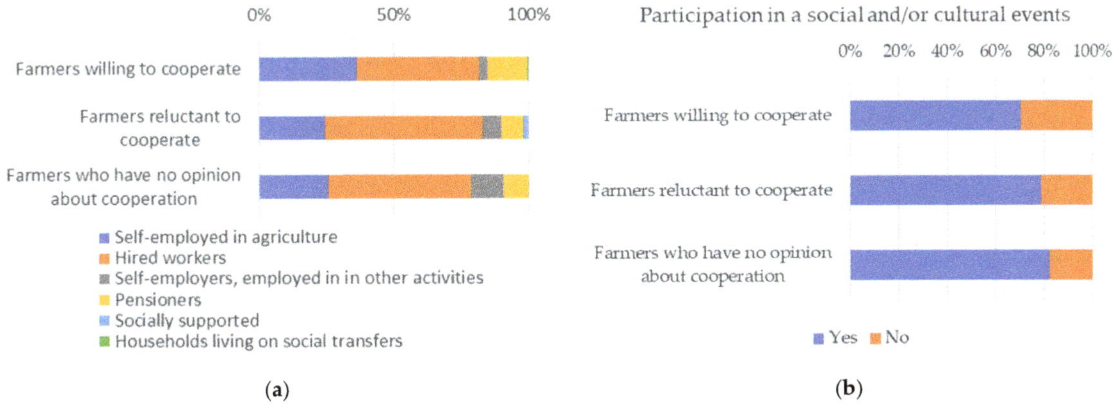

Figure 7. The socio-economic group of smallholder farmers in Lithuania (**a**) and participation in social and/or cultural events (**b**) (N = 1002).

The ratio of income from agricultural activity to total farm income is weighed in favour of income from agricultural activity. The part of the income from hired work is of high importance in total small-scale farm income. The results presuppose the conclusion that it is difficult for a small-scale farm in Lithuania to survive only from agricultural activities; therefore, farmers tend to maintain alternative sources of income, and agricultural activities are developed in parallel. Surprisingly, there are retirees among those who tend to cooperate. This means that older people with more experience are also interested in participating in joint activities and see added value in it.

Taking into consideration the level of participation in social and/or cultural events of smallholder farmers in Lithuania, it seems that the group of farmers who are willing to cooperate participate less in social activities in comparison to the other analysed groups (Figure 7b).

The next block of indicators relates to economic position of the smallholder farmers willing to cooperate. Figure 8a,b presents the farm size and the dependency of the farm on agricultural support, respectively. There is a slight difference between the analysed groups in terms of the physical size of the farm. The logical conclusion is that the larger the size of the farm, the more likely it is to cooperate. The differences between the average farm sizes of the individually analysed groups are small. The lower the share of direct payments and other support in agricultural income, the more likely farms are to cooperate. Conversely, a farm is less likely to cooperate when a higher amount of its income is from direct and other support.

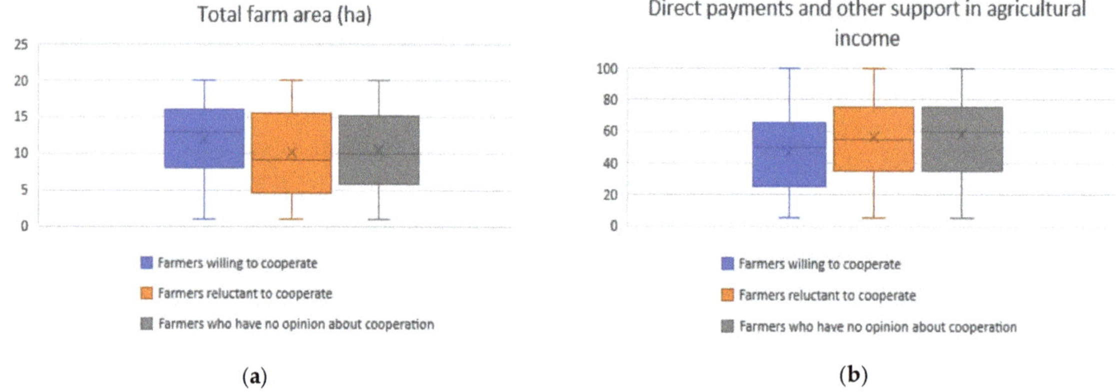

Figure 8. Small-scale farm physical size in ha (**a**) and dependency of farm on agricultural support (**b**) (N = 1002).

Farming type structures in the group of farms willing to cooperate comprise mixed farms and crop farms (Figure 9a) with obvious differences in horticulture and berry growing compared to those reluctant to cooperate (Figure 9b). Livestock farms are more likely to refuse cooperation.

The analysis of farm income structure shows that the percentage of income from different sources in total farm income differs. Farmers willing to cooperate have a higher income from agricultural activities and lower income from hired work and self-employment in comparison to those reluctant to cooperate (Figure 10a). The latter are more dependent on pensions, social transfers, and other sources of income (Figure 10b). However, it should be noted that there are also retirees who support cooperation and participate in it.

Examining the value of the output of small-scale farms, it can be seen that those with higher values of output are the more supportive of cooperation. Moreover the overall value of production is outweighed by the value of crop products (Figure 11a,b). Figure 12a,b shows the total farm assets and farm assets by type. The average value of assets is higher for those farms which are willing to cooperate (albeit very minimally); the 75% quantile is also more inclined to cooperate. Based on the sample analysed, on average, farms with higher total assets tend to cooperate. The dispersion of non-cooperators is higher.

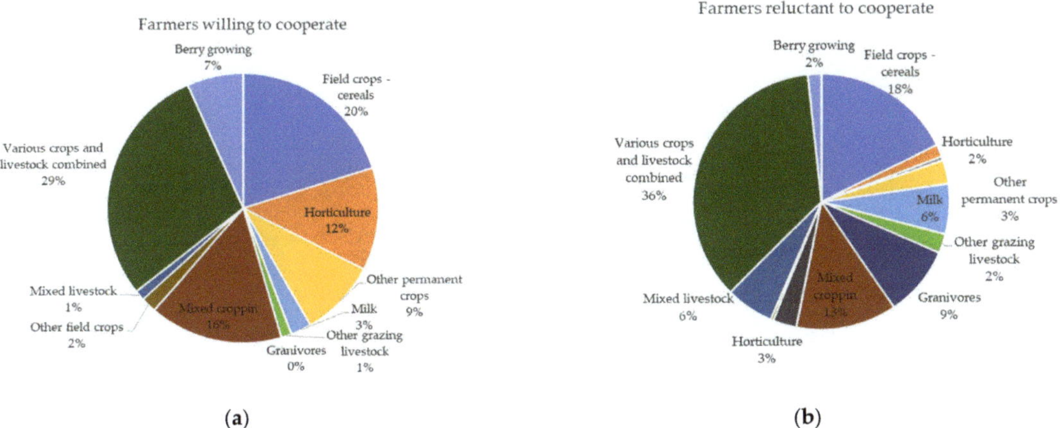

(a)   (b)

Figure 9. Farming types of the groups of small-scale farms willing to cooperate (**a**) and reluctant to cooperate (**b**) (N = 1002).

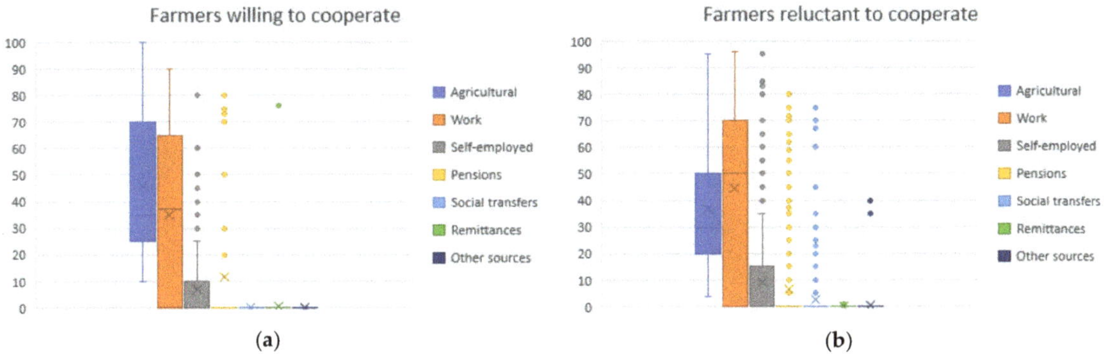

(a)   (b)

Figure 10. Small-scale farm income structure in groups of small farms willing to cooperate (**a**) and reluctant to cooperate (**b**) (N = 1002).

In summary, from the profile of the small-scale farms in Lithuania willing to cooperate, we consider that the average age of those who are willing to cooperate is lower than the age of those farm holders who are reluctant to cooperate. No significant differences can be observed when analysing the gender distribution of farm holders. A bachelor's degree education level predominates in the group where the tendency to cooperate was assessed. Smallholder farmers who are willing to cooperate are self-employed in agriculture and get a higher income from agricultural activities even when they have additional income from hired work. Taking into consideration the level of participation in social and/or cultural events of smallholder farmers in Lithuania, it seems that the group of farmers who are willing to cooperate participate less in social activities in comparison to the other analysed groups. Regarding the physical size of the farm, the larger the size of the farm, the more likely it is to cooperate. The lower the share of direct payments and other support in agricultural income, the more likely farms are to cooperate. Examining the value of the output of small-scale farms, it can be seen that higher values of output relate to more support of farm cooperation. Moreover, the overall value of production is outweighed by the value of crop products. Farms with a higher value of total assets are more willing to cooperate than those with a higher value of farm assets.

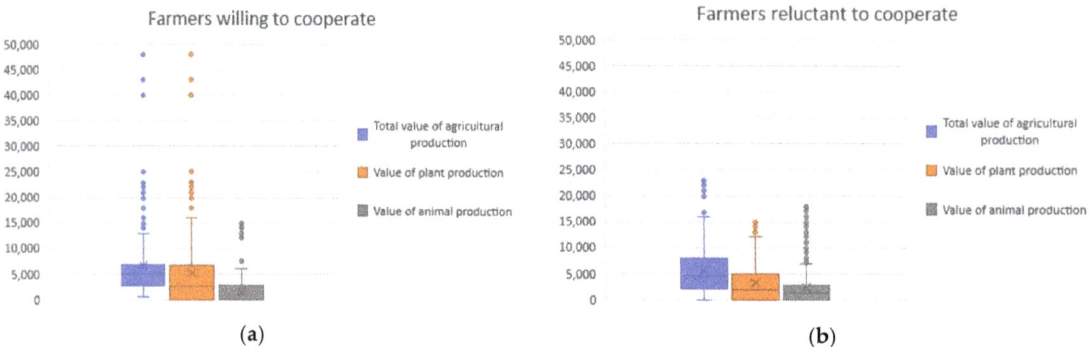

Figure 11. The value of agricultural production in groups of small-scale farms willing to cooperate (**a**) and reluctant to cooperate (**b**) (N = 1002).

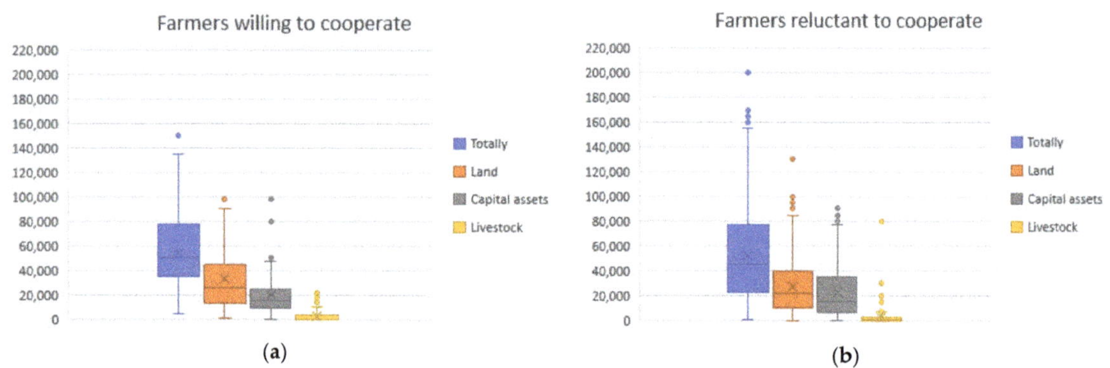

Figure 12. Farm assets in groups of small farms willing to cooperate (**a**) and reluctant to cooperate (**b**) (N = 1002).

5. Discussion

The participation of small-scale farmers in Lithuania in producer groups or cooperatives remains very low in the absence of a sufficient breakthrough compared to previous studies [13]. A specific study on dairy farmers' business strategies in Central and Eastern Europe [50] showed that Lithuanian farmers (especially milk producers) had the least interest in cooperation and chain integration. Often, this participation is only declared on paper, where farmers fictitiously join producer groups or cooperatives in order to obtain the benefits offered by the government and remain outside the joint cooperative activities. This finding proves the assumption that post-communist smallholders generally disapprove of group cooperation. Economic incentives are therefore key in forming positive intentions to join cooperatives or producer groups [51]. Even when farmers are in favour of cooperation, only one in five farmers is involved in joint activities among small farms [6]. Lithuanian researchers still have a very limited understanding of why the country's agricultural producers (especially small ones) do not engage in cooperation activities. Low involvement of small family farms in formal cooperatives might be influenced by informal mutual assistance, which is not accounted for, but has long traditions and has gained trust.

Most agricultural cooperatives are relatively small. The level of market integration of the small-scale farms in Lithuania is very low, even when the possibility to get a higher price is understood. In previous research [13], the expectation of receiving a higher purchase price as the main reason for cooperative creation was also observed. In addition, Pareigienė and Ribašauskienė [13] found that additional factors, such as EU and national support for the development of the cooperative, and access to the services provided by the cooperative are important and influence farmers' decisions to join cooperatives in Lithuania. Dendup and Aditto [52] found that the support measures were positively and strongly linked to the participation of farmers in the cooperative. Hao et al. [41] find that cooperative membership has a positive impact on selling to wholesalers and a negative impact on selling to small retailers but has no significant impact on selling to the cooperative. As products sold through cooperatives generally comply with relatively stringent food quality and safety standards, these results imply that policies promoting cooperative members to sell their products through cooperatives are likely to have a significant impact on food quality and food safety.

Smallholder farmers in Lithuania have a weak position on the market with no bargaining power; only 4% (one in twenty-four small-scale farms) sell their products within a producer group or cooperative. The results from Borychowski et al. [10] show that the resilience of farms (including Lithuania) was significantly influenced by market integration. Our study revealed that only 2% (one in fifty small-scale farms) affect the purchase prices of raw materials. More than half, i.e., 57%, of small-scale farms do not plan to join any type of cooperation and only 8% of the small-scale farms in Lithuania have intentions to join a producer group or cooperative in the future. Higher production increases the income of small-scale farms, as they are in the stage of increasing economies of scale. The effect of a higher scale of production is lower fixed unit costs, lower labour intensity, and higher bargaining power in the supply chain [10,41]. Small-scale farms have a weak market position in the input market without making a significant contribution to forming terms of contracts or negotiating more favourable input prices. Similar results were obtained in Dendup and Aditto's study [52]. The poor market, production issues like inaccessibility to inputs, and weak group cohesion among members hinder the performance of agricultural cooperatives in Bhutan. Small-scale farms in Lithuania use only the compulsory risk management tool and only 16% of the total sample are willing to cooperate.

The age of the farmers seems to be important factor in fostering the contractual integration processes. The younger a small-scale farm manager is, the more positive is his perception of cooperation activities. Dendup and Aditto's [52] study showed the positive and significant relationship of agricultural cooperatives with the age of the farmer. A literature overview shows that the age of the producer is usually associated with risk aversion and more diversified production. Together, they contribute to higher food security, and they may also stabilise agricultural income, which is an element of socio-economic sustainability. According to Gadanakis et al. [53], the relationship between a farmer's age and the farm's eco-performance was positive. This was explained by the greater experience of the manager [52,53]. Otherwise, age negatively affected resilience at a significant level (α = 0.01) in Lithuania according to a study based on the same dataset as used in this study [10].

Neves et al. [26] found that, in Brazil, higher education and smaller property sizes are associated with membership in agricultural cooperatives. In Lithuania, a bachelor's degree is predominant in the group of farms willing to cooperate, but there was no significant impact of education on resilience in Lithuania according to Borychowski et al. [10] study. The farm size analysis showed different results: the larger the size of the farm in Lithuania is, and more likely it is to cooperate. Dendup and Aditto's [52] study showed the positive and significant relationship of agricultural cooperatives and farm size. The study from Borychowski et al. [10] found that the farm area variable was significant in Lithuania, but negatively influenced the farm resilience. In the same study it was determined that animal production had a significant positive impact on farm resilience [10] in Lithuania. However, our study showed that livestock farms are reluctant to cooperate in order to strengthen their position on the market. This point was also proved in [50], which notes that dairy producers from Lithuania have the least interest in cooperation and chain integration among the analysed countries.

In the conditions of the dominance of corporations in food chains (vertical integration), reasoning would require horizontal cooperation, as the scale of production in family farms may turn out to be too small [54]. Offering large product batches is a necessity for farmers for their overall presence on the market (both domestic and foreign) as there is a growing level of concentration of the wholesale and retail commercial network in most countries in the world [55]. A structural approach to the issue of the organisation of the producer groups or cooperatives on the market and long-term financial and institutional stimulation is needed, accompanied by promotion actions.

Despite the low level of cooperation in Lithuania, farmers' willingness to cooperate should be maintained by explaining the benefits of cooperation. Kovačic et al. [6] suggested supporting small-scale farms in the joint purchase of reproduction materials, joint appearance on the market, and cross-funding (loans), detecting those which could be the initiators of cooperation activities. Generalizing the main resources, regulating work in a cooperative, promoting leadership, and informing farmers about the main benefits of cooperation are suggested as the main points in the cooperation development study carried out in Russia by Nekrasov et al. [7]. Proper agricultural and food price policies can incentivise private investment in agriculture and private banking institutions (including cooperatives) to increase their coverage in rural areas, thus strengthening farmers' resilience and risk-coping capacities [56]. Authors have even suggested international financial cooperation models. Dendup and Aditto [52] suggested promoting contractual integration through higher value-added products and improved processing, active education, and training.

Special attention should be paid to the strengthening of human capital, e.g., the training of cooperation managers, qualitative development of the network of marketing specialists, and consultants presenting and highlighting the added value generated by the cooperation. Other key variables are access to credit and extension services. Jeanneaux et al. [57] suggested implementing a strategic management approach, where cooperative members can identify their own situations with the help of a third party. The provision of governmental rural extension services mostly targets family-owned farms [26]. Providing positive examples of joining agricultural cooperatives should deliver additional incentives for farmers to cooperate in order to extract these additional benefits from cooperation [15]. Although this may be a difficult task, as young farmers in Lithuania are reluctant to use the advisory services for the development of cooperative activities [58].

New forms of information provision and inclusive cooperative activities should be encouraged in order to attract farmers, especially those of a younger age. New forms of cooperation must be offered to farmers, especially to smallholder ones. This requires the preparation of a legal framework for the establishment and development of service provision cooperatives, introducing sharing-economy principles in farming, etc. Fulton and Giannakas [59] found that procompetitive and distributional impacts of cooperatives depend critically on the sensitivity of price in the downstream retail market, the nature of the cooperative's governance structure, and the open or closed nature of cooperative membership. Ramanauskas [60] suggests that the perspective organisational form of the producer groups might be a combination of a net cooperative and a private (public) company or state capital institution. However, there is no legal basis for that in Lithuania. Šumylė and Ribašauskienė [61] emphasized that the future of the producer groups in Lithuania lies in servitisation.

6. Conclusions

This study aims to assess the level of the willingness to cooperate among smallholder farmers in Lithuania and to draw up the profiles of small-scale farms that participate in and intend to join cooperatives. The profiles of the smallholder farmers were assessed from a social point of view: the age structure, gender, education, socio-economical group, and participation in social and/or cultural events were assessed. In addition, farms' economical and financial ratios were included to draw the profile of the small-scale farm; the main indicators used in the analysis were farm size and type, income structure, share of support in total farm income, the farm market value of assets, and the level of capital resources. The research results show that participation of small-scale farms in the co-operatives or producers' groups is very low. The small-scale farms which are willing to cooperate are distinguished by the larger physical size of the farm, the lower share of direct payments and other support in agricultural income, a higher value of output, and a higher value of total assets. Additionally, younger farmers with a bachelor's degree who are self-employed in agriculture and get higher income from agricultural activities are more willing to participate in cooperative activities, regardless of gender.

In this case, the chosen analysis method (descriptive statistics) stands on its own as a research product, because the position of the small-scale farms in Lithuania and its characteristics have not previously been investigated. The analysis is based on the unique sample of small-scale farms of the Republic of Lithuania. A descriptive research method was chosen to identify the factors and characteristics of cooperative farms. The creation and description of a cooperative farm profile allows (in further research) the picture of the farm to be defined and the essential features of the farm to be identified, which are thought to form the basis for the development of farm cooperation. The direction of further research would allow for the identification of causal relationships and the establishment of a list of factors that significantly determine the participation of farms in joint activities (cooperation).

The findings of this study are important in guiding policy makers with regard to decisions on small-scale farms' cooperation development. The results of this study are also important for the farmers themselves and for other stakeholders because they can build closer relationships to develop common and sustainable future partnerships.

The presented study did not evaluate the causal relationship between the different indicators which may influence the participation in contractual integration; future research could fill in this gap which in turn could be expanded to a deeper analysis of smallholder farmers' behaviour and willingness to participate in cooperation using causal analysis and other econometric and statistical methods. Based on this research, we propose that the farmers' preferences for participating in cooperation should be assessed using qualitative research methods such as co-creation and design thinking.

Author Contributions: Conceptualization, J.D. and V.V.; methodology, J.D. and V.V.; validation, J.D. and V.V.; investigation, J.D. and V.V.; literature review, J.D. and V.V.; data curation, V.V. and L.N.; writing—original draft preparation, J.D. and V.V.; writing—review and editing, J.D., V.V. and L.N.; visualization, V.V. and L.N.; supervision, V.V.; project administration, J.D. All authors have read and agreed to the published version of the manuscript.

Funding: This research was funded under the international FAMFAR Project "The role of the small farms of the sustainable development of agri-food sector in the countries of Central and Eastern Europe" (2019–2021), financed by the Polish National Agency for Academic Exchange (NAWA), agreement no. PPI/APM/2018/1/00011/U/001.

Institutional Review Board Statement: Not applicable.

Informed Consent Statement: Not applicable.

Data Availability Statement: Data is contained within the article.

Acknowledgments: We appreciate the Chamber of Agriculture of the Republic of Lithuania for the assistance in conducting the interviews and gathering the data for the study.

Conflicts of Interest: The authors declare no conflict of interest. The views expressed are those of the author(s) and do not necessarily represent those of the Government Strategic Analysis Center.

Appendix A

Table A1. Descriptive statistics of the main quantitative variables.

Variable	Obs.	Mean	Std. Dev.	Min	Max
Age	1002	47.61	13.68	19	77
Women	490	47.73	13.67	20	77
Men	512	47.49	13.65	19	74
Total farm area (ha)	1002	10.49	5.92	1	20
Direct payments and other support in agricultural income (%)	1002	54.93	25.47	5	100
Income structure, agricultural (%)	1002	39.19	23.76	4	100
Income structure, work (%)	1002	42.04	32.02	0	100
Income structure, self-employment (%)	1002	9.72	18.19	0	95
Income structure, pensions (%)	1002	7.39	19.24	0	80
Income structure, social transfers, (%)	1002	1.42	8.11	0	75
Income structure, remittances (%)	1002	0.14	2.75	0	76
Income structure, other sources (%)	1002	0.12	2.10	0	40
Total value of agricultural production, total (euro)	1002	5614.28	4900.82	0	48,000
Total value of agricultural production, plant (euro)	1002	3287.05	4368.11	0	48,000
Total value of agricultural production, animal (euro)	1002	2330.62	3477.31	0	23,000
Estimated market value of farm, total (euro)	1002	51,308.24	37,690.97	600	205,000
Estimated market value of farm, land (euro)	1002	27,146.49	22,571.49	0	130,000
Estimated market value of farm, capital assets (euro)	1002	21,832.34	21,538.60	0	140,000
Estimated market value of farm, livestock (euro)	1002	2501.967	4559.91	0	80,000

Table A2. Descriptive statistics of the main qualitative variables.

Variable	Frequency	Variable	Frequency
Sex		Participation in a social and/or cultural events	
Female	490	*Yes*	784
Male	512	*No*	218
Education		Farm type	
No education	1	*Field crops—cereals*	185
Primary	0	*Horticulture*	38
Secondary	32	*Wine*	3
Vocational	454	*Other permanent crops*	43
General	58	*Milk*	47
Bachelor's degree	340	*Other grazing livestock*	25
Master's degree	117	*Granivores*	52
Socio-economic group		*Mixed cropping*	108
Self-employers in agriculture	268	*Horticulture*	17
Hired workers	549	*Other field crops*	34
Self-employers, employers (in in other activities)	76	*Mixed livestock*	41
Pensioners	96	*Various crops and livestock combined*	388
Socially supported	12	*Berry growing*	21
Households living on social transfers	1		

References

1. Ajates, R. Agricultural cooperatives remaining competitive in a globalised food system: At what cost to members, the cooperative movement and food sustainability? *Organization* **2020**, *27*, 337–355. [CrossRef]
2. Gonzalez, R.A. Farmers' cooperatives and sustainable food systems in Europe. *Routledge* **2018**, *11*, 45.
3. Candemir, A.; Duvaleix, S.; Latruffe, L. Agricultural cooperatives and farm sustainability–A literature review. *J. Econ. Surv.* **2017**, *35*, 1118–1144. [CrossRef]
4. Ollila, P. Farmers' cooperatives as market coordinating institutions. *Ann. Public Coop. Econ.* **2007**, *65*, 81–102. [CrossRef]

5.   Munch, D.M.; Schmit, T.M.; Severson, R.M. Assessing the value of cooperative membership: A case of dairy marketing in the United States. *J. Co-Oper. Organ. Manag.* **2021**, *9*, 100129. [CrossRef]
6.   Kovačic, D.; Juračak, J.; Žutinic, D. Willingness of farmers to cooperate. Field Study Results in the Zagreb Rural Area. *Druš. Istraž. Zagreb God.* **2001**, *6*, 1119–1129.
7.   Nekrasov, R.V.; Gusakova, E.P.; Afanaseva, E.P. Problems in Agricultural Cooperation Development in Russia (Case Study of Samara Region). *HS Web Conf.* **2019**, *62*, 08003. [CrossRef]
8.   Wolz, A.; Zhang, S.; Ding, Y. Agricultural production cooperatives and agricultural development. *Leibniz Inf. Cent. Econ. Work. Pap. CIRIEC* **2020**, *4*, 1452.
9.   Vitunskienė, V.; Drоždz, J.; Bendoraitytė, A.; Lauraitienė, L. Representing the interests of small farms in the market through cooperatives: Effect on producers' price/Mažų ūkių interesų atstovavimas rinkoje per kooperatyvus: Poveikis gamintojų kainai. *Manag. Theory Stud. Rural Bus. Infrastruct. Dev.* **2020**, *42*, 561–570. [CrossRef]
10. Borychowski, M.; Stępień, S.; Polcyn, J.; Tošović-Stevanović, A.; Ćalović, D.; Lalić, G.; Žuža, M. Socio-Economic Determinants of Small Family Farms' Resilience in Selected Central and Eastern European Countries. *Sustainability* **2020**, *12*, 10362. [CrossRef]
11. Kispál-Vitai, Z.; Regnard, Y.; Kövesi, K.; Guillotte, C.A. Cooperative case studies from three countries: Is membership a problem or a solution in the 21st century? *Soc. Econ.* **2019**, *41*, 467–485. [CrossRef]
12. Atkočiūnienė, V.; Aleksandravičius, A.; Dautartė, A.; Vitunskienė, V.; Zemeckis, R. *Farm Modernization in the Context of Markets and Rural Development: The Case of Lithuania/Ūkių Modernizacija Rinkų ir Kaimo Vystymosi Kontekste: Lietuvos Atvejis*; Aleksandras Stulginskis University: Akademija, Lithuania, 2007.
13. Pareigienė, L.; Ribašauskienė, E. Evaluation of Cooperation Development/Kooperacijos plėtros vertinimas. *Manag. Theory Stud. Rural Bus. Infrastruct. Dev.* **2008**, *1*, 127–133.
14. Kuliešis, G.; Pareigienė, L. Evaluatinion of Attitudes to Cooperate / Nuostatų kooperuotis vertinimas. *Manag. Theory Stud. Rural Bus. Infrastruct. Dev.* **2010**, *5*, 108–115.
15. Tuna, E.; Karantininis, K. Agricultural cooperatives as social capital hubs—A case in a post-socialist country. *J. Co-Oper. Organ. Manag.* **2021**, *9*, 100134. [CrossRef]
16. Livingston, M.; Erickson, K.; Mishra, A. Standard and Bayesian random coefficient model estimation of US corn—Soybean farmer risk attitudes. In *The Economic Impact of Public Support to Agriculture: An International Perspective*; Ball, V., Fanfani, R., Gutierrez, L., Eds.; Springer: New York, NY, USA, 2010. [CrossRef]
17. Stępień, S.; Polcyn, J. Risk management in small family farms in Poland. In Proceedings of the International Scientific Conference Economic Science for Rural Development, Jelgava, Latvia, 9–10 May 2019; pp. 382–388. [CrossRef]
18. Kahneman, D.; Tversky, A. Prospects theory: An analysis of decision under risk. *Econometrica* **1979**, *47*, 263–292. [CrossRef]
19. Czyzewski, B.; Sapa, A.; Kułyk, P. Human Capital and Eco-Contractual Governance in Small Farms in Poland: Simultaneous Confirmatory Factor Analysis with Ordinal Variables. *Agriculture* **2021**, *11*, 46. [CrossRef]
20. Chagwiza, C.; Muradian, R.; Ruben, R. Cooperative membership and dairy performance among smallholders in Ethiopia. *Food Policy* **2016**, *59*, 165–173. [CrossRef]
21. Mojo, D.; Fischer, C.; Degefa, T. The determinants and economic impacts of membership in coffee farmer cooperatives: Recent evidence from rural Ethiopia. *J. Rural Stud.* **2017**, *50*, 84–94. [CrossRef]
22. Zhang, B.; Fu, Z.; Wang, J.; Tang, X.; Zhao, Y.; Zhang, L. Effect of householder characteristics, production, sales and safety awareness on farmers' choice of vegetable marketing channels in Beijing, China. *Br. Food J.* **2017**, *119*, 1216–1231. [CrossRef]
23. Jitmun, T.; Kuwornu, J.; Datta, A.; Anal, A. Factors influencing membership of dairy cooperatives: Evidence from dairy farmers in Thailand. *J. Co-Oper. Organ. Manag.* **2020**, *8*, 100109. [CrossRef]
24. Martey, E.; Etwire, P.; Wiredu, A.; Dogbe, W. Factors influencing willingness to participate in multi-stakeholder platform by smallholder farmers in Northern Ghana: Implication for research and development. *Agric. Food Econ.* **2014**, *2*, 11. [CrossRef]
25. Lutz, J.; Smetschka, B.; Grima, N. Farmer Cooperation as a Means for Creating Local Food Systems—Potentials and Challenges. *Sustainability* **2017**, *9*, 925. [CrossRef]
26. Neves, M.d.C.R.; Silva, F.d.F.; Freitas, C.O.d.; Braga, M.J. The Role of Cooperatives in Brazilian Agricultural Production. *Agriculture* **2021**, *11*, 948. [CrossRef]
27. Crespi, J.M.; Saitone, T.L.; Sexton, R.J. Competition in US farm product markets: Do long-run incentives trump short-run market power? *Appl. Econ. Perspect. Policy* **2012**, *34*, 669–695. [CrossRef]
28. Melnikienė, R.; Vidickienė, D. Evaluation of the Lithuanian Agricultural Policy Based on the Analysis of Qualitative Structure. *Public Policy Adm.* **2019**, *18*, 52–67. [CrossRef]
29. Miceikienė, A.; Binkienė, D.; Savickienė, J.; Butkuvienė, V. Opportunities and problems for financing of agricultural cooperatives / Žemės ūkio kooperatyvų finansavimo galimybės ir problemos. *Sci. Stud. Account. Financ. Probl. Perspect.* **2017**, *11*, 32–40. [CrossRef]
30. Muriqi, S.; Fekete-Farkas, M.; Baranyai, Z. Drivers of cooperation activity in Kosovo's agriculture. *Agriculture* **2019**, *9*, 96. [CrossRef]
31. Vituskienė, V.; Drоždz, J.; Lauraitienė, L.; Bendoraitytė, A. 2003–2018 study (evaluation) of agricultural policy measures with a view to a more effective agricultural policy and the economic and social viability of small and medium-sized farms. In *Final Report of the R&D Project of the Research and Experimental Development Program. for Agriculture, Food, Fisheries and Rural Development 2015–2020*; Vytautas Magnus University: Kaunas, Lithuania, 2020.

32. Ciburiene, J. Farmers' Cooperative in the Context of Economic Transformations in Lithuania. In Proceedings of the 2015 International Conference "Economic Science for Rural Development", Jelgava, Latvia, 23–24 April 2015; pp. 134–141.

33. Abate, G.T. Drivers of agricultural cooperative formation and farmers' membership and patronage decisions in Ethiopia. *J. Co-Oper. Organ. Manag.* **2018**, *6*, 53–63. [CrossRef]

34. Ciliberti, S.; Frascarelli, A.; Martino, G. Drivers of Participation in Collective Arrangements in the Agri-Food Supply Chain. Evidence from Italy Using a Transaction Costs Economics Perspective. *Ann. Public Coop. Econ.* **2020**, *91*, 387–409. [CrossRef]

35. Ncube, D. The importance of contract farming to small-scale farmers in Africa and the implications for policy: A review scenario. *Open Agric. J.* **2020**, *14*, 59–86. [CrossRef]

36. Souza, A.B.D.; Fornazier, A.; Delgrossi, M.E. Local food systems: Potential for new market connections for family farming. *Ambiente Soc.* **2020**, *23*, 4422. [CrossRef]

37. Makutėnas, V.; Šukienė, A. Kooperacijos reikšmė ūkininkų ūkių veiklos efektyvumui. *Manag. Theory Stud. Rural Bus. Infrastruct. Dev.* **2017**, *39*, 203–214. [CrossRef]

38. Klepacki, B.; Krajewski, J. Wykorzystanie środków pomocowych Unii Europejskiej w rozwoju infrastruktury logistycznej grup producenckich w ogrodnictwie (The use of the European Union's aid funds in the development of logistics infrastructure of fruit producer groups in horticulture). *Rocz. Nauk. Stowarzyszenia Ekon. Rol. I Agrobiz.* **2015**, *5*, 136–140.

39. Zakić, N.; Vukotić, S.; Cvijanović, D. Organisational models in agriculture with special reference to small farmers. *Econ. Agric.* **2014**, *61*, 225–237.

40. Herbel, D.; Haddad, O.N. Successful farmer collective action to integrate food production into value chains. *Food Chain* **2012**, *2*, 164–182. [CrossRef]

41. Hao, J.; Bijman, J.; Gardebroek, C.; Heerink, N.; Heijman, W.; Huo, X. Cooperative membership and farmers' choice of marketing channels–Evidence from apple farmers in Shaanxi and Shandong Provinces, China. *Food Policy* **2018**, *74*, 53–64. [CrossRef]

42. Ortega, D.L.; Bro, A.S.; Clay, D.C.; Lopez, M.C.; Tuyisenge, E.; Church, R.A.; Bizoza, A.R. Cooperative Membership and Coffee Productivity in Rwanda's Specialty Coffee Sector. *Food Secur.* **2019**, *11*, 967–979. [CrossRef]

43. Meuwissen, M.P.; Bottema, M.J.; Ho, L.H.; Chamsai, S.; Manjur, K.; de Mey, Y. The role of group-based contracts for risk-sharing; what are the opportunities to cover catastrophic risk? *Curr. Opin. Environ. Sustain.* **2019**, *41*, 80–84. [CrossRef]

44. Vitunskienė, V. Žalio pieno vidaus rinkos struktūra ir koncentracija Lietuvoje. *Manag. Theory Stud. Rural Bus. Infrastruct. Dev.* **2019**, *41*, 576–588. [CrossRef]

45. Radzevičius, G.; Ramanauskas, J.; Contò, F. Possibilities for reduction of food loss and waste: The case study of Lithuania's producer cooperatives. *Ital. J. Food Sci.* **2015**, *14*, 99–102.

46. Agarwal, B.; Dorin, B. Group farming in France: Why do some regions have more cooperative ventures than others? *Env. Plan. A Econ. Space* **2019**, *51*, 781–804. [CrossRef]

47. Apparao, D.; Garnevska, A.; Shadbolt, N. Examining commitment, heterogeneity and social capital within the membership base of agricultural co-operatives—A conceptual framework. *J. Co-Oper. Organ. Manag.* **2019**, *7*, 42–50. [CrossRef]

48. Vitunskienė, V.; Drождz, J.; Bendoraitytė, A.; Sapa, A. Small farms in Lithuania. In *Small farms in the paradigm of sustainable development Case studies of selected Central and Eastern European countries*; Stępień, S., Maican, S., Eds.; Wydawnictwo Adam Marszałek: Toruń, Poland, 2020; Volume 14, pp. 75–99. Available online: https://marszalek.com.pl/small_farms.pdf (accessed on 28 October 2021).

49. Loeb, S.; Dynarski, S.; McFarland, D.; Morris, P.; Reardon, S.; Reber, S. *Descriptive Analysis in Education: A Guide for Researchers*; NCEE 2017-4023; Department of Education, Institute of Education Sciences, National Center for Education Evaluation and Regional Assistance: Washington, DC, USA, 2017.

50. Verhees, F.; Malak-Rawlikowska, A.; Stalgiene, A.; Kuipers, A.; Klopčič, M. Dairy farmers' business strategies in Central and Eastern Europe based on evidence from Lithuania, Poland and Slovenia. *Ital. J. Anim. Sci.* **2018**, *17*, 755–766. [CrossRef]

51. Möllers, J.; Traikova, D.; Bîrhală, B.A.M.; Wolz, A. Why (not) cooperate? A cognitive model of farmers' intention to join producer groups in Romania. *Post-Communist Econ.* **2018**, *30*, 56–77. [CrossRef]

52. Dendup, T.; Aditto, S. Performance and challenges of agriculture cooperatives in Bhutan. *Khon Kaen Agr. J.* **2020**, *48*, 1194–1205. [CrossRef]

53. Gadanakis, Y.; Bennett, R.; Park, J.; Areal, F.J. Evaluating the Sustainable Intensification of arable farms. *J. Environ. Manag.* **2015**, *150*, 288–298. [CrossRef] [PubMed]

54. Zegar, J.S. Kwestia agrarna w Polsce (The agrarian issue in Poland). *Warsaw IERiGŻ-PIB* **2018**, *14*, 575.

55. Nosecka, B. *Czynniki i Mierniki Konkurencyjności Zewnętrznej Sektora Ogrodniczego i Jego Produktów (Factors and Measures of External Competitiveness of the Horticultural Sector and its Products)*; Institute of Agricultural and Food Economics: Warsaw, Poland, 2017.

56. Calicioglu, O.; Flammini, A.; Bracco, S.; Bellù, L.; Sims, R. The Future Challenges of Food and Agriculture: An Integrated Analysis of Trends and Solutions. *Sustainability* **2019**, *11*, 222. [CrossRef]

57. Jeanneaux, P.; Capitaine, M.; Mauclair, A. PerfCuma: A framework to manage the sustainable development of small cooperatives. *Int. J. Agric. Manag.* **2018**, *7*, 68–79.

58. Baležentis, T.; Ribašauskienė, E.; Morkūnas, M.; Volkov, A.; Štreimikienė, D.; Toma, P. Young farmers' support under the common agricultural policy and sustainability of rural regions: Evidence from lithuania. *Land Use Policy* **2020**, *94*, 104542.

59. Fulton, M.; Giannakas, K. The Future of Agricultural Cooperatives. *Annu. Rev. Resour. Econ.* **2013**, *5*, 61–91. [CrossRef]

60. Ramanauskas, J. Komanditinė kooperatinė bendrovė (kooperatyvas). Socialinės inovacijos skatinant žemės ūkio gamintojų organizacijų ir kooperacijos plėtrą: Tarptautinė mokslinė konferencija: 2016 m. birželio 16-17 d.: Conference proceedings. *Vilnius LAEI* **2016**, *11*, 50–52.
61. Šumylė, D.; Ribašauskienė, E. Servitization of Lithuanian agricultural cooperatives. *Manag. Theory Stud. Rural Bus. Infrastruct. Dev.* **2017**, *39*, 510–523. [CrossRef]

Article

The Role of Cooperatives in Brazilian Agricultural Production

Mateus de Carvalho Reis Neves [1,*], Felipe de Figueiredo Silva [2], Carlos Otávio de Freitas [3] and Marcelo José Braga [1]

[1] Departamento de Economia Rural, Universidade Federal de Viçosa, Edson Potsch Magalhães Building, Purdue St., Campus Universitário, Viçosa 36570-900, Brazil; mjbraga@ufv.br
[2] Department of Agricultural Sciences, Clemson University, 235 McAdams Hall, Clemson, SC 29634, USA; fdsilva@clemson.edu
[3] Departamento de Ciências Administrativas, Universidade Federal Rural do Rio de Janeiro, BR 465 Rd., Km 07, Seropédica 23890-000, Brazil; carlosfreitas87@ufrrj.br
*   Correspondence: mateus.neves@ufv.br; Tel.: +55-031-3612-4342

Abstract: Much of the established literature on agricultural cooperatives describes their myriad contributions to farmers' economic performance. In Brazil, one of the world's leading agricultural exporters, there were more than 1500 agricultural cooperatives with 1 million members in 2020, and in 2017, 11% of all Brazilian farms were associated with one of these cooperatives. In this paper, we estimate the factors associated with the municipality share of cooperative membership (MSCM) and how municipality-level production value changes with MSCM. Our analysis is at the municipality level using aggregate data from the 2017 Agricultural Census. We find that in Brazil, higher education and smaller property sizes are associated with membership in agricultural cooperatives. To estimate how MSCM is associated with farm profits, we use a generalized propensity score and find that an increase in MSCM increases net municipal farm income, driven mostly by an increase in the value of agricultural production compared to a smaller increase in the cost of production.

Keywords: agricultural markets; generalized propensity score; cooperative organizations

JEL Classification: Q12; Q13; C31

Citation: Neves, M.d.C.R.; Silva, F.d.F.; Freitas, C.O.d.; Braga, M.J. The Role of Cooperatives in Brazilian Agricultural Production. *Agriculture* **2021**, *11*, 948. https://doi.org/10.3390/agriculture11100948

Academic Editors: Adoración Mozas Moral and Domingo Fernandez Ucles

Received: 3 September 2021
Accepted: 27 September 2021
Published: 29 September 2021

Publisher's Note: MDPI stays neutral with regard to jurisdictional claims in published maps and institutional affiliations.

1. Introduction

Brazil is one of the largest net exporters of agricultural commodities in the world [1]. However, three-quarters of its agricultural sector's income is produced by the largest farms, which represent fewer than 10% of the farms in the country [2], indicating that Brazil's agricultural sector is composed mostly of small farms. Small farmers often encounter marketing obstacles that result in lower profits [3], and this phenomenon is particularly pronounced in developing countries, such as Brazil, where marketing channels are often characterized by market failures.

Cooperatives (although the word "cooperative" can be applied to different types of collectively developed activities, we use the term to describe a democratically controlled and managed business model. In many countries, as in Brazil, cooperatives are defined legally as a specific kind of corporation and are subject to specific national legislation [4]) can serve as an alternative marketing option, especially for small farms, and may be used to narrow this wide disparity between large and small farms [5–9]. Agricultural cooperatives can help farmers obtain higher prices compared to alternatives such as capital-owned firms [10–13]. These small farmers also benefit from the support provided by cooperatives for adopting better agricultural practices and new technologies, as well as from access to inputs at lower prices [14–17]. For example, members of cooperatives in the Brazilian state of Paraná were the first farms to introduce a new technology in the 1980s, so-called "no-till farming" (Portuguese: plantio direto), which minimizes the impact of cultivation on the soil [1]. In Brazil, cooperatives can have a great impact on household income.

117

According to the Brazilian National Sample Survey of Households (PNAD) of 2014 [18], 90% of households in rural areas that reported being associated with a cooperative carried out some marketing activity through the cooperatives.

Historically, cooperatives in rural areas have been encouraged by the Brazilian government. Several articles of the Brazilian Constitution establish that the government is to encourage the creation of cooperatives; indeed, the law that regulates the operation of cooperatives highlights the role of the government in encouraging their development (Law no. 5764 of 16 December 1971 is known as the Brazilian Cooperative Law. Based on this law, the National Cooperative Policy was defined, and the legal regime of cooperative societies was instituted, in addition to other measures). In 2017, the municipality share of cooperative membership (MSCM) (The municipality share of cooperative membership (MSCM) is the ratio of farms associated with a cooperative in a municipality compared to all farms in this same municipality) was 11.42%, according to IBGE [19] data (see Table 1). Even though the number of farms is greater in the Northeastern region—home to 45% of all Brazilian farms—their contribution to the national gross value of production (11.58%) is far less than in the South (26.41%), Southeast (28.23%) and Midwest (27.07%). The number of farms associated with a cooperative follows a similar pattern, constituting almost 37% of all farms in the South. Despite these disparities, nearly 578,000 Brazilian farms are associated with a cooperative, which suggests that cooperatives may be positively related to the national gross value of production and still have room to grow (For a more comprehensive view of the heterogeneities of the Brazilian rural environment, cf. [1,20]. The paper by Neves et al. [21] conducts an initial analysis of the heterogeneity of Brazilian cooperativism in its different regions).

Table 1. Number of municipalities, population, farms, municipality share of cooperative membership (MSCM) and participation in gross value of production (GVP), Brazilian regions, 2017.

	North	Northeast	Southeast	South	Midwest	Brazil
Municipalities (#)	450	1797	1668	1191	467	5570
Population (#)	17,936,201	57,254,159	86,949,714	29,644,948	15,875,907	207,660,929
Farms (#)	580,613	2,322,719	969,415	853,314	347,263	5,073,324
Members of cooperatives (#)	20,309	33,592	165,630	313,763	46,144	579,438
MSCM (%)	3.50	1.45	17.09	36.77	13.29	11.42
Gross value of production (%)	6.72	11.58	28.23	26.41	27.07	100

Source: The authors used aggregated data from the 2017 Ag. Census made available by IBGE [19].

A number of studies have investigated the effect of cooperative membership on farmers' production and/or income for specific regions or commodities [13,22–25]. Cazzuffi [22] found that cooperative membership increased the income of milk producers in three Italian provinces, especially among small farmers and those distant from consumer centers. Jardine et al. [23] found that cooperative membership increased the income of Alaskan fishermen. Focusing on small rural banana producers in Rwanda, Verhofstadt and Maertens [13] found that participation in cooperatives increased income. Also in Rwanda, Ortega et al. [24] found that cooperative membership increased the income of coffee producers, as well as their access to inputs. Kumse et al. [25] focused on the marketing component of cooperatives for rice farmers in Thailand and found that marketing through cooperatives had a direct effect and a spillover effect on prices.

There are also a few studies that estimate the role of cooperative membership on economic and/or technical efficiency [9,26–28]. Exploring the determinants of technical efficiency among farmers in the Midwestern region of Brazil, Helfand and Levine [26] found that cooperative membership is associated with increased efficiency. For coffee producers in Costa Rica, Wollni and Brümmer [27] found no effect of cooperatives on productive efficiency. For small maize producers in Nigeria, Olagunju et al. [9] found that cooperative membership increased technical production efficiency. For Brazil, Costa et al. [28] found that cooperative membership is associated (as constructed in this paper) with higher technical efficiency among family-owned farms. In another paper concerning Brazil, Neves et al. [29] estimated an agricultural production function and found a positive effect for the share of cooperative membership on the gross value of production (GVP) in the Southern and Southeastern regions, where there are more cooperative members.

In 2018, the Organization of Brazilian Cooperatives (Organização das Cooperativas Brasileiras—OCB) [30], which has more than 6000 cooperatives as members, released a request for application (RFA) to stimulate research on cooperatives (The RFA 07/2018 was carried out and conducted by the National Council for Scientific and Technological Development (CNPq), aiming to use the wide expertise of CNPq in this type of activity. It also aimed to guarantee the impartiality and absence of conflicts of interest between researchers and the research funding entity, the Organization of Brazilian Cooperatives (OCB)). One of the priority areas of the RFA was the estimation of the economic impact of cooperatives. This is also evidence of the lack of studies on the economic impact of cooperatives. In this paper, we estimate the effect of municipality share of cooperative membership (MSCM) on net income in Brazil, breaking it down into the effect on GVP production costs (e.g., wages, fertilizers, seeds, pesticides, veterinary medicines and animal feed). To estimate how MSCM is associated with farm profits, we use a generalized propensity score that allows us to control for bias from the observable characteristics of Brazilian rural establishments.

Our study contributes to the literature in that it is one of the first to analyze the role of cooperatives for all types of farms using the most recent available data from the 2017 Census of Agriculture. Although Costa et al. [28] used the same dataset, they focused on family-owned farms, whereas we consider all farms in Brazilian municipalities. Our analysis also sheds light on the drivers of the effect of MSCM on agricultural producers' net income. Although only a few papers are studying this topic for Brazil, several studies have found a positive effect for cooperatives in other countries [13,22–25].

2. Materials and Methods

2.1. Empirical Strategy

To estimate the effect of MSCM on agricultural income in Brazil, we used the generalized propensity score (GPS) method proposed by Imbens [31]. Hirano and Imbens [32] argued that GPS accounts for the selection bias caused by observable characteristics when dealing with a continuous treatment variable. Our measure of cooperatives is the ratio of producers associated with cooperatives in each Brazilian municipality compared to all producers in the same municipality. It implies that traditional methods of treatment effect, such as propensity score matching proposed by Rosenbaum and Rubin [33], would not be adequate for this analysis, as they are applicable in situations where treatment is a dichotomous variable.

2.1.1. Generalized Propensity Score

The objective of the generalized propensity score (GPS) is to estimate an average dose-response function, which allows for the identification of the treatment effect considering different intensities of this treatment. In the context of this investigation, it allows us to identify the average effect of cooperatives for different levels of the MSCM. We first build a control group based on a vector of observable characteristics, simulating a quasi-experimental scenario.

For the application of the method, consider a sample consisting of $i = 1, \ldots, N$ municipalities. According to Hirano and Imbens [32], for each i, there is a set of potential results called $Y(t)$, with t being the level of treatment, such that $t \in T_i$. The mean dose-response function can be defined as:

$$\mu(t) = E[Y(t)], \quad \forall\, t > 0 \tag{1}$$

where $E[Y(t)]$ determines the potential response of Y (in this research: average agricultural net income, value of production and expenses), given the level of the observed MSCM ($t \in T_i$). For each municipality i, there is a set of observable characteristics (vector X_i), which are considered in the estimation of the generalized propensity score (GPS). This, in turn, is obtained by the conditional density of the treatment given the vector X, $gps(t, x) = f_{T|X}(t|x)$. Thus, Hirano and Imbens [32] define GPS as:

$$GPS = gps(T, X) \tag{2}$$

The GPS and the propensity score in the PSM method both must confirm the balance property, indicating that for municipalities with the same propensity score ($gps(t, X)$), the probability of presenting a certain intensity in the MSCM does not depend on the observable characteristics X. Thus, the model controls for bias due to observable heterogeneity when estimating the conditional expectation of the response variable, depending on the level of treatment and the GPS ($\beta(t, gps)$) and, subsequently, obtaining the dose-response function for a given treatment level ($\mu(t)$), as follows:

$$\beta(t, gps) = E[Y|T = t, GPS = gps] \tag{3}$$

$$\mu(t) = E[\beta(t, gps(t, X))] \tag{4}$$

The practical implementation of this approach has three main stages. First, under the hypothesis of normality for the treatment distribution (MSCM), the $GPS_i(t, X_i)$ is estimated by maximum likelihood:

$$G\hat{P}S_i = \frac{1}{\sqrt{2\pi\hat{\sigma}^2}} \exp\left(-\frac{1}{2\hat{\sigma}^2}(T_i - \hat{\beta}_0 - \hat{\beta}_1 'X_i)^2\right) \tag{5}$$

After testing the balancing hypothesis, the second stage consists of calculating the conditional expectation of Y_i, given the treatment level T_i and the $G\hat{P}S_i$, considering a quadratic approximation:

$$E[Y_i|T_i, G\hat{P}S_i] = \alpha_0 + \alpha_1 T_i + \alpha_2 T_i^2 + \alpha_3 G\hat{P}S_i + \alpha_4 G\hat{P}S_i^2 + \alpha_5 T_i G\hat{P}S_i \tag{6}$$

Although α's parameters have no direct interpretation, their statistical significance can be considered as evidence of the existence of the bias generated by the characteristics considered [32]. Once the parameters of the second stage were obtained, in the third stage, we estimated the average dose-response function at treatment level t, as follows:

$$E[\hat{Y}(t)] = \frac{1}{N} \sum_{i=1}^{N} (\hat{\alpha}_0 + \hat{\alpha}_1 t + \hat{\alpha}_2 t^2 + \hat{\alpha}_3 G\hat{P}S_i + \hat{\alpha}_4 G\hat{P}S_i^2 + \hat{\alpha}_1 t G\hat{P}S_i) \tag{7}$$

From (7), it becomes possible to obtain the average potential result of $\hat{Y}_i$ for each treatment level. That is, it is possible to identify the effect of different levels of MSCM on average agricultural net income, as well as on the value of production and expenses.

2.1.2. Data

We use the municipality-level dataset from the 2017 Brazilian Agricultural Census [19] (data on the Brazilian Agricultural Census 2017 can be retrieved from the IBGE Automatic Recovery System (Sistema IBGE de Recuperação Automática—SIDRA), accessible at <https://sidra.ibge.gov.br/pesquisa/censo-agropecuario/censo-agropecuario-2017>, accessed on 27 September 2021). We estimated the effect of the MSCM on three response variables: gross value of production, production cost and net income. The variable MSCM is a proxy for farmer engagement with cooperatives at the municipal level. To build this variable, we use the question, "Are you a member of a cooperative?" (in Portuguese, "Voce é associado de alguma cooperativa?"). The response or outcome variables of interest are net income, the gross value of agricultural production (GVP) and production costs (expenses), all in Brazilian reais—BRL (Brazilian currency). The GVP is the sum of the value for all Brazilian agricultural production in 2017, encompassing crops, livestock, forestry and types. The expenses are the sum of costs related to wages, soil correctives, fertilizers, pesticides, livestock medicines, seeds and seedlings, salt/feed and fuel, and is used as a proxy for agricultural inputs. Net income is obtained by subtracting production costs from the gross value of agricultural production.

Based on the previous literature, we account for several variables that determine MSCM: investments, experience, hired work (in the Brazilian Agri Census 2017, it is verified whether the farms have rural workers: (i) below the age of 14 and (ii) above the age of 14), education, access to credit (except from cooperatives), access to rural extension (technical assistance) (except from cooperatives), land ownership, farm size and Brazilian microregion dummies (see Table 2). To control for outliers, we dropped the top and bottom one percentile based on the municipal gross value of production, similarly to Helfand et al. [34]. Our sample consisted of 5252 municipalities in all regions of Brazil for 2017.

Table 2. Average of selected variables used in the model by different levels of MSCM, Brazil, 2017.

		Municipality Share of Cooperative Membership				
	Overall	**0.0**	**<0.01**	**0.01–0.05**	**0.05–0.2**	**>0.2**
Net Income ($) [1]	84.77	40.08	20.29	45.23	105.92	150.35
Gross Value of Production ($) [1]	141.93	57.88	30.96	69.94	171.66	263.43
Ag. Expenses ($) [1]	57.16	17.81	10.67	24.71	65.74	113.09
Land used (ha)	18,190.00	8669.45	13,413.73	16,751.39	23,599.24	20,157.60
Labor (Age > 14 years old) [2]	2645.13	1614.10	3701.33	3142.27	2372.89	1891.37
Capital (units) [3]	1.06	0.41	0.16	0.44	1.40	2.02
Other Investments ($) [4]	56.24	14.50	10.24	23.91	64.18	112.55
Group of Area						
0–5 ha	0.02	0.05	0.06	0.02	<0.01	<0.01
5–20 ha	0.21	0.23	0.35	0.27	0.13	0.13
20–100 ha	0.54	0.53	0.46	0.49	0.57	0.60
100–500 ha	0.21	0.18	0.13	0.18	0.26	0.25
500+ ha (base)	0.02	0.02	0.01	0.03	0.04	0.01
Landowner	0.82	0.77	0.77	0.81	0.82	0.85
Literate	0.82	0.66	0.66	0.76	0.89	0.95
College degree or higher	0.09	0.05	0.04	0.07	0.12	0.13
Experience [5]	0.01	0.02	0.02	0.02	0.01	0.01
Governmental Extension [6]	0.10	0.08	0.06	0.08	0.12	0.14
Private Extension [6]	0.17	0.04	0.04	0.07	0.17	0.36
Credit [6]	0.10	0.03	0.06	0.07	0.10	0.18
Regions						
North	0.08	0.09	0.11	0.16	0.08	<0.01
Northeast	0.32	0.74	0.74	0.48	0.10	0.01
Southeast	0.30	0.14	0.11	0.22	0.45	0.39
South (base)	0.22	<0.01	0.01	0.05	0.21	0.54
Midwest	0.08	0.03	0.02	0.10	0.16	0.06

Notes: [1] Value in millions of Brazilian reais (BRL); average exchange rate in August 2017: BRL 3.15/USD 1. [2] Average number of workers older than 14 years. [3] Units of tractors, combine harvesters, fertilizer spreaders, seeders and other tools. [4] Value (in BRL) of the livestock and buildings present on the farms. [5] Share of farms managed by farmers with 20 or more years of experience. [6] Share of farms that have access to these services. Source: The authors used aggregated data from the 2017 Ag. Census made available by IBGE [19].

It is important to consider the regional differences of Brazil when analyzing the production function for the municipalities. According to Buainain et al. [35], Neves et al. [21] and Homma et al. [20], in addition to the natural conditions, the Brazilian territory presents heterogeneities arising from other factors, such as those related to the diverse agricultural products grown in different areas of the country. With this in mind, the regression is estimated by considering fixed effects at regional levels in an attempt to control for this spatial heterogeneity. To do so, the variable of interest was being regressed on dummies for each macroregion of the country (North, Northeast, Southeast and Midwest, with the South as the base category). The dummy variables take the value of 1 when the municipality belongs to a unit of the federation, and 0 otherwise. Thus, these variables are included in the model to represent the MSCM of each region in Brazil.

3. Results and Discussion

To our knowledge, the study by Costa et al. [28] is the sole existing analysis of cooperatives using the 2017 Agricultural Census (Ag. Census). Due to the dearth of studies analyzing the results of the Ag. Census, we first discuss the results of the Ag. Census before then discussing the econometric results.

3.1. Cooperative Membership in the 2017 Agricultural Census

As of 2017, only 11.4% of Brazilian farms are associated with cooperatives. Figure 1 (left graph) shows the histogram of the MSCM. Most municipalities have low MSCM—indeed, almost one-third of the municipalities has a share equal to or less than 1%.

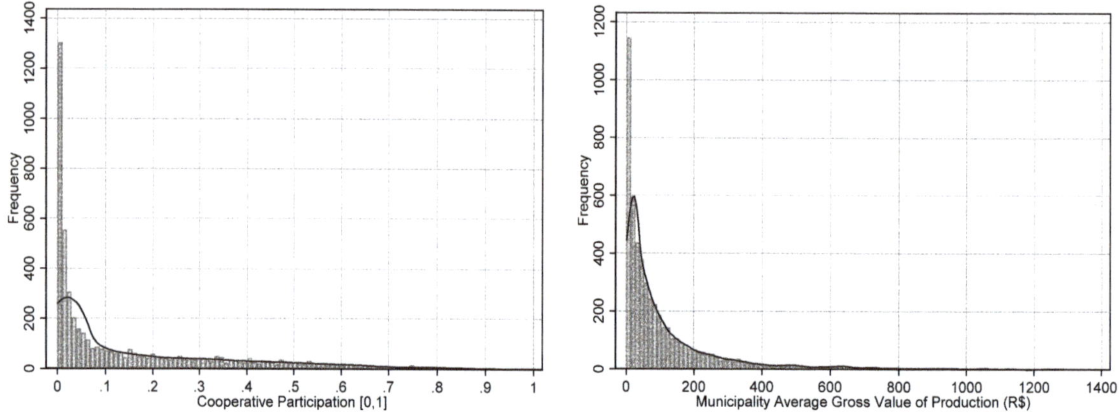

Figure 1. Histogram of MSCM (**left**) and municipality average GVP (**right**), Brazil, 2017. Source: the authors used aggregated data from the 2017 Ag. Census made available by IBGE [19].

Despite the overall low MSCM, some regions of the country stand out (see Table 1). Figure 2 displays the geographic distribution of the MSCM and the GVP; both show a similar pattern. It shows that regions with greater GVP also have a greater MSCM. The economic relevance of the Southern, Southeastern and Midwestern regions to Brazil's national agricultural production is well documented [1,2,20]. Therefore, a similar pattern in the distribution of GVP is expected. Historical factors associated with regional development of agriculture and the commodities allocation in each region play a role in the development of the Brazilian agricultural sector, which also resulted in a wide range of MSCM across the country.

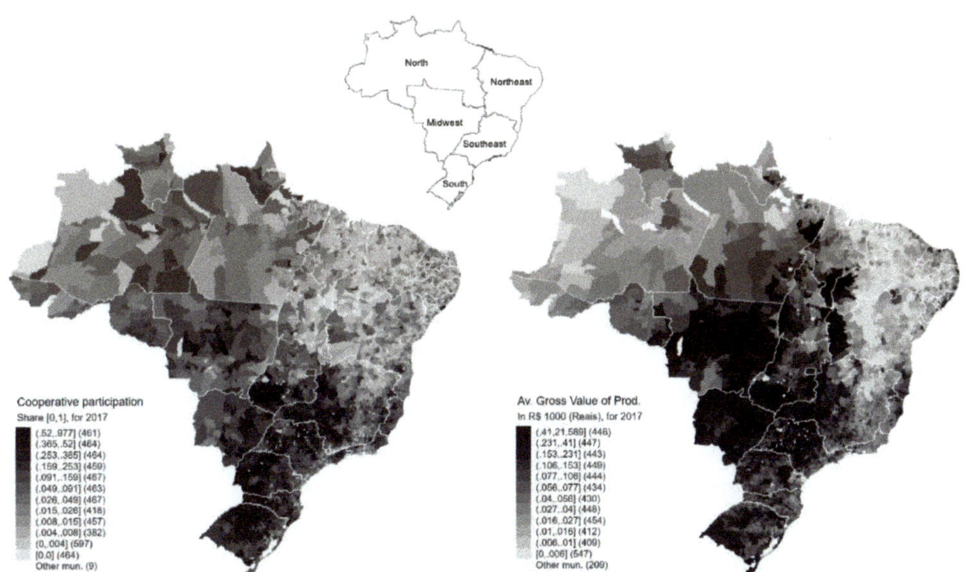

Figure 2. Geographic distribution of MSCM (left) and municipal averages of GVP (right), Brazil, 2017. Source: the authors used aggregated data from the 2017 Ag. Census made available by IBGE [19].

An exception to this pattern, also shown in Figure 2, is the newest agricultural frontier, known as the "MATOPIBA" region, a continuous area formed by the states of Maranhão (MA), Tocantins (TO), Piauí (PI) and Bahia (BA). The predominant biome in this region is the *Cerrado*, characterized by high grain productivity attributable primarily to large rural estates [36] (for a characterization of the most recent agricultural frontier of Brazil, known as "MATOPIBA", cf. Araújo et al. [36]) with a low MSCM. The Northeastern and Northern regions also show low MSCM. In the Amazon Rainforest region, even though there is no strong tendency toward cooperative membership, some cooperatives support small farms and so-called "extractivists" in their sustainable activities. These organizations have achieved good results (e.g., higher prices for members) by improving marketing capabilities and working conditions for families [37,38]. These results demonstrate the potential of using incentives such as governmental programs to increase cooperative membership in the Amazon rainforest.

The farm size, Group of Area shown in Table 2, is the municipality's average farm size. As of 2017, 54% of Brazilian municipalities have an average farm size between 20 and 100 hectares. Fewer than 2% of municipalities have an average size greater than 500 hectares. As MSCM increases, the number of municipalities with average farm sizes of 20–100 hectares and 100–500 hectares increases. Whereas 32% of the municipalities are in the Northeast, only 8% are in the South (column labeled *Overall*). This distribution is inversely correlated with the average size of municipalities in each region, given that the North has large municipalities, such as Altamira with an area of 159,533 km^2 (bigger than England). In 2017, 74% of the municipalities with MSCM equal to 0 were in the Northeast, compared to less than 1% in the South. By contrast, the South and Southeast accounted for 93% of the municipalities where more than 20% of farms are associated with a cooperative.

Historically, incentives to participate in cooperatives target rural producers with small- and medium-sized farms [28]. As described by Chaddad [1], the Southeastern and Southern regions of the country, which are marked by higher levels of cooperative membership, have a land structure dominated by medium-sized farms. Whereas the Northeastern region, with low rates of cooperative membership, is the region with the smallest average farm sizes, the Midwestern and Northern regions have the largest farms.

Table 2 also shows that average labor (average number of workers older than 14 years) grows as the MSCM increases in Brazil. However, in municipalities with the highest MSCM, average labor decreases. The opposite is found for capital (in units of tractors, combine harvester, fertilizer spreader, seeder and other tools). Studies analyzing the total productivity of factors, such as Alves et al. [39], Helfand et al. [40], Gasques et al. [41] and Rada et al. [42], found that purchased inputs and capital are responsible for the greatest productivity gains in Brazil, especially on the largest farms. These production factors replace labor on many farms, mainly in the Southern, Southeastern and Midwestern regions. Valentinov [12] and Cechin [14] argue that cooperatives facilitate their members' access to inputs and capital goods, in addition to providing access to extension and technical assistance. This may also be associated with the farmers' investments, here measured by the value of the livestock and buildings on the farms.

3.2. Econometric Results

A more adequate analysis would consider the distribution of MSCM displayed in Figure 1. The generalized propensity score (GPS) approach allows us to represent these clusters as treatment intervals [32,43]. To do so, based on the distribution of this variable (Figure 1), we split the sample into four treatment intervals: (1) observations with MSCM within the first 0.25 percentile of the distribution, which consists of values in the range (0–0.0099); (2) observations between 0.25 and 0.50 percentiles with values in the range (0.0099–0.055); (3) observations between 0.50 and 0.90 percentiles with values in the range (0.055–0.49); and (4) observations higher than 0.90 percentile with values in the range (0.49–0.97) (Table A1, in the Appendix A, displays the test of conditional means for all treatment intervals). We estimate the effect of the share of farms that are members of a cooperative within a municipality (MSCM) on response variables.

In Table 3, we display the results for the first stage of the GPS, which yield the propensity score used in the dose-response estimation (as displayed in Figure 1, the distribution of the MSCM is skewed, which led us to use a zero-skewness log transformation on this variable to estimate the propensity score). The results of Tobit in Table 3 cannot be interpreted as marginal effects. However, the sign of these estimated parameters still indicates whether the variable contributes positively or negatively to the MSCM.

Table 3. Results for the first stage of the GSP: regression on MSCM, Brazil, 2017.

Variables	Parameter	Standard Error	*t*-Test	*p*-Value
Capital	0.005	0.003	1.84	0.07
Other Investments ($)	0.001	5.12×10^{-5}	4.19	<0.01
Land used (ha)	-2.49×10^{-7}	2.86×10^{-7}	−0.87	0.38
Labor (Age > 14 years old)	7.37×10^{-6}	5.03×10^{-6}	1.46	0.14
Landowner	−0.001	0.086	−0.02	0.99
Literate	2.098	0.154	13.65	<0.01
College degree or higher	1.701	0.217	7.83	<0.01
Experience	2.023	1.096	1.85	0.065
Governmental Extension	0.716	0.106	6.74	<0.01
Private Extension	2.227	0.096	23.29	<0.01
Credit	1.776	0.140	12.69	<0.01
North	−1.221	0.063	−19.39	<0.01
Midwest	−0.716	0.060	−11.9	<0.01
Southeast	−0.599	0.040	−14.82	<0.01
Northeast	−1.299	0.062	−21.12	<0.01
0–5 ha	0.136	0.130	1.04	0.27
5–20 ha	0.360	0.102	3.53	<0.01
20–100 ha	0.407	0.096	4.22	<0.01
100–500 ha	0.428	0.091	4.69	<0.01
Constant	−4.984	0.183	−27.25	<0.01

Source: The authors.

These results indicate that municipalities with higher investments in fixed capital ("Capital" in Table 3)—tractors, combine harvesters, fertilizer spreaders and seeders—are associated with greater MSCM. We find the same results for other investments, which accounts for land rental, animal purchase, transportation and logistics costs, energy cost and others. Both associations indicate that municipalities with higher investment capabilities also have greater proportions of farms associated with cooperatives. The dummy variables for the regions (using the Southern region as the base) confirm the numbers in Table 1, indicating that most of the farmers in cooperatives are in the Southern, Southeastern and Midwestern regions. Chaddad [1] and Barros [2] state that at the beginning of the 1990s, cooperatives started to assist producers in the acquisition of inputs and machinery, as well as in better management of their rural properties.

We also find that education is associated with higher levels of cooperative membership. This finding indicates that larger shares of people in a municipality who can read and write ("Literate") and who have a college degree or higher are associated with a higher MSCM. Iliopoulos and Cook [44] and Cechin et al. [45] agreed that farmers' levels of education influence their decision-making behavior and can affect their information levels as well. According to the authors, education is a key factor in the decisions to become a member and to remain a member, as well as to take better advantage of the opportunities within these collective organizations. At the farm level, several studies have also found a positive relationship between education and farmers' cooperation, including Bernard and Spielman [46] for Ethiopia, Fischer and Quaim [6] and Abate [47] for Kenya, Cechin et al. [45] for Brazil, and Olagunju et al. [9] for Nigeria.

Other key variables associated with MSCM are access to credit and extension. The provision of governmental rural extension mostly targets family-owned farms but also reaches non-family-owned farms. Freitas et al. [48] found that extension was provided to all farm sizes in all regions of Brazil. Table 3 shows that MSCM is positively associated with governmental rural extension and access to credit. Furthermore, it has been reported that governmental extension service technicians sometimes seek to inform farmers about the benefits of cooperatives and encourage them to organize themselves into cooperatives [49].

MSCM also varies by region. All regions have a negative association compared to the base group: the Southern region. As shown in Table 2, more than 99% of the municipalities in the Southern region have at least one farmer associated with a cooperative. Neves et al. [29] emphasized the presence of cooperative clusters in the Southern and Southeastern regions of Brazil and argued that these organizations generate GVP gains. Furthermore, they conditioned these gains to the long history of cooperatives in these regions as well as improvements in management and governance throughout the decades. On the other hand, they also emphasize previous failed attempts to create cooperatives (top-down cooperative initiatives involving local elites resulted in cooperatives in which the members were mere employees and suppliers. In addition, many such endeavors had disastrous economic outcomes, with poor management leading to losses and the dissolution of cooperatives) in the Northeastern region as a factor that may discourage farmers from this kind of organization. Rios [50] and Silva et al. [51] also discussed these traumatic events.

We find a positive association of all groups by area (farm size) compared to the group of farms larger than 500 hectares (control group). Verhofstadt and Maertens [13] reported similar findings. According to an exhaustive literature review by Höhler and Kühl [52], farm size is the dimension most often analyzed in studies that investigate cooperatives and the heterogeneity of producers. Wiggins et al. [53] state that smaller farms are associated with higher external transaction costs because they cannot achieve economies of scale. They also assert that smaller farms have higher unit costs when purchasing inputs or obtaining credit, technical assistance, and certification services.

Our main result lies in the outcome of the dose response, shown in Figure 3, with Table A2 (in Appendix A) containing the results of the dose-response regression (second step of the GPS approach). Although not directly interpretable, the parameters obtained are used to calculate the average effect of membership in cooperatives for different levels of treatment (Figure 3). Figure 3a shows that as MSCM increases, net income also rises at an increasing rate up to the 0.30 percentile (30% of farms are members of cooperative(s) in the municipality). At the peak, there is an average gain of approximately BRL 150,000 per year per farm. After the peak, net income continues to increase as the treatment variable increases until reaching the 0.75 percentile.

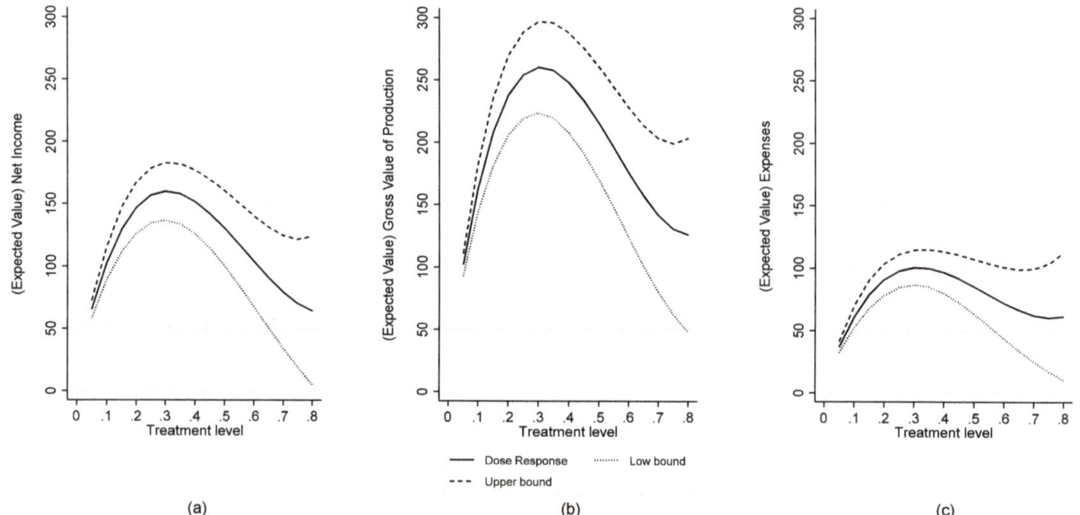

Figure 3. Estimated dose response of MSCM (treatment level) on municipality averages of agricultural (**a**) net income, (**b**) GVP and (**c**) expenses (in 1000 BRL), Brazil, 2017. Source: The authors. Notes: 95% confidence bands; average exchange rate in August 2017: BRL 3.15/USD 1.

The dose-response for net income does not indicate whether this gain occurs via changes in GVP (Figure 3b) or production costs (Figure 3c). We find a similar pattern for these two other variables, but a higher increase in the average GVP compared to the increase in the average production costs, yielding a positive impact on net income. Zhang et al. [11] and Valentinov [12] argued that cooperatives can lead to an increase in the value of production, possibly driven by an increase in production and prices obtained, which may be determined by greater access to modern inputs.

Our results on GVP are also supported by previous research using PSM at the farm level. Fischer and Qaim [6] found a positive effect on income for Kenyan banana producers. Wossen et al. [54] found an increase in the income of cassava producers in Nigeria. Hoken and Su [55] and Liu et al. [56] also found similar results for Chinese rice producers. Kumar et al. [57], using an endogenous switching regression, found that milk producers in India achieved an increase in their net income. Michalek et al. [58] also found a positive effect on net income for farmers in Slovakia.

In turn, an expansion in production is expected to increase production costs, unless it is achieved solely through an increase in productivity via costless improved management practices. Helfand et al. [34] and Gasques et al. [41] showed that there has been a gain in productivity in Brazil through better use of the land. However, this has occurred mainly through the adoption of modern inputs and equipment, which ultimately increase farmers' expenses. The optimal approach to change is to increase production costs at a slower pace than the increase in the value produced on the farms. Giannakas and Fulton [59] argued that cooperatives may promote the adoption of innovations, which can lead to both greater productivity and better use of inputs. Ortega et al. [24] documented this positive effect for coffee producers in Rwanda, where cooperative membership was linked to greater access to inputs and an increase in income.

4. Conclusions

In this paper, we estimate how municipality share of cooperative membership (MSCM) affects the value of production, production cost and net income using the generalized propensity score approach and aggregate data from the 2017 Ag. Census. We find that increases in cooperative membership share among farmers in Brazilian municipalities (MSCM) result in increases in net income, mostly driven by a greater expansion in the value of agricultural production compared to the increase in the production cost.

The findings of this study have important implications for public policies in Brazil. The results suggest that incentivizing cooperative membership may be a good instrument to improve the value of production of Brazilian farmers. Thus, public policies seeking to promote cooperatives and increase cooperative membership would have positive effects on the net income of farms. Public policies that directly and indirectly encourage and support cooperatives, such as the "Capitalization of Agricultural Cooperatives Program" (PROCAP-Agro) (program carried out by the National Bank for Economic and Social Development (BNDES). It aims to offer financing for working capital to cooperatives) [60] and the "More Cooperative Brazil Program" (Programa Brasil Mais Cooperativo) (an initiative of the Ministry of Agriculture, Livestock and Supply (MAPA), the "Brasil Mais Cooperativo Program" aims to support rural cooperatives through the provision of specialized assistance, the promotion of cooperation among cooperatives, technical training and the qualification of management, production processes and marketing) [61] can increase net income. However, farmers who are already marginalized because of low educational attainment, land access, social status and market accessibility may require additional support systems to improve their capacities, skills and resources before they can benefit from cooperative membership.

The method used in this paper does not allow us to rule out the possibility that part of these positive effects of cooperatives may be linked to spillovers caused by the presence of cooperatives' members in the Brazilian municipalities. This method also does not account for potential endogeneity issues arising from unobserved characteristics. Future research should account for these limitations. One suggestion could be to use more than one Ag. Census, with farm-level data, and the application of methodologies that correct bias of unobservable characteristics, such as Diff-Diff.

Author Contributions: Conceptualization, M.d.C.R.N., F.d.F.S., C.O.d.F. and M.J.B.; data curation, C.O.d.F.; formal analysis, M.d.C.R.N., F.d.F.S., C.O.d.F. and M.J.B.; funding acquisition, M.d.C.R.N., F.d.F.S., C.O.d.F. and M.J.B.; methodology, M.d.C.R.N., F.d.F.S. and C.O.d.F.; project administration, M.d.C.R.N.; resources, M.d.C.R.N., F.d.F.S., C.O.d.F. and M.J.B.; software, M.d.C.R.N., F.d.F.S. and C.O.d.F.; supervision, M.d.C.R.N. and M.J.B.; Validation, M.d.C.R.N., F.d.F.S., C.O.d.F. and M.J.B.; writing—original draft, M.d.C.R.N., F.d.F.S. and C.O.d.F.; Writing—review and editing, M.d.C.R.N., F.d.F.S., C.O.d.F. and M.J.B. All authors have read and agreed to the published version of the manuscript.

Funding: This research is financed by the Chamada CNPq # 01/2018, process # 427764/2018-1 and Chamada CNPq/Sescoop # 07/2018, process # 403068/2018-5. It is also supported by OCEMG. Opinions, mistakes and omissions are the authors' own.

Institutional Review Board Statement: Not applicable.

Informed Consent Statement: Not applicable.

Acknowledgments: Conselho Nacional de Desenvolvimento Científico e Tecnológico (CNPq), Serviço Nacional de Aprendizagem do Cooperativismo (Sescoop) e Organização das Cooperativas do Estado de Minas Gerais (OCEMG).

Conflicts of Interest: The authors declare no conflict of interest.

Appendix A

Table A1. Test of conditional means of the pretreatment variables under the generalized propensity score, Brazil, 2017.

	Interval 1		Interval 2		Interval 3		Interval 4	
	Mean Difference	*t*-Value	Mean Difference	*t*-Value	Mean Difference	*t*-Value	Mean Difference	*t*-Value
Capital	0.751	3.753	0.415	2.269	−0.250	−1.842	−0.764	−2.277
Other Investments ($)	45.439	3.846	30.718	2.842	−51.770	−6.157	−20.581	−1.176
Land used (ha)	7644.200	4.449	1804.300	1.196	−2811.300	−1.730	3330.300	0.911
Labor (Age >14 years old)	215.360	2.354	149.100	1.721	264.760	2.808	1062.100	4.757
Landowner	0.028	4.825	0.005	1.001	−0.006	−1.104	−0.049	−4.132
Literate	0.107	38.464	0.034	7.106	−0.063	−15.909	−0.121	−10.030
College degree or higher	0.037	10.886	0.001	0.376	−0.027	−11.625	−0.012	−2.189
Experience	−0.002	−4.733	−0.0002	−0.499	0.002	3.833	0.005	4.874
Governmental Extension	0.020	3.531	0.001	0.240	0.011	2.807	−0.070	−7.712
Private Extension	0.129	17.736	0.074	11.748	−0.012	−2.183	−0.106	−12.485
Credit	0.058	13.330	0.047	12.343	−0.009	−2.515	−0.047	−7.484
North	−0.010	−1.245	−0.013	−1.965	−0.024	−2.499	0.081	3.523
Midwest	0.032	2.504	−0.041	−4.088	−0.063	−7.132	0.027	1.200
Southeast	0.062	3.158	0.002	0.099	−0.181	−13.044	0.015	0.427
Northeast	−0.293	−33.780	−0.055	−4.455	0.200	15.918	0.289	7.941
0–5 ha	−0.017	−4.092	−0.006	−1.329	0.017	3.522	0.018	1.600
5–20 ha	−0.050	−3.229	−0.058	−4.036	0.135	9.732	−0.062	−1.912
20–100 ha	0.019	0.931	0.030	1.631	−0.037	−2.197	0.002	0.062
100–500 ha	0.039	2.219	0.039	2.523	−0.104	−7.693	0.029	0.954

Source: The authors.

Table A2. Results for the dose-response regression for the municipality averages of net income, gross value of production, and agricultural expenses, Brazil, 2017.

Variables	Av. Mun. Net Income	Av. Mun. Gross Value of Production	Av. Mun. Agricultural Expenses
MSCM	814.31 *** (90.1)	1386.25 *** (120.53)	571.94 *** (65.05)
Squared *MSCM*	−1979.42 *** (286.15)	−3346.27 *** (382.81)	−1366.85 *** (206.59)
Cubic *MSCM*	1176.76 *** (248.73)	2044.54 *** (332.75)	867.78 *** (179.57)
GPS	−732.09 *** (264.94)	−948.79 *** (354.44)	−216.7 (191.28)
Squared *GPS*	2801.36 ** (1160.11)	3668.12 ** (1552)	866.76 (837.57)
Cubic *GPS*	−3135.1 ** (1461.54)	−4089.49 ** (1955.26)	−954.39 (1055.2)
Interaction *MSCM—GPS*	413.21 *** (110.36)	728.88 *** (147.65)	315.67 *** (79.68)
Constant	70.66 *** (17.49)	90.08 *** (23.4)	19.42 (12.63)

Note: Standard errors are reported in parentheses. *** significant at 1%, ** at 5%.

References

1. Chaddad, F. *The Economics and Organization of Brazilian Agriculture*, 1st ed.; Elsevier: London, UK, 2016; ISBN 9780128016954.
2. Barros, G.S.C. Agriculture, industry and the economy: From extensive farming to a global agro-food power. In *Agricultural Development in Brazil: The Rise of a Global Agro-Food Power*; Series: Routledge Studies in Agricultural Economics, 2019; Buainain, A.M., Lanna, R., Navarro, Z., Eds.; Routledge: New York, NY, USA, 2019; pp. 12–29. ISBN 9781351029742.
3. Jayne, T.S.; Zulu, B.; Nijhoff, J.J. Stabilizing Food Markets in Eastern and Southern Africa. *Food Policy* **2006**, *31*, 328–341. [CrossRef]
4. Zeuli, K.A.; Radel, J. Cooperatives as a Community Development Strategy: Linking Theory and Practice. *J. Reg. Anal. Policy* **2005**, *35*, 43–54.
5. Bernard, T.; Taffesse, A.S.; Gabre-Madhin, E. Impact of Cooperatives on Smallholders' Commercialization Behavior: Evidence from Ethiopia. *Agric. Econ.* **2008**, *39*, 147–161. [CrossRef]
6. Fischer, E.; Qaim, M. Linking Smallholders to Markets: Determinants and Impacts of Farmer Collective Action in Kenya. *World Dev.* **2012**, *40*, 1255–1268. [CrossRef]
7. Bizikova, L.; Nkonya, E.; Minah, M.; Hanisch, M.; Turaga, R.M.R.; Speranza, C.I.; Karthikeyan, M.; Tang, L.; Ghezzi-Kopel, K.; Kelly, J.; et al. A Scoping Review of the Contributions of Farmers' Organizations to Smallholder Agriculture. *Nat. Food* **2020**, *1*, 620–630. [CrossRef]
8. Batzios, A.; Kontogeorgos, A.; Chatzitheodoridis, F.; Sergaki, P. What Makes Producers Participate in Marketing Cooperatives? The Northern Greece Case. *Sustainability* **2021**, *13*, 1676. [CrossRef]
9. Olagunju, K.O.; Ogunniyi, A.I.; Oyetunde-Usman, Z.; Omotayo, A.O.; Awotide, B.A. Does Agricultural Cooperative Membership Impact Technical Efficiency of Maize Production in Nigeria: An Analysis Correcting for Biases from Observed and Unobserved Attributes. *PLoS ONE* **2021**, *16*, e0245426. [CrossRef]
10. Merrett, C.D.; Walzer, N. *A Cooperative Approach to Local Economic Development*, 1st ed.; Quorum Books: Westport, CT, USA, 2001; ISBN 9781567203950.
11. Zhang, J.; Goddard, E.; Lerohl, M. Estimating Pricing Games in the Wheat-handling Market in Saskatchewan: The Role of a Major Cooperative. In *Cooperative Firms in Global Markets*; Novkovic, S., Sena, V., Eds.; Emerald Group Publishing Limited: West Yorkshire, UK, 2007; ISBN 9780762313891.
12. Valentinov, V. Why Are Cooperatives Important in Agriculture? An Organizational Economics Perspective. *J. Inst. Econ.* **2007**, *3*, 55–69. [CrossRef]
13. Verhofstadt, E.; Maertens, M. Can Agricultural Cooperatives Reduce Poverty? Heterogeneous Impact of Cooperative Membership on Farmers' Welfare in Rwanda. *Appl. Econ. Perspect. Policy* **2015**, *37*, 86–106. [CrossRef]
14. Cechin, A. Cooperativas brasileiras nos mercados agroalimentares contemporâneos. In *O Mundo Rural no Brasil do Século 21: A Formação de um Novo Padrão Agrário e Agrícola*; Buainain, A.M., Alves, E., da Silveira, J.M., Navarro, Z., Eds.; Embrapa: Brasília, Brazil, 2014; pp. 479–507.
15. Herrera, G.P.; Lourival, R.; da Costa, R.B.; Mendes, D.R.F.; Moreira, T.B.S.; de Abreu, U.G.P.; Constantino, M. Econometric Analysis of Income, Productivity and Diversification among Smallholders in Brazil. *Land Use Policy* **2018**, *76*, 455–459. [CrossRef]
16. Ma, W.; Abdulai, A.; Goetz, R. Agricultural Cooperatives and Investment in Organic Soil Amendments and Chemical Fertilizer in China. *Am. J. Agric. Econ.* **2018**, *100*, 502–520. [CrossRef]
17. Zhang, S.; Sun, Z.; Ma, W.; Valentinov, V. The Effect of Cooperative Membership on Agricultural Technology Adoption in Sichuan, China. *China Econ. Rev.* **2020**, *62*, 101334. [CrossRef]
18. IBGE. Instituto Brasileiro de Geografia e Estatística. Microdados da Pesquisa Nacional por Amostra de Domicílios (PNAD). Available online: https://www.ibge.gov.br/estatisticas/sociais/populacao/9127-pesquisa-nacional-por-amostra-de-domicilios.html?edicao=9451&t=downloads (accessed on 22 May 2021).
19. IBGE. Instituto Brasileiro de Geografia e Estatística. SIDRA - Sistema IBGE de Recuperação Automática: Censo Agropecuário. 2017. Available online: https://sidra.ibge.gov.br/pesquisa/censo-agropecuario/censo-agropecuario-2017 (accessed on 22 April 2020).
20. Homma, A.K.O.; Lima, J.R.F.; Vieira, P.A. Structural heterogeneity in rural Brazil: Three regional cases. In *Agricultural Development in Brazil: The Rise of a Global Agro-Food Power*; Buainain, A.M., Lanna, R., Navarro, Z., Eds.; Routledge: New York, NY, USA, 2019; pp. 189–207.
21. Neves, M.C.R.; Castro, L.S.; Freitas, C.O. Cooperatives and Brazilian Agricultural Production: A Spatial Analysis. In Proceedings of the 46th Brazilian Economics Meeting, Rio de Janeiro, Brazil, 11–14 December 2018. ANPEC—Associação Nacional dos Centros de Pós-graduação em Economia [Brazilian Association of Post-Graduate Programs in Economics].
22. Cazzuffi, C. Small Scale Farmers in the Market and the Role of Processing and Marketing Cooperatives: A Case Study of Italian Dairy Farmers. Ph.D. Thesis, University of Sussex, Brighton, UK, 2012.
23. Jardine, S.L.; Lin, C.Y.C.; Sanchirico, J.N. Measuring Benefits from a Marketing Cooperative in the Copper River Fishery. *Am. J. Agric. Econ.* **2014**, *96*, 1084–1101. [CrossRef]
24. Ortega, D.L.; Bro, A.S.; Clay, D.C.; Lopez, M.C.; Tuyisenge, E.; Church, R.A.; Bizoza, A.R. Cooperative Membership and Coffee Productivity in Rwanda's Specialty Coffee Sector. *Food Secur.* **2019**, *11*, 967–979. [CrossRef]
25. Kumse, K.; Suzuki, N.; Sato, T.; Demont, M. The Spillover Effect of Direct Competition between Marketing Cooperatives and Private Intermediaries: Evidence from the Thai Rice Value Chain. *Food Policy* **2021**, *101*, 102051. [CrossRef]
26. Helfand, S.; Levine, E. Farm Size and the Determinants of Productive Efficiency in the Brazilian Center-West. *Agric. Econ.* **2004**, *31*, 241–249. [CrossRef]

27. Wollni, M.; Brümmer, B. Productive Efficiency of Specialty and Conventional Coffee Farmers in Costa Rica: Accounting for Technological Heterogeneity and Self-Selection. *Food Policy* **2012**, *37*, 67–76. [CrossRef]

28. Costa, R.A.; Vizcaino, C.A.C.; Costa, E.M. Participação em Cooperativas e Eficiência Técnica entre Agricultores Familiares no Brasil. In *Uma Jornada Pelos Contrastes do Brasil: Cem anos de Censo Agropecuário*; Vieira Filho, J.E.R., Gasques, J.G., Eds.; IPEA: Brasília, Brazil, 2020; pp. 243–255. ISBN 9786556350110.

29. Neves, M.D.C.R.; Castro, L.S.D.; Freitas, C.O.D. O Impacto das Cooperativas na Produção Agropecuária Brasileira: Uma Análise Econométrica Espacial. *Rev. Econ. Sociol. Rural.* **2019**, *57*, 559–576. [CrossRef]

30. OCB. *Organização das Cooperativas Brasileiras. Anuário do Cooperativismo Brasileiro 2020*; OCB: Hochiminh City, Vietnam, 2020.

31. Imbens, G. *The Role of the Propensity Score in Estimating Dose-Response Functions*; Biometrika: Cambridge, MA, USA, 1999.

32. Hirano, K.; Imbens, G.W. The Propensity Score with Continuous Treatments. In *Applied Bayesian Modeling and Causal Inference from Incomplete-Data Perspectives*; John Wiley & Sons: Hoboken, NJ, USA, 2004.

33. Rosenbaum, P.R.; Rubin, D.B. The Central Role of the Propensity Score in Observational Studies for Causal Effects. *Biometrika* **1983**, *70*, 41–55. [CrossRef]

34. Helfand, S.M.; Magalhães, M.M.; Rada, N.E. *Brazil's Agricultural Total Factor Productivity Growth by Farm Size*; IDB Working Paper Series; Inter-American Development Bank: Washington, DC, USA, 2015.

35. Buainain, A.M.; Gonzáles, M.G.; Souza Filho, H.M.; Vieira, A.C.P. *Alternativas de Financiamento Agropecuário: Experiências no Brasil e na América Latina*; Buainain, A.M., Gonzáles, M.G., Eds.; IICA: Brasília, Brazil, 2007.

36. de Araújo, M.L.S.; Sano, E.E.; Bolfe, E.L.; Santos, J.R.N.; dos Santos, J.S.; Silva, F.B. Spatiotemporal Dynamics of Soybean Crop in the Matopiba Region, Brazil (1990–2015). *Land Use Policy* **2019**, *80*, 57–67. [CrossRef]

37. Silva, L.D.J.D.S.; Pinheiro, J.O.C.; Dos Santos, E.M.; Da Costa, J.I.; Meneghetti, G.A. O Cooperativismo Como Instrumento Para a Autonomia de Comunidades Rurais da Amazônia: A Experiência dos Agricultores Extrativistas do Municipio de Lábrea, AM. *Boletín Asoc. Int. Derecho Coop.* **2019**, *55*, 199–226. [CrossRef]

38. Futemma, C.; de Castro, F.; Brondizio, E.S. Farmers and Social Innovations in Rural Development: Collaborative Arrangements in Eastern Brazilian Amazon. *Land Use Policy* **2020**, *99*, 104999. [CrossRef]

39. Alves, E.; Souza, G.D.S.; Rocha, D.D.P. Lucratividade da Agricultura. *Rev. Política Agrícola* **2012**, *21*, 45–63.

40. Helfand, S.M.; Rada, N.E.; Magalhães, M.M. L'agriculture Brésilienne: S'agit-Il Uniquement de Grandes Exploitations? *EuroChoices* **2017**, *16*, 17–22. [CrossRef]

41. Gasques, J.G.; Bacchi, M.R.P.; Bastos, E.T. Nota Técnica IV-Crescimento e Produtividade da Agricultura Brasileira de 1975 a 2016. *Carta Conjunt.* **2018**, 1–9.

42. Rada, N.; Helfand, S.; Magalhães, M. Agricultural Productivity Growth in Brazil: Large and Small Farms Excel. *Food Policy* **2019**, *84*, 176–185. [CrossRef]

43. Bia, M.; Mattei, A. A Stata Package for the Estimation of the Dose-Response Function through Adjustment for the Generalized Propensity Score. *Stata J. Promot. Commun. Stat. Stata* **2008**, *8*, 354–373. [CrossRef]

44. Iliopoulos, C.; Cook, M.L. The Efficiency of Internal Resource Allocation Decisions in Customer-Owned Firms: The Influence Costs Problem. In Proceedings of the Annals of 3rd Annual Conference of the International Society for New Institutional Economics, Washington, DC, USA, 16–18 September 1999.

45. Cechin, A.; Bijman, J.; Pascucci, S.; Zylbersztajn, D.; Omta, O. Drivers of Pro-Active Member Participation in Agricultural Cooperatives: Evidence from Brazil. *Ann. Public Coop. Econ.* **2013**, *84*, 443–468. [CrossRef]

46. Bernard, T.; Spielman, D.J. Reaching the Rural Poor through Rural Producer Organizations? A Study of Agricultural Marketing Cooperatives in Ethiopia. *Food Policy* **2009**, *34*, 60–69. [CrossRef]

47. Abate, G.T. Drivers of Agricultural Cooperative Formation and Farmers' Membership and Patronage Decisions in Ethiopia. *J. Co-Oper. Organ. Manag.* **2018**, *6*, 53–63. [CrossRef]

48. de Freitas, C.O.; Silva, F.D.F.; Braga, M.J.; Neves, M.D.C.R. Rural Extension and Technical Efficiency in the Brazilian Agricultural Sector. *Int. Food Agribus. Manag. Rev.* **2021**, *24*, 215–232. [CrossRef]

49. Costa, B.A.L.; Junior, P.C.G.A.; Da Silva, M.G. As Cooperativas de Agricultura Familiar e o Mercado de Compras Governamentais Em Minas Gerais. *Rev. Econ. Sociol. Rural.* **2015**, *53*, 109–126. [CrossRef]

50. Rios, G.S.L. Pré-cooperativismo: Etapa queimada. In *A Problemática Cooperativista no Desenvolvimento Econômico*; Uwe, J., Ed.; Fundação Friedrich Naumann: São Paulo, Brazil, 1973; pp. 315–347.

51. Silva, E.S.; Salomão, I.L.; McIntyre, J.P.; Guerreiro, J.; Pires, M.L.L.S.; Albuquerque, P.P.; Bergonsi, S.; Vaz, S.C. Panorama do Cooperativismo Brasileiro: História, Cenários e Tendências. *Rev. uniRcoop* **2003**, *1*, 75–102.

52. Höhler, J.; Kühl, R. Dimensions of member heterogeneity in cooperatives and their impact on organization—A literature review. *Ann. Public Coop. Econ.* **2018**, *89*, 697–712. [CrossRef]

53. Wiggins, S.; Kirsten, J.; Llambí, L. The Future of Small Farms. *World Dev.* **2010**, *38*, 1341–1348. [CrossRef]

54. Wossen, T.; Abdoulaye, T.; Alene, A.; Haile, M.G.; Feleke, S.; Olanrewaju, A.; Manyong, V. Impacts of Extension Access and Cooperative Membership on Technology Adoption and Household Welfare. *J. Rural. Stud.* **2017**, *54*, 223–233. [CrossRef] [PubMed]

55. Hoken, H.; Su, Q. Measuring the Effect of Agricultural Cooperatives on Household Income: Case Study of a Rice-Producing Cooperative in China. *Agribusiness* **2018**, *34*, 831–846. [CrossRef]

56. Liu, Y.; Ma, W.; Renwick, A.; Fu, X. The Role of Agricultural Cooperatives in Serving as a Marketing Channel: Evidence from Low-Income Regions of Sichuan Province in China. *Int. Food Agribus. Manag. Rev.* **2019**, *22*, 265–282. [CrossRef]
57. Kumar, A.; Saroj, S.; Joshi, P.K.; Takeshima, H. Does Cooperative Membership Improve Household Welfare? Evidence from a Panel Data Analysis of Smallholder Dairy Farmers in Bihar, India. *Food Policy* **2018**, *75*, 24–36. [CrossRef]
58. Michalek, J.; Ciaian, P.; Pokrivcak, J. The Impact of Producer Organizations on Farm Performance: The Case Study of Large Farms from Slovakia. *Food Policy* **2018**, *75*, 80–92. [CrossRef]
59. Giannakas, K.; Fulton, M. Process Innovation Activity in a Mixed Oligopoly: The Role of Cooperatives. *Am. J. Agric. Econ.* **2005**, *87*, 406–422. [CrossRef]
60. BNDES. Banco Nacional de Desenvolvimento Econômico e Social. PROCAP-Agro. Available online: https://www.bndes.gov.br/wps/portal/site/home/financiamento/produto/procap-agro (accessed on 3 February 2021).
61. MAPA. Ministério da Agricultura, Pecuária e Abastecimento. Programa Brasil Mais Cooperativo. Available online: https://www.gov.br/agricultura/pt-br/assuntos/agricultura-familiar/brasil-mais-cooperativo (accessed on 3 February 2021).

MDPI

Article

Popularity in Social Networks. The Case of Argentine Beekeeping Production Entities

Jimena Andrieu [1,*], Domingo Fernández-Uclés [2], Adoración Mozas-Moral [2] and Enrique Bernal-Jurado [3]

[1] National Institute of Agricultural Technology (INTA), EEA San Juan-Argentina/National University of San Juan (UNSJ), San Juan 5400, Argentina
[2] Faculty of Social and Legal Sciences, Department of Business Organization, Marketing and Sociology, University of Jaén, 23071 Jaén, Spain; dfucles@ujaen.es (D.F.-U.); amozas@ujaen.es (A.M.-M.)
[3] Faculty of Social and Legal Sciences, Department of Economics, University of Jaén, 23071 Jaén, Spain; ebernal@ujaen.es
* Correspondence: andrieu.jimena@inta.gob.ar; Tel.: +54-92644921079 (ext. 144)

Abstract: The context of the COVID pandemic has accelerated the pace of the digitalization of society, especially of its business fabric. Among the various applications offered by the Internet, social networking platforms have been identified as powerful tools that organizations have at their disposal for the development of their online business activities. This is due to the closeness and trust generated by word-of-mouth communication. In this context, the aim of this article is to identify which organizational characteristics are directly related to popularity on social networks, measured by the number of followers on these accounts. In order to achieve this objective, the Argentinean beekeeping organizations have been taken as a case study and the fuzzy set Qualitative Comparative Analysis method has been used. The results obtained allow us to validate the different organizational factors which, beyond the use of Facebook itself, lead to better results for the organizations in their social network strategies. These factors include their cooperative nature, localization, environmental sensitivity and presence on other digital platforms.

Keywords: Facebook; cooperatives; beekeeping; fuzzy set qualitative comparative analysis

Citation: Andrieu, J.; Fernández-Uclés, D.; Mozas-Moral, A.; Bernal-Jurado, E. Popularity in Social Networks. The Case of Argentine Beekeeping Production Entities. *Agriculture* **2021**, *11*, 694. https://doi.org/10.3390/agriculture11080694

Academic Editor: Manuel García-Herreros

Received: 23 June 2021
Accepted: 21 July 2021
Published: 23 July 2021

Publisher's Note: MDPI stays neutral with regard to jurisdictional claims in published maps and institutional affiliations.

1. Introduction

Argentina plays a central role in beekeeping production, both regionally and globally. Thus, in 2018 it was ranked as the first honey-producing country in America and the third in the world, after China and Turkey. This importance is maintained not only in volume produced, but also in volume marketed. Thus, in 2018, on a world export volume of approximately 650 thousand tons of honey, Argentina contributed 10.5%, being only surpassed by China, with 19% of the total. Between the two countries, they concentrated almost a third of the international sales of this product [1].

Despite Argentina's global importance in the world beekeeping market, the sector has been facing important difficulties, both in production and marketing. On the one hand, primary production has been suffering a drop in average yields since the beginning of the 21st century, due to a series of factors that negatively affect the beehive environment and beekeeping practice. Examples include the loss of biodiversity, the advance of agriculture over pastures and natural forests, homogenization of landscapes and the increased presence of diseases, among others [2,3]. On the other hand, with respect to the structure of the production system, this is characterized by a high atomization in primary production, a low relative size of the productive units and, as a consequence, a strong dependence on distribution. Argentina has a total of 13,722 beekeeping registries located throughout the country, most of which have fewer than 500 hives (The same negative trend can be observed in sales of honey certified as organic. Since 2008, Argentina's share in the world organic market has been declining despite its favorable evolution in both demand and world prices) [4–6].

In addition to these problems present in the field of production, there are others related to the marketing of these products, such as the significant dependence on foreign markets and their evolution, the lack of differentiation derived from the preferential sale in bulk and the concentration of sales from a few exporting companies. Indeed, in 2018, 85% of domestic production was destined to international markets, mainly to the United States, Germany and Japan [1,7]; a figure that has shown a downward trend if we consider that it had reached up to 95% in previous years [8]. Another commercial problem is related to the pursuit of a commercial strategy based on the exchange of large volumes of honey (in drums). This form of sale has an impact on the average prices per kilo exported, causing Argentine prices to be below the world average (An analysis of the evolution of prices for the Americas throughout the 21st century shows that they are, on average, intermediate prices between those of Europe (mainly Western Europe) and Asia. Oceania, based on the differentiation of "Manuka" honey, shows considerably higher prices.) [1,8].

This problem tends to worsen as exports in this large-volume format become more and more frequent, making product differentiation difficult. The sector also loses a large part of the added value generated in the packaging and distribution stages (The same negative trend can be observed in sales of honey certified as organic. Since 2008, Argentina's share in the world organic market has been declining despite its favorable evolution in both demand and world prices.) [9–11]. In turn, it is estimated that only three entities control approximately half of the exported volume [10,12]. The way of organizing such commercialization has traditionally been structured on the basis of the figure of the "stockpiler" as the main intermediary. However, during the last few years there has been an advance in backward integration in the chain by exporting companies through the implementation of direct contracts with producers [10]. This situation tends to reinforce a distancing between the primary production sector and the final consumer, and with it the loss of participation of the first sector in the honey valorization process [8,13].

This is taking place in a context in which, in general terms, honey consumption is showing a favorable evolution (In 2018, the apparent honey consumption data implies a value of 243 grams per person per year. This figure is higher than the average value for the period 2011–2016 that implies 156 grams per person per year (p. 21). However, these values still represent a low level of national domestic consumption.) [1,7,14,15]. Thus, honey has managed to consolidate itself within the group of natural products related to healthy eating [16,17], due to its characteristic as a natural sweetener, in addition to other properties that expand its potential use [18,19]. In the Argentine case, its consumption will be favored in the future to the extent that strategies tending to improve the diet of the local population are developed based on a reduction in the consumption of ultra-processed products and an increase in the consumption of natural products with an intersectional look. This is coupled with a commitment to the implementation of production and distribution systems that are more sustainable [20–22].

The above problems suggest the need for the sector to adopt measures to improve the supply of this food product, not only from the perspective of production but also along the distribution chain [23]. Such measures should include those aimed at improving exchange networks in order to achieve greater differentiation of production and more direct communication with the consumer [13,24]. In this sense, several authors have highlighted the determining role that can be played by the use of virtual applications, such as company websites or social networks [25–27] especially in the case of natural products, as is the case of honey [26,28]. Thus, certain "experiential" agricultural products, due to their intrinsic characteristics, are particularly suitable for marketing through the Web [29]. For example, the purchase and consumption of products such as honey or wine are based on an intensive exchange of information on tangible aspects of the product as well as on symbols, tradition, culture, tourism and gastronomy; all aspects that can significantly enhance the value perceived by the consumer [30]. Thus, social networks, due to their closeness and interactivity with users, are positioned as an ideal communication channel for transmitting such information [31].

In line with the above, some authors have defended the need to address the link between the beekeeping subsector and ICT (Information and Communications Technology) [32]. However, the literature review shows that there are no studies focused on the beekeeping sector that have tried to investigate the degree of use of social networks and which factors are associated with greater success in the use of these technologies. Indeed, studies focused on the beekeeping sector can be divided into three groups: first, those that have focused on the analysis of honey markets and their potential at the local scale [12,33,34]; a second group has focused on the analysis of production, its socio-economic structure and the analysis of technical efficiency [10,35]; and, finally, a third group of studies has focused on the transformations of the digital era, paying more attention to production than to interactions between producers and consumers [36,37]. Thus, the focus of these latter studies tends to be on the analysis of the digital impact on machinery, equipment and other inputs needed to increase production efficiency, with less frequent studies addressing the benefits of a comprehensive use of the Internet in terms of communication and management improvements [38].

Based on the premise that social networks are tools that can provide answers to the problematic situation described above, especially the commercial one, the objective of this paper is to analyze which characteristics of beekeeping organizations are directly related to greater success in the use of these technologies. To achieve this objective, the Argentine beekeeping organizations have been taken as a case study and their use of the social network Facebook and the factors related to greater popularity in this network have been analyzed using the fuzzy set Qualitative Comparative Analysis method. The paper is structured as follows: after this introduction, the contextual framework detailing the study propositions is presented, followed by the technical characteristics of the research in the methodology section, after which the results are presented and finally, the conclusions and reflections are derived from the data analysis.

2. Materials and Methods

2.1. Contextual Framework

In this sense, transaction cost theory has often been used as a basis for analyzing and highlighting the potential of information and communication technologies (ICT) for business, especially in the commercial sphere [39]. Social media, as the main example of this phenomenon, brings multiple benefits in terms of business performance and reduction of different types of costs. Thus, information costs decrease thanks to the informational potential of online social networks, which facilitate, improve and speed up information exchanges [40]. Negotiation costs decrease because online media can improve customers' access to the organization and enable them to receive more personalized offers [40]. Finally, assurance costs are minimized because users are offered reliable and good quality information and feedback [41].

However, experience indicates that not all companies have the same ability to take advantage of the benefits offered by the Internet. Rather, several studies indicate that adaptation to the Internet and social media in which one operates in this medium depends on several factors, which differ according to the sector or region in which the company operates [42]. In the specific case of agricultural markets, McFarlane el al. [43]. found that the characteristics of the distribution chain, the scope of the company and the type of product it sells (organic) influence the intensity of adoption of the ICT strategy.

The literature has been concerned with investigating the challenges and organizational characteristics of cooperative entities that delay or could delay the adoption of technology and the use of ICT [44]. This happens despite the fact that aspects shared between Social Economy entities and Web 2.0 technologies and the different tools that integrate them are highlighted. Particular reference is made to the affinity of these types of tools with cooperative principles, such as their participatory and democratic nature and the predominance of the social component of capital [45]. Thus, it is identified that the benefits of ICT are increased in cooperative societies due to their ability to coordinate activities, people and processes [46]. In short, the potential of ICT use within cooperative societies for information exchange and communication is recognized as a key factor for their management [46]. In turn, this potential is identified in terms of improving aspects that determine the economic viability of organizations [47]. Thus, the following assertion is established:

Proposition 1. *Having the legal form of a cooperative is directly related to popularity in social networks.*

The literature points out the importance of agglomeration economies for the circulation of knowledge and innovation of firms located in the same geographical space. Thus, the role of the network of linkages will be key to the extent that it is also structured beyond the local level [48]. In this line, several studies in Argentine companies account for the importance of connections with third parties and cooperation links for the circulation of information and on the innovation process [10,12,49]. Thus, the literature makes it clear how clusters are an element that enhances competitiveness, improving the innovative performance of the organization and its commercial actions [50]. Based on these arguments, the following proposition is put forward:

Proposition 2. *The location of the entity in a central area for beekeeping production favors greater popularity in social networks.*

On the other hand, a growing social concern for environmental issues and environmental conservation can be observed. An increasing number of consumers are seeking information in this regard and are considering environmental aspects in their purchasing decision process [51]. In this sense, these consumers are increasingly resorting to the use of digital media as an alternative purchasing channel [52]. In addition, consumers of organic products tend to be more active on the Internet, in part because of their greater need for information [53]. Precisely, among the various tools offered by the Internet, social networks are the ideal platforms for acquiring trusted information and mitigating existing misinformation when purchasing organic food [54]. From the business point of view, Mozas et al. [55] identify that the organic character of the organization positively affects its innovative character. Moreover, Fernández-Uclés et al. [52] show how organizations of a greener character are likely to achieve higher performance in the use of social networks. Thus, the following proposition is established:

Proposition 3. *The environmental sensitivity of companies favors the increase of their popularity in social networks.*

Virtual social networks are a key communication channel in the commercial strategy of organizations to increase their notoriety and improve their performance [56]. To do this, it is necessary to make a solid commitment to these tools, which will give the organization greater competitiveness and better business results [57]. Furthermore, when an organization integrates a technology, a learning process begins that will lead to a better use of the technology [58]. This know-how, the result of the experience of using an innovation, makes it possible to improve business performance in an innovative environment, favoring an increase in performance and even the ability to obtain sustainable competitive advantages [59]. Therefore, experience is going to be a factor that will go hand in hand with an increasingly efficient use of social networks [60]. This line of argument leads us to put forward the following proposition:

Proposition 4. *The experience of using social networks favors the greater popularity of the entities in them.*

A key aspect in organizational performance is the integration and combination of different social media [61]. Having a corporate website when accompanied by virtual social networks facilitates contact, information exchange and interaction with consumers, improving the company's positioning on the Internet [62]. The interconnectivity of the different online platforms gives them greater visibility and therefore better results in this medium [63]. The existence of an increasingly sophisticated audience requires companies to increase the amount of company information on the internet and its presence on different platforms [64]. Thus, the optimization of social networks requires an effective strategy based on the interaction with the different platforms where the organization is present [65]. In this line, we find the strategies of so-called Inbound Marketing, based on the interconnection of all the virtual platforms in which the company has a presence and aimed at the consumer, which will supposedly increase the performance of these technologies and the company itself [66]. Thus, we put forward the following propositions:

Proposition 5. *The interconnectivity of social networks with the website leads to greater popularity in social networks.*

Proposition 6. *Presence on digital platforms other than the entities' websites favors greater popularity in social networks.*

2.2. Population and Methodology

2.2.1. Population

In order to determine the organizational structure of the Argentine beekeeping sector, information was obtained on the population and basic commercial characteristics of the set of legal entities taxed under the category "beekeeping production" during the fiscal year 2019/2020 [67]. In July 2020, a total of 228 legal entities registered in the category "beekeeping production" in the Argentine territory were identified [67]. The population under study in this study will be only those entities present on the social network Facebook. Precisely, the population thus defined implies a total of 65 entities, of which 43% had the legal form of cooperative. It should be noted that this group of 65 entities concentrates approximately half (47%) of the employment generated within the category analyzed and contains 67% of the entities with a high-income level, 42% of those with a medium income and 27% of those with a low income within the Argentine beekeeping sector.

2.2.2. Methodology

With respect to the methodology used, the Qualitative Comparative Analysis (QCA) technique was employed, using the fuzzy sets approach (fsQCA), in order to establish technological and organizational variables that are jointly associated with a higher level of efficiency. The QCA technique, based on Boolean algebra, uses a verbal, conceptual and mathematical language that configures it as a qualitative and quantitative approach, useful for small samples by combining the main advantages of both [68]. Thus, by applying QCA it is possible to systematically analyze a set of cases to determine causal patterns in the form of necessity and sufficiency relationships between a set of conditions and an outcome [69]. This method has the advantage over a regression technique of establishing relationships between subsets of variables in order to explain relationships. Specifically, QCA has three main variations: crisp-set QCA (csQCA), multi-valued QCA (mvQCA) and fuzzy-set QCA (fsQCA). Fuzzy set (fsQCA) is positioned as one of the most widely used QCA variants, as it resolves one of the main drawbacks and criticisms of the initial approach called csQCA, namely its strictly dichotomous approach [70].

Thus, fsQCA will provide as a result one or more antecedent combinations sufficient for obtaining a particular result, such as: $X1*{\sim}X2*X3$ sufficient for a result (Y). Making use of the symbology of this technique $(X1*{\sim}X2*X3{\rightarrow}Y)$. Being: X1, X2 and X3, antecedents; Y, the result; * the union and ~ the absence or negation, in this case the opposite value to X2 (1 − X2). Thus, this technique makes it possible to identify logically simplified statements that describe different combinations (or configurations) of conditions that indicate a specific result [68].

The fsQCA technique was developed for small sample or population environments [68], so it is not an inconvenience for this research, in which the study universe was small. For the correct execution of this technique, the phases recommended in the literature were followed: (1) data calibration (transform variables into fuzzy sets), (2) simplify the multiple solutions, (3) interpret the results [69]. Next, a necessity analysis of the efficiency scores on the different causal conditions was carried out to verify that none of the values obtained exceeded the threshold recommended in the literature of 0.9, established by Ragin [71], and this was corroborated.

In this study, the number of followers of the different organizational accounts on Facebook was used as the outcome (dependent variable). In turn, as conditions (independent variables), the different variables shown below (Table 1) were used.

Table 1. Description of the variables used in this study.

Variable	Description	Type of Variable
Followers (dependent variable)	Number of followers on Facebook	Continuous [1]
Coop	Organization is a cooperative society	Dichotomous [2]
Location	Location in a central productive region	Dichotomous [2]
Environmental	Degree of environmental sensitivity [3]	Continuous [1]
Experience	Days of use of the social network Facebook	Continuous [1]
Social web	Website interfaced with social network	Continuous [1]
Other sites	Presence on other digital platforms	Dichotomous [2]

[1] The continuous variables were calibrated using the fsQCA 3.0 software. [2] Dichotomous variables (1: yes; 0: no). [3] This variable is constructed by evaluating both the presence and frequency with which different environmental aspects appear in the network. The information was structured along four axes: (i) explanation of the contribution of beekeeping to sustainability, (ii) manifestation of environmental concern, (iii) characteristics of the type of product (organic/ecological/agro-ecological) and (iv) indication of the health benefits of consuming the main product of the activity, honey. Source: own compilation.

3. Results and Discussion

3.1. Descriptive Analysis

In a first approach to the study, Table 2 shows the average descriptive values of the variables considered in this study.

Table 2. Descriptive values of the variables used.

Variable	Description
Coop	43% of the companies have a cooperative legal form
Location	75% of the companies are located in a central production region for beekeeping.
Environmental	70% have some degree of environmental sensitivity.
Experience	The average company has been on this social network for 3.8 years (1397 days).
Social web	32% of companies have a website that leads to user interaction, linking social networks and including comments and ratings.
Other sites	63% of the organizations are present on other online platforms, different from the website and social networks.

Source: own compilation.

The information included in Table 2 is here interpreted according to the trends observed in the international literature. Firstly, several studies have shown the potential use of ICTs by agri-food cooperatives [44–47]. Our data reveals that the cooperative nature becomes more important when the population is analyzed by its presence on social media (28/65 over 85/228). Secondly, there is evidence that the problems with generating websites interconnected with social networks still persist for an important group of entities in the agri-food sector [52,55,62]. Table 2 shows that only one of three entities has this interconnectivity. Thirdly, there are other studies that show that the concern for sustainability aspects is more frequently found in products classified as natural [25–31]. Here we can see how environmental sensitivity is present in almost two of three cases. Finally, it is recognized in the literature that the presence in social media is not isolated, it is rather connected with other networks and platforms [63,66]. Table 2 highlights that most of the entities have a multi-platform presence.

3.2. fsQCA Analysis

Table 3 identifies which of the factors listed in previous sections are positively related to the level of followers on the Facebook social network. The results obtained after applying fsQCA are shown in Table 3. The combination of the parsimonious and intermediate solution is used, which can provide a more detailed and aggregated view of the findings [72].

Table 3. fsQCA analysis results.

Configurations	1	2	3	4	5
Coop				●	●
Location	●	●	●	●	
Environmental		●	●		●
Experience	●	●		●	●
Social web	●	●	●	●	●
Other sites	●		●		●
Raw coverage	0.438469	0.414685	0.397449	0.278525	0.225784
Unique coverage	0.032402	0.053439	0.044812	0.330921	0.042399
Consistency	0.880886	0.934732	0.905735	0.898777	0.953421
Model coverage	0.62082				
Model consistency	0.86337				

Source: own compilation. Black circles (●) denote the presence of a condition, and a blank space represents the "do not care" condition. The distinction between core condition and peripheral condition is made by using large and small circles, respectively [73].

The results obtained show that the first configuration presents a gross coverage of 43.84 percent. This configuration establishes that, as a whole, the relationship between the variables of the organization's experience in the use of Facebook, its location, the link between Facebook and the organization's website and its presence on other online platforms explain a greater popularity on Facebook, measured by the number of followers on this social network. Similarly, it is worth highlighting the other configurations. Overall, this model presents a total coverage of 62 percent, which denotes the proportion of organizations that are explained by the six variables considered, and a total consistency of 86 percent of the cases. This value far exceeds the minimum consistency level recommended in the literature of 0.74, which strengthens the validity of the model proposed [74,75].

The results obtained are in line with the results of others research. The social economy as well as organic agriculture sector agree that marketing should be carried out through short channels [76]. ICT are a fundamental tool for improving organizational results, as they have the possibility of bringing producers closer to the final market [77]. This approach is crucial, based on empirical evidence in LA, to increasing the participation of producers in the honey value chain [8:156]. In addition, greater popularity and acceptance is expected for those organizations that are in line with the Sustainable Development Goals, which include linkage to the territory, innovation as a transversal axis, environmental commitment and those of attachment to the territory shared by cooperativism and the social economy [78]. Furthermore, there are studies in which the location dimension does not end up being a discriminatory variable [57]. The importance of localization found here is interpreted in the context of the production environment, therefore we invite further research to explore how to incorporate this dimension into studies on ICT.

This study is also significant in terms of the local and regional evidence that points to the importance of the use of social networks, especially Facebook, in the agricultural sector [79,80]. Other studies have also revealed the importance of the use of these networks in the marketing and promotion of products. However, this evidence is still of a sectoral and spatial nature [81,82]. Furthermore, it is noted that both for Argentina and for Latin America the volume of electronic commerce is lower than that recorded in other regions of the world [83]. In this way, it is necessary to examine in more detail both the use of ICT and its impact within the entities related to the beekeeping sub-sector and the agricultural sector in Latin America [32,38].

4. Conclusions

The Argentine beekeeping sector, the third largest honey producer in the world, faces important problems that challenge the sustainability of the activity. Among them are the atomization and the scale of work in the primary sector, which implies a strong dependence on distribution. This situation is intensified by the concentration of the commercial export sector and the sale of the product, preferably in bulk, thus losing much of the added value generated in the packaging and marketing stages. In response to this situation, it is necessary to improve the exchange networks in order to achieve greater differentiation of production and more direct communication with the consumer. Several authors have highlighted the decisive role that the use of social networks can play as a communication and information channel due to the trust they generate among users as a result of their closeness and interactivity [45,56]. Accordingly, the aim of this paper is to analyze which characteristics of beekeeping organizations are directly related to greater success in the use of these technologies, taking as a case study the third largest honey producer in the world: The Argentine beekeeping sector.

The results obtained offer empirical evidence to accept the premise that all the propositions analyzed here are relevant to explaining the popularity of organizations in the most important worldwide social networks, such as Facebook. In this way, and beyond aspects related to the use of the network itself, the cooperative nature, environmental sensitivity, appropriate location and the linkage of the company with different online platforms are reasons that lead companies to improve their positioning in social networks. It is identified that the variables of location of the entity and the interconnection between the network and its own website become necessary conditions in most of the configurations. It is also noted that the variables of environmental sensitivity and presence on other platforms will be of particular importance, especially for cooperatives that are not located in a central area.

The results obtained can help to make beekeeping sector entities aware of the potential of social networks to address the commercial problems they face and their best position to take advantage of it, given the specific characteristics of this market. At the same time, they should serve as an incentive for both public and private organizations to take measures to correct in time the possible lags that may occur with respect to other sectors in terms of the use of online social networks for commercial purposes. The relevance and topicality of the beekeeping sector, as well as the presence of a society that is increasingly technological and demanding of natural products, makes it relevant to continue delving into this line of research. As a proposal for future developments, it is interesting to analyze other sectors or to quantify economically the impact of these tools on the organizational structure of beekeeping organizations.

At this point, it is necessary to point out the main limitations of this study. On the one hand, it is worth mentioning that this research has been directed especially at beekeeping sector entities, although we believe that these contributions can be extrapolated to a large part of the agri-food sector, which presents, in general terms, a similar basic problem in terms of marketing. On the other hand, we also note as a limitation that this study has focused on the national level. In this sense, although Argentina occupies a privileged position in honey production, it might be interesting to contrast its situation with that of other producing countries.

Author Contributions: Conceptualization, methodology and formal analysis, D.F.-U., A.M.-M., and E.B.-J. Investigation and resources, J.A. All authors have read and agreed to the published version of the manuscript.

Funding: This research received no external funding.

Institutional Review Board Statement: Not applicable.

Informed Consent Statement: Not applicable.

Data Availability Statement: This study did not report any data.

Conflicts of Interest: The authors declare no conflict of interest. The funders had no role in the design of the study; in the collection, analyses, or interpretation of data; in the writing of the manuscript, or in the decision to publish the results.

References

1. Food and Agriculture Organization of the United Nations. FAOSTAT- Producción y Comercio Agropecuario. Available online: http://www.fao.org/faostat/es/#data/QL (accessed on 26 March 2020).
2. Andrieu, J. De tranqueras y candados. Las mujeres, su acceso a la tierra y bienes comunes. In *Economía Feminista Para la Sostenibilidad de la Vida*; Friedrich-Ebert-Stiftung: Buenos Aires, Argentina, 2020; pp. 18–24. Available online: https://www.fes-argentina.org/publicaciones/ (accessed on 10 December 2020).
3. De-Groot, G.S.; Aizen, M.A.; Sáez, A.; Morales, C.L. Large-scale monoculture reduces honey yield: The case of soybean expansion in Argentina. *Agric. Ecosyst. Environ.* **2021**, *306*, 107203. [CrossRef]
4. MAGYP. Cadena Apícola: Informe Coyuntura Mensual N186. Área Apícola—Ministerio de Agroindustria: Capital Federal, Argentina. 2019. Available online: http://www.alimentosargentinos.gob.ar/HomeAlimentos/Apicultura/documentos/SintesisApic186.pdf (accessed on 10 April 2020).
5. MAGYP. Alimentos Argentinos. Apicultura. RENAPA. MAGyP: Capital Federal, Argentina. 2020. Available online: http://www.alimentosargentinos.gob.ar/HomeAlimentos/Apicultura/renapa.php (accessed on 23 April 2021).
6. MAGYP. Alimentos Argentinos. Apicultura. Síntesis Apícola. MAGyP: Capital Federal, Argentina. 2020. Available online: http://www.alimentosargentinos.gob.ar/HomeAlimentos/Apicultura/documentos/Sintesis-Apicola-Julio2020.pdf (accessed on 23 April 2021).
7. TRADEMAP. Estadísticas del Comercio Para el Desarrollo Internacional de las Empresas. Available online: http://www.trademap.org (accessed on 26 March 2020).
8. Tonantzi, K. Mercado y Comercialización de la Miel: Un Acercamiento a Oaxaca. Master's Thesis, Ciencias en Estrategia Agroempresarial, Universidad Autónoma Chapingo, Chapingo, Mexico, 2020.
9. MAGYP. Alimentos Argentinos. Apicultura. Tendencias Producción Orgánica. MAGYP: Capital Federal, Argentina. 2019. Available online: http://www.alimentosargentinos.gob.ar/HomeAlimentos/Organicos/documentos/TendenciaORGANICO12.pdf (accessed on 10 April 2020).
10. Estrada, M.E. Rasgos de la Territorialización en Complejos Productivos no Tradicionales Basados en Recursos Naturales: La Apicultura en el Sudoeste Bonaerense. Ph.D. Thesis, Departamento de Geografía y Turismo, Universidad Nacional del Sur, Bahía Blanca, Argentina, 2015.
11. MAGYP. Alimentos Argentinos. Apicultura. Argentina Exportó Miel Fraccionada a Brasil Después de Más de Una Década. MAGYP: Capital Federal, Argentina. 2018. Available online: http://www.alimentosargentinos.gob.ar/HomeAlimentos/Noticias/nota/1124/argentina-exporto-miel-fraccionada-a-brasil-despues-de-mas-de-una-decada (accessed on 10 April 2020).
12. Goslino, M. Apicultura en el Sudoeste Bonaerense: Una Propuesta de Eficiencia y Sustentabilidad en Esquemas de Comercialización Conjunta. Master's Thesis, Economía Agraria y Administración Rural, Universidad Nacional del Sur, Bahía Blanca, Argentina, 2017.
13. Malak-Rawlikowska, A.; Majewski, E.; Wąs, A.; Borgen, S.O.; Csillag, P.; Donati, M.; Freeman, R.; Hoang, V.; Lecoeur, J.-P.; Mancini, M.C.; et al. Measuring the Economic, Environmental, and Social Sustainability of Short Food Supply Chains. *Sustainability* **2019**, *11*, 4004. [CrossRef]
14. World Bank. Datos Por País. Argentina. Población. Available online: https://datos.bancomundial.org/indicador/SP.POP.TOTL?locations=AR (accessed on 26 March 2020).
15. Sanchez, C.; Castignani, H.; Rabaglio, M. *Mercado Apícola Internacional*; INTA Ediciones: Buenos Aires, Argentina, 2018; pp. 1–23.
16. Instituto Nacional de Tecnología Agropecuaria. Alimentos: El Consumo Responsable Cambia Paradigmas. 2018. Available online: https://intainforma.inta.gob.ar/alimentos-el-consumo-responsable-cambia-paradigmas/ (accessed on 7 May 2021).
17. Pippinato, L.; Blanc, S.; Mancuso, T.; Brun, F. A Sustainable Niche Market: How Does Honey Behave? *Sustainability* **2020**, *12*, 678. [CrossRef]
18. Garry, S.; Parada Gómez, Á.M.; Salido Marcos, J. Incorporación de mayor valor en la cadena de la miel y productos derivados de la colmena en el Pacífico Central, Costa Rica. Ciudad de México. 2017. Available online: https://repositorio.cepal.org/bitstream/handle/11362/42232/1/S1700970_es.pdf (accessed on 5 May 2021).
19. Urquiza-Jozami, J.; Berges, M.; Casellas, K.; De-Greef, G.; Gil, J.M.; Liseras, N. Preferencias del consumidor y canales cortos de comercialización de miel en Mar del Plata. *Doc. Trab. CIEP* **2019**, *2*, 1–40.
20. Sebillotte, C. L'Argentine dans le contexte latino-américain: Consommations alimentaires, santé et politiques nutritionnelles. *OCL Oilseeds Fats Crop. Lipids EDP* **2018**, *25*, 1–20. [CrossRef]

21. Blacha, L. El menú del agronegocio: Monocultivo y malnutrición del productor al consumidor (1996–2019). *Rev. Hist. Debates E Tend.* **2020**, *20*, 9–24. [CrossRef]
22. Aguirre, P. Feeding, cooking, sharing: A brief social history of food. *Mètode Sci. Stud. J. Annu. Rev.* **2021**, *11*. [CrossRef]
23. Potts, J.; Lynch, M.; Wilkings, A.; Hupp, G.; Cunningham, M.; Voora, V. *The Sate of Sustainability Initiatives Review 2014: Standards and the Green Economy*; International Institute for Sustainable Development (IISD): Winnipeg, MB, Canada; International Institute for Environment and Development (IIED): London, UK, 2014; pp. 1–345.
24. Filippi, M.; Chapdaniel, A. Sustainable demand-supply chain: An innovative approach for improving sustainability in agrifood chains. *Int. Food Agribus. Manag.* **2020**, *24*, 1–16. [CrossRef]
25. Scuderi, A.; Bellia, C.; Foti, V.T.; Sturiale, L.; Timpanaro, G. Evaluation of consumers' purchasing process for organic food products. *AIMS Agric. Food* **2019**, *4*, 251–265. [CrossRef]
26. Kim, J. Analysis for Growth Potential in Response to Changes in the Online Food Market. *Sustainability* **2020**, *12*, 4386. [CrossRef]
27. Cristobal-Fransi, E.; Montegut-Salla, Y.; Ferrer-Rosell, B.; Daries, N. Rural cooperatives in the digital age: An analysis of the Internet presence and degree of maturity of agri-food cooperatives' e-commerce. *J. Rural. Stud.* **2020**, *74*, 55–66. [CrossRef]
28. Lyubenov, L. Beekeeping Markets outside the Honey Category. *J. Mt. Agric. Balk.* **2020**, *23*, 73–92. Available online: http://www.rimsa.eu/images/stockbreeding_vol_23-2_2020.pdf (accessed on 10 April 2021).
29. Stricker, S.; Mueller, R.A.; Sumner, D.A. Marketing Wine on the Web. *Choices Mag.* **2007**, *22*, 31–34. Available online: http://www.choicesmagazine.org/2007_1/foodchains/2007_1_06.htm (accessed on 26 March 2020).
30. Canavari, M.; Regazzi, D.; Spadoni, R. *Origin Labels on the World Wide Web*; Working Paper; Università di Bologna, Dipartimento di Economia e Ingegneria Agrarie: Bologna, Italy, 2002.
31. Ribeiro-Soriano, D. Small business and entrepreneurship: Their role in economic and social development. *Entrep. Reg. Dev.* **2017**, *29*, 1–3. [CrossRef]
32. Gallacher, M.; Justo, A. *Uso de TICs en el Sector Agropecuario con Énfasis en el Subsector Apícola de Argentina, Uruguay, República Dominicana y Costa Rica*; INTA CIEP: Buenos Aires, Argentina, 2016; pp. 1–22.
33. Scaglione, A. Percepción del Consumidor de Miel de Abejas en la Ciudad de La Plata. Tesis de Grado, Facultad de Ciencias Agrarias y Forestales, Universidad Nacional de La Plata, La Plata, Argentina. 2016. Available online: http://sedici.unlp.edu.ar/handle/10915/56144 (accessed on 10 April 2021).
34. Mouteira, M.C.; Dedomenici, A.C.; Alberto, C.M.; Pérez, R.C.; Paradela, M.P. Productos agroalimentarios: Hábito de compra y consumo de miel en la localidad de La Plata. In Proceedings of the IV Reunión Transdisciplinaria en Ciencias Agropecuarias, La Plata, Argentina, 11–12 December 2019; Available online: http://sedici.unlp.edu.ar/handle/10915/88223 (accessed on 5 May 2021).
35. ESIC Market Submitter; Bragulat, T.; Angón, E.; Giorgis, A.; Perea, J. Typology and Characterization of the Pampean Beekeeping Systems. *ESIC Mark. Econ. Bus. J.* **2020**, *51*, 299–318. Available online: https://ssrn.com/abstract=3609956 (accessed on 5 May 2021). [CrossRef]
36. Lavarello, P.; Bil, D.; Vidosa, R.; Langard, F. Reconfiguración del oligopolio mundial y cambio tecnológico frente a la agricultura 4.0: Implicancias para la trayectoria de la maquinaria agrícola en Argentina. *Ciclos* **2019**, *XXVI*, 163–193. Available online: http://hdl.handle.net/11336/123424 (accessed on 10 April 2021).
37. Di-Bartolo, A.L.; Martin, M. Evaluación de Potencial Innovador y de Mercado de Plataforma de Gestión Para Empresas Agropecuarias. Tesis de Grado, Facultad de Ingeniería, Universidad Nacional de Mar del Plata, Mar del Plata, Argentina. 2020; pp. 1–102. Available online: http://rinfi.fi.mdp.edu.ar/xmlui/handle/123456789/447 (accessed on 10 April 2021).
38. Bellini-Saibene, Y.; Caldera, J.; Ramos, L. Cosechando Datos. Desarrollos para la agricultura en la era digital. *Electron. J. Sadio EJS* **2020**, *19*, 64–95. Available online: https://ojs.sadio.org.ar/index.php/EJS/article/view/158 (accessed on 10 April 2021).
39. Powell, T.C.; Dent-Micallef, A. Information technology as competitive advantage: The roles of human, business and technology resources. *Strateg. Manag. J.* **1997**, *18*, 375–405. [CrossRef]
40. Karoui, M.; Dudezert, A.; Leidner, D.E. Strategies and symbolism in the adoption of organizational social networking systems. *J. Strateg. Inf. Syst.* **2015**, *24*, 15–32. [CrossRef]
41. Laroche, M.; Habibi, M.R.; Richard, M.O. To be or not to be in social media: How brand loyalty is affected by social media. *Int. J. Inf. Manag.* **2013**, *33*, 76–82. [CrossRef]
42. Vlachvei, A.; Notta, O.; Diotallevi, F.; Marchini, A. Web marketing strategies in Agro Food SMES—Evidence from Greek and Italian wine SMEs. In *E-Innovation for Sustainable Development of Rural Resources during Global Economic Crisis*; Andreopoulou, Z., Samathrakis, V., Louca, S., Vlachopoulou, M., Eds.; IGI Global: Hershey, PA, USA, 2014; pp. 199–220.
43. Befecadu, J.; Chembezi, D.M.; McFarlane, D. Internet Adoption and Use of E-Commerce Strategies by Agribusiness Firms in Alabama. In Proceedings of the Southern Agricultural Economics Association Annual Meeting, Mobile, AL, USA, 1–5 February 2003. [CrossRef]
44. Mozas, A.; Bernal, E. Posibilidades y aplicaciones de la Web 2.0: Un caso de estudio aplicado a la economía social. *CIRIEC Esp. Rev. Econ. Pública Soc. Coop.* **2012**, *74*, 261–283.
45. Fernández-Uclés, D.; Mozas-Moral, A.; Bernal-Jurado, E.; Medina-Viruel, M.J. Uso y eficiencia de la social media. Un análisis desde la economía social. *CIRIEC Esp. Rev. Econ. Pública Soc. Coop.* **2016**, *88*, 4–27. [CrossRef]
46. Montegut, Y.; Cristóbal, E.; Gómez, M.J. La implementación de las TIC en la gestión de las cooperativas agroalimentarias: El caso de la provincia de Lleida. *REVESCO Rev. Estud. Coop.* **2013**, *110*, 223–253. [CrossRef]

47. López, E.I.; Arcas, N.; Alcón, F. Uso y calidad de los sitios Web: Evaluación en las empresas agroalimentarias murcianas. *Rev. Esp. Estud. Agrosoc. Pesq.* **2014**, *237*, 155–179.

48. Boschma, R.A.; Ter Wal, A.L. Knowledge networks and innovative performance in an industrial district: The case of a footwear district in the South of Italy. *Ind. Innov.* **2007**, *14*, 177–199. [CrossRef]

49. Arza, V.; López, E. Obstacles to innovation and firm size. Obstacles of innovation and firm size: A quantitative study for Argentina. *Int. Dev. Bank.* 2018. Available online: https://publications.iadb.org/publications/english/document/Obstacles-to-Innovation-and-Firm-Size-A-Quantitative-Study-for-Argentina.pdf (accessed on 3 August 2020).

50. Claver-Cortés, E.; Marco-Lajara, B.; Sánchez-García, E.; Seva-Larrosa, P.; Manresa-Marhuenda, E.; Ruiz-Fernández, L.; Poveda-Pareja, E. A Literature Review on the Effect of Industrial Clusters and the Absorptive Capacity on Innovation. *Int. J. Ind. Manuf. Eng.* **2020**, *14*, 489–498.

51. Hilverda, F.; Kuttschreuter, M.; Giebels, E. Social media mediated interaction with peers, experts and anonymous authors: Conversation partner and message framing effects on risk perception and sense-making of organic food. *Food Qual. Prefer.* **2017**, *56*, 107–118. [CrossRef]

52. Fernández-Uclés, D.; Mozas-Moral, A.; Bernal-Jurado, E.; Medina-Viruel, M.J. The importance of websites for organic agri-food producers. *Econ. Res. Ekon. Istraž.* **2019**, *33*, 2867–2880. [CrossRef]

53. Lee, H.J.; Yun, Z.S. Consumers' perceptions of organic food attributes and cognitive and affective attitudes as determinants of their purchase intentions toward organic food. *Food Qual. Prefer.* **2015**, *39*, 259–267. [CrossRef]

54. Evelyn, C.; Qiang, L.; Rohan, M. Social media, customer relationship management, and consumers' organic food purchase behavior. In *Strategic Customer Relationship Management in the Age of Social Media*; Khanlari, A., Ed.; IGI Global: Hershey, PA, USA, 2015; pp. 198–215.

55. Mozas-Moral, A.; Bernal-Jurado, E.; Fernández-Uclés, D.; Medina-Viruel, M.J. Innovation as the Backbone of Sustainable Development Goals. *Sustainability* **2020**, *12*, 4747. [CrossRef]

56. Shiau, W.L.; Dwivedi, Y.K.; Yang, H.S. Co-citation and cluster analyses of extant literature on social networks. *Int. J. Inf. Manag.* **2017**, *37*, 390–399. [CrossRef]

57. Jorge-Vázquez, J.; Cebolla, C.; Peana, M.; Salinas Ramos, F. La transformación digital en el sector cooperativo agroalimentario español: Situación y perspectivas. *CIRIEC Esp. Rev. Econ. Pública Soc. Coop.* **2019**, *95*, 39–70. [CrossRef]

58. Fagerberg, J.; Mowery, D.C.; Nelson, R.R. *The Oxford Handbook of Innovation*; Oxford University Press: New York, NY, USA, 2005.

59. Jiménez, D.; Sanz, R. Innovation, organizational learning, and performance. *J. Bus. Res.* **2011**, *64*, 408–417. [CrossRef]

60. Mozas, A.; Bernal, E.; Medina, M.J.; Fernández, D. Factors for success in online social networks: An fsQCA approach. *J. Bus. Res.* **2016**, *69*, 5261–5264. [CrossRef]

61. Kaplan, A.M.; Haenlein, M. Users of the world, unite The challenges and opportunities of Social Media. *Bus. Horiz.* **2010**, *53*, 59–68. [CrossRef]

62. Tian, Y.; Robinson, J.D. Media use and health information seeking: An empirical test of complementarity theory. *Health Commun.* **2008**, *23*, 184–190. [CrossRef]

63. Ghulam, A.; Depar, M.H.; Ali, S.; Rahu, S. On-Page Search Engine Optimization (SEO) Techniques Model: A Use Case Scenario of a Business Entity Website. *Int. J. Res. Appl. Sci. Eng. Technol.* **2017**, *11*, 3076–3083.

64. Revilia, D.; Irwansyah, I. Social Media Optimization: Promotional Strategy towards Library 4.0 Era. *Proc. Int. Conf. Doc. Inf.* **2019**, *2*, 37–71. [CrossRef]

65. Sahai, S.; Goel, R.; Malik, P.; Krishnan, C.; Singh, G.; Bajpai, C. Role of social media optimization in digital marketing with special reference to Trupay. *Int. J. Eng. Technol.* **2018**, *7*, 52–57. [CrossRef]

66. Patrutiu, L. Inbound Marketing-the most important digital marketing strategy. *Bull. Transilv. Univ. Brasov Econ. Sci. Ser. V* **2016**, *9*, 61–68.

67. NOSIS. Investigación y Desarrollo. Available online: http://www.nosis.com.ar (accessed on 14 August 2020).

68. Ragin, C. *The Comparative Method: Moving beyond Qualitative and Quantitative Strategies*; University of California Press: Berkeley, CA, USA, 1987.

69. Schneider, C.Q.; Wagemann, C. Standards of good practice in qualitative comparative analysis (QCA) and fuzzy-sets. *Comp. Sociol.* **2010**, *9*, 397–418. [CrossRef]

70. Sehring, J.; Korhonen, K.; Brockhaus, M. *Qualitative Comparative Analysis (QCA): An Application to Compare National REDD+, Policy Processes*; CIFOR: Bogor, Indonesia, 2013. [CrossRef]

71. Ragin, C.C. Set relations in social research: Evaluating their consistency and coverage. *Political Anal.* **2006**, *14*, 291–310. [CrossRef]

72. Fiss, P.C. A set-theoretic approach to organizational configurations. *Acad. Manag. Rev.* **2007**, *32*, 1180–1198. [CrossRef]

73. Pappas, I.O.; Woodside, A.G. Fuzzy-set Qualitative Comparative Analysis (fsQCA): Guidelines for research practice in Information Systems and marketing. *Int. J. Inf. Manag.* **2021**, *58*, 102310. [CrossRef]

74. Ragin, C.C. *Redesigning Social Inquiry: Fuzzy Sets and beyond*; University of Chicago Press: Chicago, IL, USA, 2008.

75. Woodside, A.G. Moving beyond multiple regression analysis to algorithms: Calling for adoption of a paradigm shift from symmetric to asymmetric thinking in data analysis and crafting theory. *J. Bus. Res.* **2013**, *66*, 463–472. [CrossRef]

76. Schwab, F.; Calle-Collado, A.; Muñoz, R. Economía social y solidaria y agroecología en cooperativas de agricultura familiar en Brasil como forma de desarrollo de una agricultura sostenible. *CIRIEC Esp. Rev. Econ. Pública Soc. Coop.* **2020**, *98*, 189–211. [CrossRef]

77. Mozas, A.; Bernal, E.; Fernández, D.; Medina, M.J.; Puentes, R. Cooperativismo de segundo grado y adopción de las TIC. *CIRIEC Esp. Rev. Econ. Pública Soc. Coop.* **2020**, *100*, 67–85. [CrossRef]
78. Alarcón, M.Á.; Álvarez, J.F. El balance social y las relaciones entre los objetivos de desarrollo sostenible y los principios cooperativos mediante un análisis de redes sociales. *CIRIEC Esp. Rev. Econ. Pública Soc. Coop.* **2020**, *99*, 57–87. [CrossRef]
79. Santini, S.; Ghezan, G.; Bontempo, M. Uso de las TIC por parte de Agricultores Familiares en el Sudeste de la provincia de Buenos Aires. In Proceedings of the X Jornadas Interdisciplinarias de Estudios Agrarios y Agroindustriales Argentinos y Latinoamericanos, Buenos Aires, Argentina, 7–10 November 2017; Available online: http://hdl.handle.net/20.500.12123/4172 (accessed on 5 May 2021).
80. Urcola, M. Articulación de las TIC en el sector agrícola pampeano: La apropiación de la telefonía celular, las computadoras e Internet entre los productores de una localidad del sur santafesino. *Rev. Temas Debates* **2012**. [CrossRef]
81. Santini, S.; Ghezan, G. Uso y resignificación de las TIC en una red de comercio electrónico de alimentos agroecológicos. In Proceedings of the II Jornadas de Sociología Facultad de Humanidades-UNMP, Mar del Plata, Argentina, 28–29 March 2019; Available online: http://hdl.handle.net/20.500.12123/4780 (accessed on 5 May 2021).
82. Jones, C.; Motta, J.; Alderete, M.V. Gestión estratégica de tecnologías de información y comunicación y adopción del comercio electrónico en Mipymes de Córdoba, Argentina. *Estud. Gerenc.* **2016**, *32*, 4–13. [CrossRef]
83. Perdigón, R.; Pérez, M. Holistic analysis of the social impact of electronic business in Latin America, from 2014 to 2019. Análisis holístico del impacto social de los negocios electrónicos en América Latina, de 2014 a 2019. *Paakat Rev. Tecnol. Soc.* **2020**, *10*, 1–23. [CrossRef]

agriculture

Article

Does the Self-Identity of Chinese Farmers in Rural Tourism Destinations Affect Their Land-Responsibility Behaviour Intention? The Mediating Effect of Multifunction Agriculture Perception

Xingping Cao [1], Zeyuan Luo [1], Manli He [2], Yan Liu [1,*] and Junlin Qiu [1]

[1] Business and Tourism School, Sichuan Agricultural University, No. 288, Jianshe Road, Dujiangyan 611830, China; cxp@sicau.edu.cn (X.C.); luozy@stu.sicau.edu.cn (Z.L.); 2020227007@stu.sicau.edu.cn (J.Q.)

[2] School of Economics and Management, Fuzhou University, No. 2, Wulongjiang North Avenue, Xue Yuan Road, University Town, Fuzhou 350108, China; 200710023@fzu.edu.cn

[*] Correspondence: liuyan@sicau.edu.cn

Abstract: Farmers are the heart of rural tourism destinations, and their land-responsibility behaviours affect sustainable development. In this study, four rural tourist sites in the suburbs of Chengdu were selected, and the structural equation model was used to analyse the influence of farmers' self-identity on their land-responsibility behaviours intention under the condition of agricultural multifunction perception as a mediation variable. The results show that, in rural tourism destinations of suburban districts of China, farmers' self-identity is an important variable that affects their land-responsibility behaviour intention. Agricultural economic function perception mediates the relationship between farmers' self-identity and land-responsibility behaviour intention. Agricultural non-economic function perception positively affects their agricultural economic function perception.

Keywords: self-identity; agricultural non-economic function perception; agricultural economic function perception; land-responsibility behaviour intention

Citation: Cao, X.; Luo, Z.; He, M.; Liu, Y.; Qiu, J. Does the Self-Identity of Chinese Farmers in Rural Tourism Destinations Affect Their Land-Responsibility Behaviour Intention? The Mediating Effect of Multifunction Agriculture Perception. *Agriculture* **2021**, *11*, 649. https://doi.org/10.3390/agriculture11070649

Academic Editors: Domingo Fernandez Ucles and Adoración Mozas Moral

Received: 20 May 2021
Accepted: 28 June 2021
Published: 9 July 2021

Publisher's Note: MDPI stays neutral with regard to jurisdictional claims in published maps and institutional affiliations.

1. Introduction

In recent years, China's sustainable rural development has faced many challenges, some of which involve farmers' land decisions such land-responsibility behaviour intention. For example, farmers are confronted with decisions on non-point source pollution caused by the overuse of pesticides, the loss of traditional farming culture caused by the abandonment of peasants, as well as vegetable and food safety problems caused by pesticide residues. Farmers are the main subjects of the countryside and the principal force promoting the revitalisation of China's rural areas. Therefore, understanding farmers' land behaviour could help to clarify the role of policies in development of rural areas, and to aid the integrated and coordinated development of multiple industries, such as agriculture, forestry, and tourism, as well as encourage the protective use of environmental and cultural factors in land-planning.

Researchers have primarily used the research framework of the theory of planned behaviour (TPB) [1] to investigate farmers' land behaviour and policy interactions. The Theory of Planned Behaviour (TPB) is an extension of the theory of reasoned action [2]. The theory uses factors such as attitudes, social norms, and personal abilities to predict individual farmers' behavioural intentions and directions [3–6]. Farmers' decision making in agricultural practice can be rooted in certain social and cultural backgrounds, and many behaviours cannot be explained by the theoretical framework of rational behaviour alone. Indeed, the included variables in the TPB cannot fully explain the large variances in individual behavioural intentions [7,8]. Therefore, increasingly, researchers have attempted

145

to combine the rational logic of TPB with farmers' personal emotions, feelings, and other psychological factors to more accurately predict and manage farmers' land behaviours and policy responses [9,10]. To enhance the predictive power of TBP in farmers' decision making, social psychologists have renewed their interests recently in articulating how self-identity can play an important role [10,11]. Self-identity has a strong correlation with behavioural intention across a wide range of public health areas [12–14], consumer behaviour [15,16], and environmental behaviour [17–19]

The role of self-identity to farmers' behavioural intentions has been illustrated empirically in many studies [20,21]. Based on Self-Identity Theory, self is envisaged as a social construct in which a distinctive self-component represents each of the roles we occupy in different social settings, internally generated role-expectation [22]. It views self as a societal role and incorporates the meanings and expectations associated with that role. For farmers, when their self-perception positively matches with the behavioural outcomes, they would have more intention to undertake the action. To summarise, self-identity is one of the most important non-rational factors [23] for predicting farmers' land-responsibility behaviour.

At present, the process of urbanisation in China is quickly advancing to the countryside, and rural tourism around large cities is developing rapidly. This rapid development can be expected to affect the thinking, especially in terms of how to re-recognise themselves, agriculture and land. The development of rural tourism has changed the way that rural residents and urban populations interact, forming a new social interaction context. On the one hand, more and more Chinese urban residents flock to the countryside to experience rural life. This may cause farmers to re-examine their environment and self-identity, and thus generate a new self-identity. On the other hand, in these rural areas, part of the land is dedicated to rural tourism, and agriculture has gradually developed from having a single function (food and vegetable production) to multiple functions, such as natural landscape provision, cultural atmosphere creation, and education; furthermore, these non-economic functions are linked to economic functions through the tourism industry. At the same time, farmers' land decisions such as land-responsibility behaviour can also be affected by farmers' perception of internal self and external environment.

Self-Identity Theory emphasises that self-identity affects the cognitive style of individuals and can predict the direction of individual perception and cognitive process. Lee et al. [24] used it within the context of the Technology Acceptance Model (TAM), introducing an extension of a validated framework, and proved that self-identity influences the adoption behaviour of WebCT through perceived usefulness. From this perspective, how does farmers' self-identity affect their cognition of agricultural function? Additionally, whether it determines their land responsibility behaviour? This has not been discussed in the literature before.

The environmental protection behaviour of farmers and the issue of inheritance of traditional culture are related to future rural sustainable development. The study defined the land decision-making behaviours that are conducive to sustainable rural development as land responsibility behaviour intention. In rural tourism text, farmers' land-responsibility behaviour intention refers to their intention of behaviour aimed the environmental protection, social profit, and inheritance of traditional culture that are related to future rural sustainable development. Land-responsibility behaviour is the most important factor related to sustainable development of rural areas. The aim of sustainable development is to achieve a balance between the complementary goals of providing environmental, economic, and social opportunities for the benefit of present and future generations, while also maintaining and enhancing the quality of the land resource [25].

From a farm level, contributing to the preservation of the landscape's character, strengthening the landscape's quality, and sustainable development, the study presents the relations of farmers self-identity, agricultural multifunction perception (including agricultural non-economic function perception and agricultural economic function perception), and their land responsibility behaviour. This empirical study sought to answer how farmers' self-identity affects their perception of agricultural economic function, non-economic function and land responsibility behaviour. The findings could assist policymakers and land use planners in decision making related to sustainable rural tourism in China.

2. Literature Review and Research Hypotheses

2.1. Farmers' Self-Identity and Its Impact on Farmers' Land-Responsibility Behaviour Intention

Self-identity, also known as role identity, is derived from Identity Theory [26]. Identity Theory is an important theory in the field of sociology that focuses on the social structural attributes of people's connection with others. Self-identity is defined by the specific role that a person plays, or considers themselves to play, in the existing social structure, such as social roles (parent or child), professional roles (farmer, student), and group roles (manager, employee). These different role classifications lead to the formation of different behavioural intentions, and people spontaneously behave according to their role expectations [27,28]. Scholars introduced the concept of self-identity to modify the traditional theoretical model of planned behaviour. They pointed out that the role positioning of self-identity triggered a habitual behaviour that supports self-concept verification [29]. In this way, people try to establish self-identity consistency between attitudes and behaviours [30,31]. Self-identity also reflects the enduring characteristics of individual self-cognition [32], and the prediction of individual behaviour through self-identity is stable. Therefore, the socio-psychological factor of self-identity is a key influencer of individual behavioural intentions [23]. The link between self-identity and the behavioural intentions of farmers has been confirmed by many studies. For example, farmers' pro-environmental behaviours, environmental protection behaviours under non-economic subsidies, as well as land decisions not only depend on rational decisions [10,33], but also variables such as perception of farmers' job independence, pride [30,34], and farmers' lifestyles [35], which have been reported to have a direct or indirect effect on farmers' individual behavioural intentions. Research from Lokhorst et al. [33] has shown that farmers' self-identity affects their pro-environmental behaviours by affecting their connectedness to nature.

Self-identity can be a direct or intermediary condition that affects farmers' land behaviour by affecting their cognition and other socio-psychological factors (such as attitudes and social norms). Therefore, based on the perspective of farmers' self-identity, this study explored the impact path of farmers' land-responsibility behaviour intention in suburban rural tourism destinations. As such, we made the following hypothesis:

Hypothesis (H1). *Farmers' self-identity has a positive significant impact on farmers' land-responsibility behaviour intention.*

2.2. Farmers' Perception of Agricultural Multifunction: Agricultural Non-Economic Function Perception and Agricultural Economic Function Perception

The concept of "agricultural multifunction" or "agricultural versatility" describes the multiple non-productive benefits of agricultural systems and land. It has been highlighted that, in addition to food production, agriculture also exerts unique production functions such as economy, society, environment, and culture, and is the result of the joint production of economic and non-economic products (Millennium Ecosystem Assessment, 2005).

With the continuous changes in agriculture and land functions in modern society, as well as the development of agricultural tourism and rural tourism, the theme of agricultural multifunction has received more and more academic attention. Much research on agricultural multifunction has focused on the national and regional levels, mainly investigating the value judgement of agricultural multifunction, land multifunction planning, agricultural compensation system design, urban agricultural development positioning, and agricultural multifunction technical practice. Practical technical methods, such as rural governance and village planning, need to be studied from multiple dimensions, such as subject and function. Multifunctional agriculture is the basic unit of decision making and a direct expression of multifunctional agriculture in families, farms, and rural communities; in particular, the agricultural multifunction at the family farm level is closely related to farmers' attitudes, ideas, and identity [36].

From the perspective of European land management practices, there is reportedly a profound interaction between the policy of agricultural multifunction and farmers [37]. However, in the context of China's tourism development, previous research on the multifunction of agriculture in China has overlooked the micro-levels of communities, businesses, families, and individual farmers. Instead, it has mainly focused on human geography, environmental science and resource utilisation, and landscape planning, involving the evaluation of the multifunction of agricultural landscapes, spatial identification, and planning management. In response to this gap in the literature, this study took farmers who are most closely connected to the land as the research object. From the perspective of the perception of agricultural "multifunction", we focused on the micro-expression of the "agricultural multifunction" of rural tourism destinations; additionally, we combined the self-identity of farmers in rural tourism destinations, the perception of the importance of agricultural multifunction, and farmers' land-responsibility behaviour intention, and in such a way, a micro-perspective for multifunctional agricultural research is provided.

Social logic and economic studies have traditionally studied agriculture as a means of making money and improving the livelihoods of farmers. At the same time, agricultural activities are also affected by various irrational factors, such as culture, family, and lifestyle preferences [38]. Therefore, we believe that farmers' perception of agricultural multifunction can be divided into two dimensions: agricultural economic function perception and agricultural non-economic function perception. Some studies have shown that there are large differences between farmers' perceptions and practices of multifunctional agriculture. For example, a survey results of Norwegian farmers' showed that the respondents identified themselves as producing not only high-quality food, but also public goods such as cultural landscapes and cultural heritage, and income maximisation is less important [39]. Another study found that residents in the state of Maine felt protecting farmland was important, but that protecting natural resources/wild landscapes was more important [40]. Hence, this study puts forward the following research hypotheses:

Hypothesis (H2a). *Farmers' self-identity positively affects their agricultural non-economic function perception.*

Hypothesis (H2b). *Farmers' self-identity positively affects their agricultural economic function perception.*

Farmers' perception of agricultural multifunction may influence their land-responsibility behaviour intention. Previous studies have shown that, at the farm level, the perception and practice of multifunctional agriculture affects farmers' environmental protection behaviours (such as reducing the use of pesticides), and farmers' protection and inheritance of traditional culture [41]. The awareness of agricultural landscape values enable farmers to create new strategies [42].

Kontogeorgos et al. [43] mentioned that the farmers' perceptions towards land environment impact their responsibility to protect it. While, the farmers were beginning to realise the importance of landscape culture and tourism and leisure functions of cultivated land. Taking tourism and leisure functions as an example, in order to create a good atmosphere and to create their own farm characteristics to attract tourists, farmers will learn how to improve the ecological environment protection of their cultivated land. Then, we believe that the farmers' agricultural non-economic perception will affect their willingness to take land responsibility behaviour positively.

Previous study has investigated farmers' attitudes about farming, the results indicated that land is always closely related to farmers' income and livelihoods, then they view themselves in a caretaker role for the land and showed their greater concern for the soil as a resource [44]. Emerton and Snyder [45] identified characteristics such as the ability to generate higher crop yields, better food supplies influence farmers' sustainable land management choices. Hence, we consider that farmers' economic perception of agricultural may influence their land-responsibility behavioural intention.

Then, the study puts forward the following research hypotheses:

Hypothesis (H3a). *Farmers' agricultural non-economic function perception positively affects their land-responsibility behavioural intention.*

Hypothesis (H3b). *Farmers' agricultural economic function perception positively affects their land-responsibility behavioural intention.*

In tourism context, the agricultural production functions of social, environmental, and cultural products have met the market demand for rural tourism products [46]. Accordingly, the non-economic function of agriculture has been transformed into economic functions through the tourism industry under the multifunctional system of agriculture [47].

Based on the previous findings outlined above, we made the following research hypothesis:

Hypothesis (H4). *Farmers' agricultural non-economic function perception positively affects their agricultural economic function perception.*

The proposed research framework is shown in Figure 1.

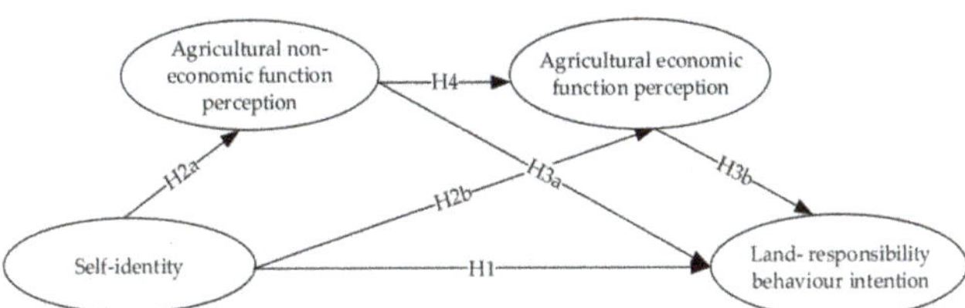

Figure 1. Research framework.

3. Methodology

3.1. Research Area

Over nearly 10 years of development, a more mature suburban rural tourism industry cluster has formed around Chengdu, Sichuan Province, which is a region in China that well reflects multifunctional agriculture. In this study, four rural tourist sites around Chengdu were selected as our survey sites (Nongke Village in Pi County, Mingyue Village in Pujiang County, Taohuagou in Longquanyi District, and Sansheng Township in Jinjiang District).

Rural tourism has been developing in Sansheng Township, Jinjiang District, since the early 1990s. The "Five Golden Flowers", which has an area of about 12 km^2, is a typical representative area of rural tourism; it is a tourism and leisure area that integrates business, leisure, vacation, culture, and creativity. Taohuagou, Longquanyi District, is a representative area of multifunctional agriculture; it is an agricultural tourist attraction with peach and pear trees as the main rural landscape resources. Compared with Sansheng Township, the degree of government participation is relatively low, and most of initiatives are farmers' independent development. Nongke Village, which is located in Youai Town, Pidu District, is a national agricultural tourism site; its main tourism interest is agritainment, which originated in the 1980s, whereby farmers attract citizens by opening their Sichuan bonsai nurseries for tourists to visit. The tourism industry started later in Mingyue Village, Pujiang County, which relies on local resources such as Phyllostachys praecox, ecological tea gardens, and ancient kilns, with the theme of pottery culture and home to the Mingyue International Ceramic Art Industry Cultural and Creative Park. With the creation of a humanistic ecological resort that integrates ceramic art production and sales, cultural display, creative experience, leisure sports, and rural vacations, this area is now a well-known rural tourism destination and a prime example of rural construction in Chengdu.

3.2. Sampling Procedures

From March to June 2018, random sampling method was used to select farmers from village household lists provided by local government authorities and institutions in the selected four rural tourist sites, resulting in a total of 393 famers overall.

After 46 farmers were removed from the dataset because of missing and inconsistent answers, a total of 347 valid questionnaires were obtained. Of these, 92 were from Mingyue Village, 88 from Taohuagou Village, 91 from Nongke Village, and 76 from Sansheng Township. The socioeconomic characteristics of the study sample are shown in Table 1.

Table 1. Socio-economic characteristics of sample.

Characteristics of Participants	Frequency (*n*)	Percentage (%)
Gender		
Male	167	48.1
Female	180	51.9
Age		
18–29 years old	52	15
30–39 years old	72	20.7
40–59	167	48.1
Over 60 years old	55	15.9
Education level		
Under 9 years	276	79.5
High school	47	13.5
College	19	5.5
Bachelor's degree and above	3	0.9
Land area		
<1 mu	152	43.8
1–5 mu	154	44.4
5–10 mu	40	11.5
10 mu	1	0.3
Status of land management		
Always cultivate by own	156	45.0
Lease to others	85	24.5
Hand over to the government for unified management	88	25.4
Other	18	5.2
Farming time per year		
None	64	18.4
1–3 months	109	31.4
4–6 months	65	18.7
7–9 months	52	15.0
10–12 months	42	12.1
Other forms of economic sources		
Go out for work	167	48.1
Tourism industry services	91	26.2
Other	79	22.7
None	10	3

Note: 1 mu = 0.07 acres. (*n* = 347, with some missing).

3.3. Measures

3.3.1. Farmers' Self-Identity

The three items of farmers' self-identity were measured using the self-identity scale by Lee et al. [24]. Amendments to this scale were made with consideration to the characteristics of farmer identity, mainly from the perspective of professional identity, such as perception of farmers' job independence, pride (Christensen & P., 2004; Key, 2005), and lifestyles (Howley, 2014). Based on previous interviews and the identity characteristics of Chinese farmers, three measurement items were used, including "I enjoy the lifestyle of being a farmer" (F1), "Being a farmer is an honest profession" (F2), and "I have freedom and independence as a farmer" (F3). All variables were scored on a Likert-type scale that ranged from totally disagree (1) to totally agree (7).

3.3.2. Multifunctional Agriculture Perception: Agricultural Non-Economic Function Perception and Agricultural Economic Function Perception

The multifunctional agriculture perception scale was developed by Kvakkestad et al. [39]. In that original study, the authors designed a 16 item land multifunctional perception survey that was completed by farmers in Norwegian agricultural cultural heritage sites. In this study, to ensure the validity of the measurement items, we randomly selected 15 farmers in the surveyed area and conducted one-to-one in-depth interviews. The main question of the interview was "As a farmer, what function of the agriculture is important to you?" According to the interview results and specific items of multifunctional agriculture from Kvakkestad et al. [39], nine items were put forward by the farmers (e.g., keeping the land or countryside tidy), seven items were not suitable for the research situation (e.g., securing the workplace of myself and my family). Finally, nine items were used to measure multifunctional agriculture perception including agricultural non-economic function perception and agricultural economic function perception in this study [48].

In these items, agricultural non-economic function perceptions are as follows: carrying forward production knowledge and a lifestyle that had been passed down from ancestors (NEF1); keeping the land or countryside tidy (NEF2); maintaining the rural cultural landscape (NEF3); conserving the natural environment (e.g., by minimising pollution) (NEF4); taking care of the land and other resources left by seniors (NEF5). Farmers were asked to rate their importance on the five items, ranging from 1 to 7, representing the importance from not being important at all to being very important.

Agricultural economic function perceptions are as follows: receiving a higher income through agriculture on the basis of constant land area (EF1); obtaining the maximum economic benefits (EF2); receiving a satisfactory income (EF3); ensuring that there are sufficient food and vegetable supplies in the event of an emergency (such as a natural disaster) (EF4). Farmers were asked to rate their importance on the four items, ranging from 1 to 7, representing the importance from not being important at all to being very important.

3.3.3. Land-Responsibility Behaviour Intention

Considering three typical land decision-making behaviours of farmers that affect sustainable rural tourism, such as food production, farmland landscape and cultural inheritance [11,49,50], farmers' land-responsibility behaviour intention was measured by the degree of attention paid by farmers to the three following aspects during land disposal: social benefits (such as food security and reducing pesticide use) (P1), environmental benefits (such as reducing pollution and protection of farmland landscape) (P2), and cultural conservation benefits (such as teaching agricultural knowledge to future generations) (P3). All variables were scored on a Likert-type scale that ranged from totally disagree (1) to totally agree (7).

3.4. Pre-Test

To ensure the validity of the questionnaire, five farmers in Sansheng Township were asked to complete a pre-interview, and the words and expressions that appeared in the pre-interview were used to form a pre-test questionnaire. The pre-test questionnaire was then used in a small sample survey of 40 farmers. Analysis of pre-test sample scores was performed to identify variables that passed the reliability and validity tests. The corrected item-total correlation and the internal consistency reliability index (Cronbach's α coefficient) were used to test the reliability of the four variables measured by the questionnaire, and SPSS 25.0 was used. The corrected item-total correlation values of all items retained exceeded 0.6 [51], and the Cronbach's α reliability coefficients of all variables exceeded the recommended value of 0.6. After deleting any item, the overall Cronbach's α reliability coefficient of each variable did not increase significantly. Thus, we concluded that the scale had good internal consistency, reliability, and stability, and ideal internal reliability. Furthermore, the validity of the construction of the measurement scale was tested using factor analysis.

As shown in Table 2, the Kaiser–Meyer–Olkin values of the four latent variables (self-identity, non-economic function perception, economic function perception, and land-responsibility behaviour intention) all exceeded 0.6, which was greater than the recommended value of 0.5. The significance of the Bartlett's test of sphericity was 0.000. The original hypothesis of the Bartlett's test of sphericity was rejected, and so the questionnaire measurement scale and the construct validity of each variable could be considered as good.

Table 2. Reliability estimations for the questionnaire.

Variables	Items	Cronbach's α	KMO Test	Bartlett's (SIG)
Self-identity	3	0.780	0.656	0.000
Agricultural non-economic function perception	5	0.799	0.855	0.000
Agricultural economic function perception	4	0.771		
Land-responsibility behaviour intention	3	0.763	0.634	0.000

4. Results and Discussion

4.1. Measurement Model Estimation

We used Mplus8.0 to perform confirmatory factor analysis to detect the structural validity of the measurement scale, including factors self-identity, agricultural non-economic function perception, agricultural non-economic function perception, and land-responsibility behaviour intention. The various inspection indicators after deleting "taking care of the land and other resources left by seniors is important to me." (NEF5) for correction are shown in Table 3; the revised scale had a better composition reliability and structural validity. Table 3 presents all constructs' factor loadings, Construct Reliability (CR), and Average Variance Extracted (AVE), and Table 4 presents the relationships between the constructs.

Table 3. Structural validity test of the measurement scale.

Constructs	Items	Unstd.	S.E.	P	Std.	SMC	CR	AVE
	F1	1			0.897	0.805		
SI	F2	0.847	0.067	0.000	0.747	0.558	0.802	0.581
	F3	0.647	0.060	0.000	0.616	0.379		
	NEF1	1		0.000	0.548	0.300		
ANEFP	NEF2	1.069	0.114	0.000	0.754	0.569	0.808	0.517
	NEF3	1.097	0.118	0.000	0.748	0.560		
	NEF4	1.064	0.111	0.000	0.800	0.640		
	EF1	1		0.000	0.793	0.629		
AEFP	EF2	0.940	0.069	0.000	0.731	0.534	0.859	0.605
	EF3	1.105	0.068	0.000	0.842	0.709		
	EF4	1.037	0.073	0.000	0.740	0.548		
	P1	1		0.000	0.849	0.721		
LRBI	P2	1.003	0.072	0.000	0.839	0.704	0.810	0.593
	P3	0.833	0.082	0.000	0.595	0.354		

Note: ANEFP = agricultural non-economic function perception, AEFP = agricultural economy function perception, LRBI = land-responsibility behaviour intention, SI = self-identity.

Table 4. Differential validity tests of the measurement scale.

	SI	**ANEFP**	**AEFP**	**LRBI**
SI	**0.762**			
ANEFP	0183	**0.718**		
AEFP	0.450	0.517	**0.778**	
LRBI	0.399	0.157	0.563	**0.770**

Note: The diagonal values are the value of the AVE root sign, and the value of the lower triangle is the correlation coefficient between the variables. ANEFP = agricultural non-economic function perception, AEFP = agricultural economy function perception, LRBI = land-responsibility behaviour intention, SI = self-identity.

The results revealed that the model fit was $\chi^2 = 212.604$, df = 71, $\chi^2/\text{df} = 2.9$ (less than the recommended value of 3), CFI = 0.942, TLI = 0.926, RMSEA = 0.068, and SRMR = 0.046. The factor loadings of most items were >0.7, the composition reliability of each factor was >0.8, the convergence validity was >0.5, and the fit degree of the measurement model reached an ideal value

4.2. Structural Model Estimation

4.2.1. Path Analysis

First, we adopted Mplus8.0 to estimate the regression coefficient between variables. Given that the data were non-normally distributed, we used the maximum likelihood estimation method provided by Mplus8.0 to verify the relationships between the variables, and the standard error and mean-variance corrected chi-square test (MLMV) as the estimation method. The results revealed that RMSEA = 0.076, SRMR = 0.048 (recommended value < 0.08), CFI = 0.942, TLI = 0.915 (recommended value > 0.9), and $\chi 2$ (162) = 411; these results show that the data fit the model well. We further tested the hypotheses. Self-identity had a significant positive influence on land-responsibility behaviour intention ($\beta = 0.224$, $p < 0.01$), agricultural non-economy function perception ($\beta = 0.128$, $p < 0.01$) and agricultural economy function perception ($\beta = 0.319$, $p < 0.001$), thus supporting H1, H2a and H2b. Agricultural non-economic function perception positively and significantly affected agricultural economy function perception ($\beta = 0.557$, $p < 0.001$), supporting H4, but agricultural non-economic function perception negatively and significantly affected land-responsibility behaviour intention ($\beta = -0.319$, $p = 0.016$), not supporting H3a. Similarly, agricultural economy function perception was found to significantly influence land-responsibility behaviour intention ($\beta = 0.866$, $p < 0.001$), which supported H3b. Results of the hypotheses tests are summarised in Table 5.

Table 5. Results of the hypothesis tests.

DV	IV	Std. Est.	S.E.	Est./S.E.	*p*-Value	R-Square	Hypothesis
ANEFP	SI	0.183	0.046	3.978	0.005	0.034	Supported (H2a)
AEFP	SI	0.368	0.051	7.216	0.000	0.398	Supported (H2b)
	ANEFP	0.450	0.086	5.233	0.000		Supported (H4)
LRBI	SI	0.172	0.084	2.048	0.008	0.364	Supported(H1)
	ANEFP	−0.171	0.132	−1.295	0.016		Not supported (H3a)
	AEFP	0.574	0.124	4.629	0.000		Supported (H3b)

Note: ANEFP = agricultural non-economic function perception, AEFP = agricultural economy function perception, LRBI = land-responsibility behaviour intention, SI = self-identity.

4.2.2. Mediating Effects Estimation

Then, we used Mplus8.0 bootstrapping to test the mediating role of ANEFP and AEFP in the relationship between SI and farmers' LRBI. The bootstrapping method has more advantages than the traditional mediation analysis method because it can statistically calculate the significance of direct effects, indirect effects, and total effects within a certain confidence interval (CI) [52,53]. The results are shown in the Table 6.

Table 6. Mediation effect test (standardisation).

Model Test	β	S.E.	*p*	CI (95%)	Results
Total effect	0.521	0.103	0.000	[0.335,0.743]	Supported
Direct effect	0.224	0.097	0.020	[0.039,0.416]	Supported
Total indirect effect	0.297	0.075	0.000	[0.169,0.467]	Supported
SI→AEFP→LRBI	0.224	0.071	0.000	[0.152,0.427]	Supported
SI → ANEFP → LRBI	−0.041	0.023	0.076	[−0.110, −0.008]	Not supported
SI → ANEFP → AEFP → LRBI	0.062	0.030	0.039	[0.017,0.144]	Supported

Note: ANEFP = agricultural non-economic function perception, AEFP = agricultural economy function perception, LRBI = land-responsibility behaviour intention, SI = self-identity.

We found that farmers' self-identity had a significant positive impact on farmers' land-responsibility behaviour intention (total effect: β = 0.521, 95% CI = (0.335, 0.743)). The total indirect effect was β = 0.297, 95% CI = (0.169, 0.467). The ratio of total indirect effects to total effects was 0.297/0.521 = 0.692. In other words, 69.2% of the impact of farmers' self-identity on land-responsibility behaviour intention was affected by agricultural non-economic function perception and agricultural economic function perception. Results also indicated that self-identity directly or through intermediary conditions affected land-responsibility behaviour intention.

In addition, the mediation test of agricultural economic function perception revealed that the significant impact of self-identity on farmers' land-responsibility behaviour intention was mediated by agricultural economic function perception (β = 0.224, 95% confidence interval (CI) = (0.152, 0.427)). However, the mediation test for agricultural non-economic function perception was not supported. This means that farmers' agricultural non-economic function perception mediates the relationship between farmers' self-identity and land-responsibility behaviour intention (H2a) was not supported. The results of each hypothesis test are shown in Table 6.

4.3. Discussion

This study investigated the impact of farmers' self-identity on land-responsibility behaviour intention from individual perspective of local farmers and examined the mediating effect of agricultural multifunctional perception on the relationship between these two variables. Two main research conclusions were obtained.

First, in rural tourism destinations in suburban districts of China, farmers' self-identity is an important variable that affects farmers' land-responsibility behaviour intention, whereby the higher the level of farmers' self-identity, the more likely they are to adopt land-responsibility behaviours. On the one hand, farmers' self-identity can directly affect farmers' land-responsibility behaviour intention. On the other hand, self-identity can also positively influence farmers' land-responsibility behaviour intention through the mediating effect of agricultural multifunctional perception, which means that farmers' self-identity will further initiate rational behaviour decision making through the functional evaluation of agriculture. That is, self-identity support land-responsibility behaviour, directly and indirectly supporting the perceived economic function of agriculture.

Second, agricultural non-economic function perception negatively and significantly affected land-responsibility behaviour intention. It was found that some non-economic benefits of agriculture, such as environmental protection and social culture, may come at the cost of individual farmers' interests, which supports previous findings [54,55]. Therefore, although farmers know that adopting certain technologies or programs can improve the non-economic functions of agriculture, they may not adopt corresponding land behaviours [56,57]. In other words, although farmers can envision the non-economic functions that their land-responsibility behaviour may bring about, they may not adopt it, even oppose this kind of behaviour. The main reason for this could be that the farmers bear the additional economic costs for land-responsibility behaviour.

Third, the analysis of the mediating effect shows that the perception of the importance of agricultural multifunction is an important mediating condition for the influence of farmers' self-identity on farmers' land-responsibility behaviour intention. However, the utility of importance perception of agricultural economic function and that of non-economic function are different. Self-identity can influence farmers' land-responsibility behaviour intention through the mediating role of agricultural economic function perception. However, agricultural non-economic function perception does not have a direct mediating effect; that is, although farmers perceive the non-economic functions that their land-responsibility behaviour may bring, this does not directly stimulate them to generate land-responsibility behaviour intention. Farmers' agricultural non-economic functions perception can only significantly influence their land-responsibility behaviour intention through the intermediary effect of agricultural economic functions of land. That is, only when farmers perceive that non-economic function is positively related to economic function does the corresponding land-responsibility behaviour intention occur. Furthermore, the likelihood of adopting land-responsibility behaviours will increase if farmers feel that the non-economic functions of agriculture are accompanied by economic functions that can offset the perceived costs.

5. Conclusions

The results indicate that self-identity is a vital factor that affects Chinese farmers' land-responsibility behaviour intention in rural tourism areas. What is more, "not well respected, rather perceived as a low-rank profession" and "the low social status" are primary factors discourage youths from getting involved with farming [58]. Therefore, the government should understand how farmer perspective the value of farming and consider its role in growing the rural economy and rural development.

Second, the conclusions of this study highlight the path dependence of farmers on the economic functions of agriculture. The income of farmers is generally low in China and obtaining economic benefits through agriculture remains the most important motivation for farmers to take land-responsibility behaviour. Therefore, only when farmers get the economic benefits of the tourism industry caused by the non-economic functions of agriculture, will farmers' land-responsibility decisions be effectively stimulated.

Third, farmers' agricultural non-economic function perception positively affects their agricultural economic function perception. That is, farmers can perceive the transformation of agricultural non-economic functions into economic functions, thereby increasing their understanding and support for sustainable land policies. For the sustainable development of villages, our findings are consistent with those of Ahnström et al. [56] and Reimer et al. [57], among others. We believe that investing in the non-economic functions of agriculture will be beneficial to rural communities and their sustainable development. The multifunctional development of agriculture is of great significance to the protection of the rural landscape, ecology, the cultural environment, and the protection of biodiversity and cultural heritage; these factors form the basis for agriculture to generate economic benefits in tourism and other industries. The tourism industry is an important means by which to transform agricultural non-economic functions into economic functions. When farmers in these areas understand that agricultural non-economic functions can achieve economic functions through tourism and other industries, they could be more likely to adopt land behaviours that are more conducive to sustainable development. For agricultural heritage sites and remote rural tourism sites with abundant tourism resources, agriculture and farming culture are important tourist attractions, and rural tourism can be developed with the help of agricultural non-economic functions. Therefore, rural tourism can not only be used as a means by which to alleviate poverty through the development of rural areas but could also enable farmers to experience the transformation of agricultural non-economic functions into agricultural economic functions, thereby stimulating local farmers to adopt more sustainable land decisions.

6. Limitations and Future Research

Due to differences in economic development levels, land policies, and location relationships between these sites and central cities, the conclusion from the four destinations cannot fully represent other rural tourist destinations in China. All the variables selected in this study were assessed using cross-sectional data, and future work could further track variables such as farmers' self-identity, agricultural multifunction perception, to better understand how these variables impact farmers' land awareness, and land-responsibility behaviour intention in the context of rural tourism development. In addition, farmers' attitudes, values, and land behaviour were different [59]. Therefore, future research should acknowledge the heterogeneity between "farmers", and strengthen the localisation and differentiation of cultural and social factors. For example, the large wave of people returning home to start a business in China has resulted in huge, ongoing alterations in the structure of farmers in rural tourism destinations. Future work should also consider further analysing the psychology and land decision-making behaviours of different types of agricultural practitioners.

Author Contributions: Conceptualization, X.C.; Z.L.; Y.L. and J.Q.; methodology, X.C.; software, X.C.; validation, X.C.; formal analysis, Z.L. and J.Q.; investigation, X.C.; Z.L. and M.H.; resources, X.C.; data curation, Z.L.; writing—original draft preparation, X.C. and Z.L.; writing—review and editing, Y.L. and X.C.; visualization, J.Q.; supervision, X.C. and Y.L.; project administration, Y.L.; funding acquisition, X.C. and Y.L. All authors have read and agreed to the published version of the manuscript.

Funding: This research was funded by the Education Department of Sichuan Province in China [Grant No. 15sb0002]; System Science and Enterprise Development Research Center (Grant No. Xq20C05).

Conflicts of Interest: The authors declare no conflict of interest.

References

1. Ajzen, I. The Theory of Planned Behavior. *Organ. Behav. Hum. Decis. Process.* **1991**, *50*, 179–211. [CrossRef]
2. Ajzen, I.; Fishbein, M.J.E.C. *Understanding Attitudes and Predicting Social Behavior;* Prentice-Hall, Inc.: Englewood Cliffs, NJ, USA, 1980; p. 278.
3. Lokhorst, A.M.; Staats, H.; Dijk, J.V.; Dijk, E.V.; Snoo, G.D. What's in it for Me? Motivational Differences between Farmers' Subsidised and Non-Subsidised Conservation Practices. *Appl. Psychol.* **2011**, *60*, 337–353. [CrossRef]
4. Rob, J.F.; Burton, U.H.P. Creating culturally sustainableagri-environmental schemes. *J. Rural Stud.* **2011**, *27*, 95–104.
5. Rob, J.F.; Burton, C.; Kuczera, G. Exploring Farmers' Cultural Resistance to Voluntary Agri-environmental Schemes. *Sociol. Ruralis* **2008**, *48*, 16–37.
6. Van Dijk, W.F.A.; Schaffers, A.P.; Leewis, L.; Berendse, F.; De Snoo, G.R. Temporal effects of agri-environment schemes on ditch bank plant species. *Basic Appl. Ecol.* **2013**, *14*, 289–297. [CrossRef]
7. Dean, M.; Raats, M.M.; Shepherd, R. The Role of Self-Identity, Past Behavior, and Their Interaction in Predicting Intention to Purchase Fresh and Processed Organic Food. *J. Appl. Soc. Psychol.* **2012**, *42*, 669–688. [CrossRef]
8. Sparks, P.; Shepherd, R. Self-Identity and the Theory of Planned Behavior: Assessing the Role of Identification with "Green Consumerism". *Soc. Psychol. Q.* **1992**, *55*, 388–399. [CrossRef]
9. Defrancesco, E.; Gatto, P.; Runge, F.; Trestini, S. Factors affecting farmers' participation in agri-environmental measures: A northern Italian perspective. *J. Agric. Econ.* **2008**, *59*, 114–131. [CrossRef]
10. Van Dijk, W.F.A.; Lokhorst, A.M.; Berendse, F.; de Snoo, G.R. Factors underlying farmers' intentions to perform unsubsidised agri-environmental measures. *Land Use Policy* **2016**, *59*, 207–216. [CrossRef]
11. Aliabadi, V.; Gholamrezai, S.; Ataei, P. Rural people's intention to adopt sustainable water management by rainwater harvesting practices: Application of TPB and HBM models. *Water Supply* **2020**, *20*, 1847–1861. [CrossRef]
12. Karimy, M.; Niknami, S.; Hidarnia, A.R.; Hajizadeh, I. Intention to start cigarette smoking among Iranian male adolescents: Usefulness of an extended version of the theory of planned behaviour. *Heart Asia* **2012**, *4*, 120–124. [CrossRef]
13. Leske, S.; Strodl, E.; Hou, X.-Y. Predictors of dieting and non-dieting approaches among adults living in Australia. *BMC Public Health* **2017**, *17*, 214–233. [CrossRef] [PubMed]
14. Zerbini, C.; Luceri, B.; Vergura, D.T. Leveraging consumer's behaviour to promote generic drugs in Italy. *Health Policy* **2017**, *121*, 397–406. [CrossRef]
15. O'Connor, E.L.; Sims, L.; White, K.M. Ethical food choices: Examining people's Fair Trade purchasing decisions. *Food Qual. Prefer.* **2017**, *60*, 105–112. [CrossRef]

16. Pyo, L.K.; Kwon, S.M. The Effects of Fair-tourism' Perceptions on Buying Intention and Willingness to Pay Premium Price: An application of the Extended Theory of Planned Behavior. *J. Tour. Leis. Res.* **2016**, *28*, 23–39.

17. Ates, H. Merging Theory of Planned Behavior and Value Identity Personal norm model to explain pro-environmental behaviors. *Sustain. Prod. Consumption* **2020**, *24*, 169–180. [CrossRef]

18. Gkargkavouzi, A.; Halkos, G.; Matsiori, S. Environmental behavior in a private-sphere context: Integrating theories of planned behavior and value belief norm, self-identity and habit. *Resour. Conserv. Recycl.* **2019**, *148*, 145–156. [CrossRef]

19. Anastasia, G.; George, H.; Steriani, M. Environmental behavior in a private-sphere context: Integrating theories of planned behavior and value belief norm, self-identity and habit. *Sustain. Prod. Consumption* **2019**, *148*, 145–156.

20. Fielding, K.S.; Terry, D.J.; Masser, B.M.; Hogg, M.A. Integrating social identity theory and the theory of planned behaviour to explain decisions to engage in sustainable agricultural practices. *Br. J. Soc. Psychol.* **2008**, *47*, 23–48. [CrossRef]

21. Borges, J.A.R.; Domingues, C.H.D.F.; Caldara, F.R.; da Rosa, N.P.; Senger, I.; Guidolin, D.G.F. Identifying the factors impacting on farmers' intention to adopt animal friendly practices. *Prev. Vet. Med.* **2019**, *170*, 104718. [CrossRef]

22. Thoits, P.A.; Virshup, L.K. Me's and we's: Forms and functions of social identities. In *Self and Identity: Fundamental Issues*; Oxford University Press: Oxford, UK, 1997; pp. 106–133.

23. Paquin, R.S.; Keating, D.M. Fitting identity in the reasoned action framework: A meta-analysis and model comparison. *J. Soc. Psychol.* **2016**, *157*, 47–63. [CrossRef] [PubMed]

24. Lee, Y.; Lee, J.; Lee, Z. Social influence on technology acceptance behavior. *ACM Sigmis Database* **2006**, *37*, 60–75. [CrossRef]

25. Dumanski, J. Criteria and indicators for land quality and sustainable land management. *ITC J.* **1997**, *1997*, 216–222.

26. Stryker, S.; Serpe, R.T. *Commitment, Identity Salience, and Role Behavior: Theory and Research Example*; Springer: New York, NY, USA, 1982.

27. De Bruijn, G.J.; Verkooijen, K.; De Vries, N.K.; Bas, V.D.P. Antecedents of self identity and consequences for action control: An application of the theory of planned behaviour in the exercise domain. *Psychol. Sport Exerc.* **2012**, *13*, 771–778. [CrossRef]

28. Hagger, M.; Chatzisarantis, N. Self-identity and the theory of planned behaviour: Between- and within-participants analyses. *Br. J. Soc. Psychol.* **2011**, *45*, 731–757. [CrossRef] [PubMed]

29. Burke, S.P.J. Identity Theory and Social Identity Theory. *Soc. Psychol. Q.* **2000**, *63*, 224–237.

30. Christensen, P.N.; Rothgerber, H.; Wood, W.; Matz, D.C. Social norms and identity relevance: A motivational approach to normative behavior. *Pers. Soc. Psychol. Bull.* **2004**, *30*, 1295–1309. [CrossRef] [PubMed]

31. Laverie, D.A.; Arnett, D.B. Factors Affecting Fan Attendance: The Influence of Identity Salience and Satisfaction. *J. Leis. Res.* **2000**, *32*, 225–246. [CrossRef]

32. Sparks, P. Subjective Expected Utility-Based Attitude-Behavior Models: The Utility of Self-Identity. In *Attitudes, Behavior and Social Context: The Role of Norms and Group Membership*; Psychology Press: London, UK, 2000.

33. Lokhorst, A.M.; Hoon, C.; Le Rutte, R.; De Snoo, G. There is an I in nature: The crucial role of the self in nature conservation. *Land Use Policy* **2014**, *39*, 121–126. [CrossRef]

34. Key, N. How much do farmers value their independence? *Agric. Econ.* **2005**, *33*, 117–126. [CrossRef]

35. Howley, P.; Dillon, E.; Hennessy, T. It's not all about the money: Understanding farmers' labour allocation decisions. *Agric. Hum. Values* **2014**, *31*, 261–271. [CrossRef]

36. Wilson, G.A. The spatiality of multifunctional agriculture: A human geography perspective. *Geoforum* **2009**, *40*, 269–280. [CrossRef]

37. Morgan, S.L.; Marsden, T.; Miele, M.; Morley, A. Agricultural multifunctionality and farmers' entrepreneurial skills: A study of Tuscan and Welsh farmers. *J. Rural Stud.* **2010**, *26*, 116–129. [CrossRef]

38. Van der Ploeg, J.D.; Roep, D. Multifunctionality and rural development: The actual situation in Europe. In *Multifunctional Agriculture: A New Paradigm for European Agriculture and Rural Development*; Ashgate: Aldershot, UK, 2003.

39. Kvakkestad, V.; Rorstad, P.K.; Vatn, A. Norwegian farmers' perspectives on agriculture and agricultural payments: Between productivism and cultural landscapes. *Land Use Policy* **2015**, *42*, 83–92. [CrossRef]

40. Walker, A.J.; Ryan, R.L. Place attachment and landscape preservation in rural New England: A Maine case study. *Landsc. Urban Plan.* **2008**, *86*, 141–152. [CrossRef]

41. Ravikumar, R.K.; Thakur, D.; Choudhary, H.; Kumar, V.; Kumar, V. Social engineering of societal knowledge in livestock science: Can we be more empathetic? *Vet. World* **2017**, *10*, 86–91. [CrossRef] [PubMed]

42. Slamova, M.; Kruse, A.; Belcakova, I.; Dreer, J. Old but Not Old Fashioned: Agricultural Landscapes as European Heritage and Basis for Sustainable Multifunctional Farming to Earn a Living. *Sustainability* **2021**, *13*, 4650. [CrossRef]

43. Kontogeorgos, A.; Tsampra, M.; Chatzitheodoridis, F. Agricultural Policy and the Environment Protection through the Eyes of New Farmers: Evidence from a Country of Southeast Europe. *Procedia Econ. Financ.* **2015**, *19*, 296–303. [CrossRef]

44. Shannon, S.; Mccann, E.; De Young, R.; Erickson, D. Farmers' attitudes about farming and the environment: A survey of conventional and organic farmers. *J. Agric. Environ. Ethics* **1996**, *9*, 123–143.

45. Emerton, L.; Snyder, K.A. Rethinking sustainable land management planning: Understanding the social and economic drivers of farmer decision-making in Africa. *Land Use Policy* **2018**, *79*, 684–694. [CrossRef]

46. Cawley, M.; Gillmor, D.A. Integrated rural tourism: Concepts and Practice. *Ann. Tour. Res.* **2008**, *35*, 316–337. [CrossRef]

47. Vanclay, F. Multifunctional Agriculture: A Transition Theory Perspective—By Geoff A. Wilson. *Geogr. Res.* **2009**, *47*, 221–222. [CrossRef]

48. Peng, J.; Liu, Z.C.; Liu, Y.X. Research progress on assessing multi-functionality of agriculture. *Chin. J. Agric. Resour. Reg. Plan.* **2014**, *35*, 1–8.

49. Raymond, C.M.; Bieling, C.; Fagerholm, N.; Martin-Lopez, B.; Plieninger, T. The farmer as a landscape steward: Comparing local understandings of landscape stewardship, landscape values, and land management actions. *Ambio* **2016**, *45*, 173–184. [CrossRef]

50. Vuillot, C.; Coron, N.; Calatayud, F.; Sirami, C.; Mathevet, R.; Gibon, A. Ways of farming and ways of thinking: Do farmers' mental models of the landscape relate to their land management practices? *Ecol. Soc.* **2016**, *21*, 35–58. [CrossRef]

51. Shimp, A.T.; Sharma, S. Consumer Ethnocentrism: Construction and Validation of the cetscale. *J. Mark. Res.* **1987**, *24*, 280–289. [CrossRef]

52. Kenny, D.A.; Judd, C.M. Power anomalies in testing mediation. *Psychol. Sci.* **2014**, *25*, 334–339. [CrossRef]

53. Zhao, X.; Lynch, J.G., Jr.; Chen, Q. Reconsidering Baron and Kenny: Myths and Truths about Mediation Analysis. *J. Consum. Res.* **2010**, *37*, 197–206. [CrossRef]

54. Kabii, T.; Horwitz, P. A review of landholder motivations and determinants for participation in conservation covenanting programmes. *Environ. Conserv.* **2006**, *33*, 11–20. [CrossRef]

55. Ryan, R.; Erickson, D.; De Young, R. Farmers' Motivations for Adopting Conservation Practices along Riparian Zones in a Mid-western Agricultural Watershed. *J. Environ. Plan. Manag.* **2003**, *46*, 19–37. [CrossRef]

56. Ahnström, J.; Höckert, J.; Bergeå, H.L.; Francis, C.A.; Skelton, P.; Hallgren, L. Farmers and nature conservation: What is known about attitudes, context factors and actions affecting conservation? *Renew. Agric. Food Syst.* **2009**, *24*, 38–47. [CrossRef]

57. Reimer, A.P.; Thompson, A.W.; Prokopy, L.S. The multi-dimensional nature of environmental attitudes among farmers in Indiana: Implications for conservation adoption. *Agric. Hum. Values* **2012**, *29*, 29–40. [CrossRef]

58. Raj, K.G.C.; Hall, R.P. The Commercialization of Smallholder Farming-A Case Study from the Rural Western Middle Hills of Nepal. *Agriculture* **2020**, *10*, 143–158.

59. Darnhofer, I.; Schneeberger, W.; Freyer, B. Converting or not converting to organic farming in Austria:Farmer types and their rationale. *Agric. Hum. Values* **2005**, *22*, 39–52. [CrossRef]

 *agriculture*

Article

Would Kazakh Citizens Support a Milk Co-Operative System?

Samal Kaliyeva [1,*], **Francisco Jose Areal** [2] **and Yiorgos Gadanakis** [1]

1   School of Agriculture Policy and Development, University of Reading, Reading RG6 6AR, UK; g.gadanakis@reading.ac.uk
2   Centre for Rural Economy, School of Natural and Environmental Sciences, Newcastle University, Newcastle upon Tyne NE1 7RU, UK; francisco.areal-borrego@newcastle.ac.uk
*   Correspondence: s.kaliyeva@pgr.reading.ac.uk

Abstract: We estimate the monetary value of a policy aimed at increasing rural co-operative production in Kazakhstan to increase milk production. We analyse the drivers associated with public support for such policy using the contingent valuation method. The role of individuals' psychological aspects, based on the reasoned action approach, along with individuals' views on the country's past regime (i.e., to the former Soviet Union), their awareness about the governmental policy, their sociodemographic characteristics, and household location on their willingness to pay (WTP) for the policy is analysed using an interval regression model. Additionally, we examine changes in individuals' WTP before and during the COVID-19 pandemic. The estimated total economic value of the policy is KZT 1335 bn for the length of the program at KZT 267 bn per year, which is approximately half the total program budget, which includes other interventions beyond the creation of production co-operatives. The total economic value of the policy would equal the cost of the whole program after 10 years, indicating public support for this policy amongst Kazakh citizens. Psychological factors, i.e., attitude, perceived social pressure, and perceived behavioural control, and the respondents' awareness of the policy and views on the Soviet Union regime are associated with their WTP. Sociodemographic factors, namely, age, income, and education, are also statistically significant. Finally, the effect of the shocks of COVID-19 is negatively associated with the respondents' WTP.

Keywords: co-operative creation policy; contingent valuation; reasoned action approach; Kazakhstan; COVID-19

Citation: Kaliyeva, S.; Areal, F.J.; Gadanakis, Y. Would Kazakh Citizens Support a Milk Co-Operative System?. *Agriculture* **2021**, *11*, 642. https://doi.org/10.3390/agriculture11070642

Academic Editors: Adoración Mozas Moral and Domingo Fernandez Ucles

Received: 18 June 2021
Accepted: 6 July 2021
Published: 8 July 2021

Publisher's Note: MDPI stays neutral with regard to jurisdictional claims in published maps and institutional affiliations.

1. Introduction

Prior to Kazakhstan joining the World Trade Organisation in 2015, Kazakhstan joined Belarus and Russia in 2014 to create the Eurasian Economic Union (EAEU), a free trade zone. Later Armenia, Belarus, and Kyrgyzstan also joined the EAEU. The opening of Kazakhstan's economy to international markets challenged its agricultural competitiveness, which was detrimental to the rural economy, highly dependent on agricultural production [1]. Hence, improving agriculture productivity is key for the development of the rural economy of Kazakhstan. Consequently, the government decided to stimulate the production of agricultural products by allocating a significant part of its governmental budget, 2374.2 billion tenges (KZT) for 5 years, for the development of the country's agricultural sector, part of which also considers the creation of agricultural co-operatives. This is a relatively large budget, accounting for 9% of the revenue of the state, republican, and local budgets in 2017. To compare, 1868.4 billion tenges (KZT) was budgeted under the state program for the development of education and science for the period 2016–2019; 1385.6 billion was budgeted for the development of tourism for the period 2019–2025, and 1762.5 billion tenges was budgeted for regional development for the period 2015–2020 [2].

Amongst agricultural products produced in Kazakhstan, dairy is one of the key agricultural sectors, representing 16% of the total agricultural production of the country [3]. Milk production has increased by 16% in the last 5 years reaching a total of 5,820,000 t

159

of cow's milk produced in 2019 compared to 5,020,000 t in 2014. However, the domestic supply of dairy products is insufficient to meet the internal demand. Specifically, dairy product exports amounted to USD 53,517,500 (1 US dollar (USD) is equal to 426.84 tenges (KZT) as of 27 June 2021), whereas the imports were USD 252,450,400 in 2019, indicating a 198,932,900 trade deficit. Hence, a transformation of the structure of the dairy sector seems key to reduce this gap.

Currently, the structure of Kazakh's dairy is dominated by small-scale producers, such as rural households and individual/peasant farms, representing 93% of total production (of which rural households are 78% and individual/peasants are 22%), whereas only 7% of the milk was produced by agricultural enterprises. Thus, due to the prevailing of small-scale production, dairy factories face a deficit of milk for processing, and consequently, the country experiences a low supply of processed dairy products [4,5]. In 2019, a total of 262,000 t of milk went to the processing factories in Kazakhstan, only 4.5% of the 5,820,000 t produced that year. Considering therefore the status of the agricultural sector, the government's intervention plan aimed at reducing the number of agricultural activities conducted by small farm/household with the objective of expanding agricultural production (including dairy) in enterprises through the creation of co-operatives in rural areas. It is worth noting that although there are other supply chain pathways to reach dairy factories (e.g., peasant and small farms, merchants), more than 70% of milk is produced by rural households, consequently making them the main body in the dairy supply chain.

The legislative basis of co-operatives is set out in the law "On Agricultural Cooperation", adopted in 2015. The policy on creating co-operatives was introduced in 2017. However, the initial government plan was revised in July 2018, and is no longer aiming to create more co-operatives under the Programme (the reason of which remains unclear). Despite this fact, the idea of creating co-operatives is still relevant and it has been included in the Strategic Plan of noncommercial organization "Atameken" for 2018–2023; thus, in 2019, the number of rural households involved in co-operative production was 27.2 thousand whereas the production of cow's milk by co-operatives was 65.4 thousand tonnes (the country's total production was 5820.1 thousand tonnes of milk in the same year).

According to the law, an agricultural co-operative is created when there are at least three members. All members of the co-operative are obliged to pay an entrance fee, in accordance with the charter of the co-operative. If necessary, members of the co-operative can make additional contributions (on a voluntary basis). In addition, the founders and members of the co-operative can also make a material (share) contribution. The basic principles of the creation and functioning of co-operatives are expected to comply with the international principles specified in the International Co-operative Alliance (ICA). According to the ICA, there are seven main international co-operative principles: (1) voluntary and open membership; (2) democratic member control; (3) member economic participation; (4) autonomy and independence; (5) provision of education, training, and information; (6) co-operation among co-operatives; and (7) concern for the community.

Unlike the Soviet Union where production output and all assets (productive and social, except land) were owned jointly by the collective (i.e., kolkhozes) and by the state (i.e., sovkhozes/state farms) [6], under the current policy, the individuals do not own the means of production and share the means of production to produce an output. Nevertheless, access to technologies, equipment, feeding, and subsidies are expected to be facilitated through co-operatives.

Although co-operatives can potentially be organised in many forms, e.g., service co-operatives, the main focus of the policy and therefore of this study is focused on production co-operatives. Rural households are expected to be engaged in the supply chain to facilitate constant milk supply to dairy factories via co-operatives. Members of production co-operatives, i.e., rural households and individual/peasant farms, are expected to supply the co-operatives with fresh milk that goes directly to the dairy processing industry. As there are no intermediates, rural households (and individual/peasant farms) will be paid from the dairy processing units directly. In turn, co-operatives receive KZT 10 per litre of milk in the form of subsidies from the government [7].

Co-operatives can contribute to uplifting livelihoods by reducing poverty and food insecurity in rural areas through the improved use of technology, share of knowledge between members, and distribute income from a market-oriented output [8–11].

We estimate the consumer's willingness to pay (WTP) for a Kazakh's government intervention to create production co-operatives in rural areas to obtain the total economic value of the policy. We also analyse the heterogeneity in WTP and investigate whether the COVID-19 pandemic affected consumer WTP for the government's policy. Estimating the total economic value of agricultural policies, or any other policy for that matter, is paramount for policy decision-making under constrained budgets. As Price [12] points out, an "unbiased and focused evaluation of unpriced benefits is an important pre-condition for needed policy interventions". The estimation of monetary value of agricultural policies, such as conservation of agricultural genetic resources [13], safe vegetables [14], and agri-environment schemes [15] has been previously studied. Although the attitudes of Kazakh rural households towards joining and creating co-operatives was previously studied [16], to the best of our knowledge, no study has estimated the total economic value of a policy aimed at increasing milk production through co-operative creation. More specifically, we contribute to the literature in three ways: (1) by estimating the total economic value a of the transformation of the milk production system from small-scale production to industrial production through a policy aiming at creating co-operatives; (2) to our knowledge, this is the first paper that has used and expanded the reasoned action approach to gaining an understanding of how the total economic value for the policy is moderated by a number of elements. These include individual psychological aspects based on the reasoned action approach (RAA), views on the past regime (i.e., to the former Soviet Union), awareness concerning the governmental policy, sociodemographic characteristics, and geographical location; and (3) by analysing whether a pandemic shock such as COVID-19 may be associated with changes in individuals' WTP for the policy.

2. Materials and Methods

We used the contingent valuation (CV) method to elicit the total economic value of the policy through the respondents' WTP for a premium price on a litre of milk in order to support the government policy. The program allows farmers to receive support from government and other co-operatives, such as a subsidy in the amount of KZT 10 per litre of milk and discounted animal feed products. This information was provided to respondents along with the policy objective of supporting dairy producing households to expand dairy production in Kazakhstan. We used the RAA to analyse how psychological factors may be associated to respondents' WTP. We extend the RAA to integrate the respondents' (a) views on the past regime (i.e., to the former Soviet Union), (b) their sociodemographic characteristics and the location, (c) awareness about the governmental policy, and (d) COVID-19 into our framework to investigate the role of these elements on respondents' WTP.

2.1. Contingent Valuation Method

The total value associated with the implementation of governmental policies includes not only the provision of market goods, but the provision of nonmarket goods and services, too (i.e., those that cannot be traded in the marketplace, and consequently do not have a market price). The policy might provide substantial benefits for the society, such as increasing milk production whilst supporting rural development and allowing farmers to increase their livelihoods as a result of receiving higher returns for their products. Co-operative production promotes sustainable agriculture, enhancing not only the environment but also the social sustainability of local communities [17]. The stated preferences method is employed as a double-bounded dichotomous choice contingent valuation (CV) to elicit the total value of the policy. Although the majority of the stated preference research focuses on the demand for environmental benefits, the use of this technique has spread to evaluating other type of goods, including farmers' WTP for crop insurance [18], animal welfare [19], agricultural genetic resources [13], and the provision of production services [20].

Preferences of the respondents are explained by the random utility theory (RUT) since it is the theoretical basis for the CV method [21,22]. Thus, the utility of a good is expressed as follows:

$$U_{iq} = V_{iq} + \varepsilon_{iq} \tag{1}$$

where U is the utility of good i for individual q, V_{iq} is the expected value of U, and ε is the error term.

Two main approaches are used to elicit the value of a good using CV: (a) single-bounded (take-it-or-leave-it) and (b) double-bounded (take-it-or-leave-it with follow-up) dichotomous choice techniques. However, the single-bounded approach has been criticized due to the limitation in revealing the true WTP [23,24]. The double-bounded dichotomous choice approach was used to deal with the limitations of a single-bound approach. The singularity of this approach is that participants are simply asked if they would pay a certain amount of money for the good and if the answer is "Yes" ("No"), the monetary amount can be raised (or decreased) with follow-up questions according to Yes/No answers [23,25–27]. Consequently, by follow-up questions, four possible outcomes can be derived [28]:

1. Respondent answers YES for both the main bid P^I and the higher bid P^H (YES–YES), in this case, WTP $\geq P^H$
2. Respondent answers YES for the main bid P^I and NO for the higher bid P^H (YES–NO), in this case, $P^I \leq$ WTP $> P^H$
3. Respondent answers NO for the main bid P^I and YES for the lower bid P^L (NO–YES), in this case, $P^L \leq$ WTP $< P^I$
4. Respondent answers NO for both the main bid P^I and the lower bid P^L (NO–NO), in this case, WTP $< P^L$

A common issue that researchers face while applying the CV method is the identification and treatment of protest WTP responses [29]. In CV studies, protest responses can account for 50% of WTP [30,31].

The most common treatment of protest bids is the exclusion of them from the sample [31,32]. However, some researchers argue that only deleting is not an option, it is important to investigate protest responses to define the motivation behind protest bids [29,30]. Thus, several reasons have already been identified in the literature. Namely, possible subjects of protest might be (a) need in more information or (b) a conviction that the government is responsible for payment, while (c) "I cannot afford it" is defined as a true WTP of zero [29,30].

2.2. Reasoned Action Approach

We use the reasoned action approach (RAA) to assess the level of influence that psychological factors may have on Kazakh citizens' valuation of the government policy aimed at increasing milk production through co-operatives. How psychological factors may underlie individual's behaviour was stated by Fishbein and Ajzen [33] in their theory of reasoned action (TRA), where beliefs, attitudes, intentions, and behaviour were identified as its main elements. The TRA was extended by adding perceived behavioural control in the theory of planned behaviour (TPB) [34], which was defined as a determinant of behavioural intention and behaviour [35]. RAA is a continuation of the TPB, where behaviour is assumed to consist of four elements—action, target, time, and context [36]. Hence, the generality of behaviour can be controlled by making those elements more or less specific. Following the RAA, individuals construct (a) behavioural belief b_i, which is weighted by evaluation e_i of its outcome, (b) normative beliefs n_i that are evaluated by the motivation to comply m_i with a referent, and (c) control beliefs c_i assessed by the power p_i of that belief. Together they compose attitude (i.e., $A = \sum b_i e_i$), social norms (i.e., $SN = \sum n_i m_i$), and perceived behavioural control (i.e., $PBC = \sum c_i p_i$), which underly the intention to perform the given behaviour (Figure 1). Thus, constructed and weighted *A*, *SN*, and *PBC* are combined to formulate the behavioural intention (BI).

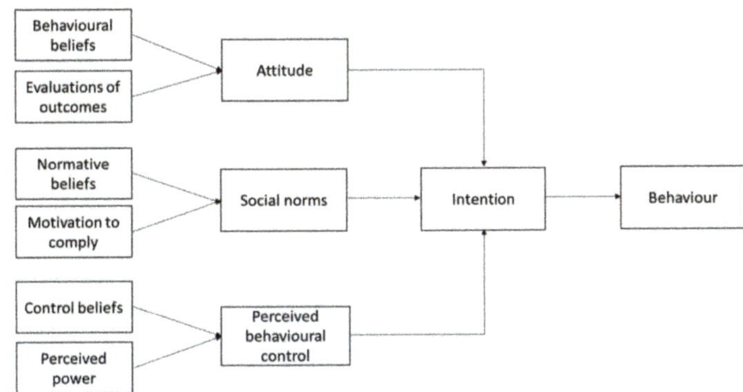

Figure 1. The reasoned action approach. The figure was drawn by authors following the model described in the text.

2.3. Other Constituents of the Model

We expand the RAA framework to include other contextual elements that may be relevant in the respondent's valuation of the policy in our framework (Figure 2).

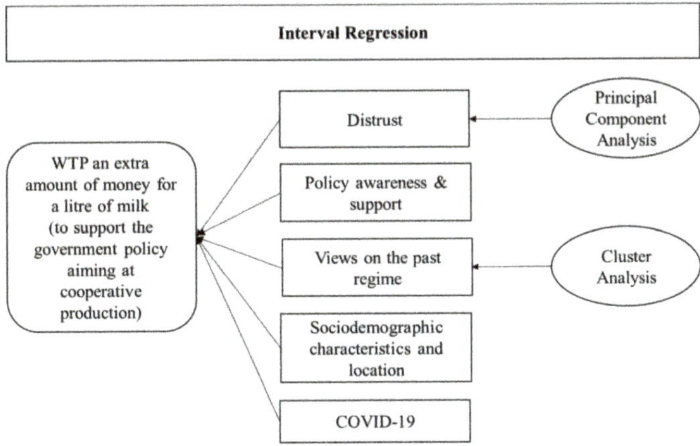

Figure 2. The conceptual framework of the study.

Prior to announcing independence in 1991, Kazakhstan was a part of the Soviet Union and regarding the collectivist–communist regime, agricultural production was organised mostly on the basis of collective farming, i.e., kolkhozes and sovkhozes [6,37,38]. Even though almost 30 years have passed since the collapse of collective farms, the transition from centrally planned to market economy may have left some impact on individuals' views towards the current government and its policies. Although numerous studies tried to shed a light on implications of the transition economy on post-Soviet countries' development [39–42], the influence of post-communist regime on the policy in question is not yet clear. Thus, we investigate how individuals' views on the past regime may be associated with their valuation of a policy aimed at increasing production co-operatives. Several main associations are possible. Individuals who miss the Soviet Union may (a) be supportive of the policy that reminds them of the previous regime (the structure and function of kolkhozes as agricultural production systems), but they may also (b) be sceptical about the current regime delivering the policy on co-operative production as one in the past, and as a result, may be less likely to support it. Thus, the mistrust of the current regime and unattractiveness of current policies compared with the Soviet Union regime might lead to less support of the current regime by the general public.

Moreover, we investigate the association of (a) sociodemographic characteristics, (b) the location where a respondent resides, (c) awareness of the policy in question, and (d) COVID-19 relationship with respondents' intention to pay extra money for a litre of milk.

2.4. Survey and Questionnaire

A snowball sampling technique was used to contact Kazakh citizens to voluntarily take part in the study, i.e., by using an already existing network of contacts via social media platforms to distribute the link to the questionnaire. The questionnaire was approved by the Ethics Committee of the University of Reading (protocol code/ethical clearance application number 001151P, approved on 2 December 2019).

The instrument used to collect information was a questionnaire survey using Qualtrics XM (Version 12, Qualtrics, Provo, UT, USA). All participants were provided with an information sheet and consent form containing information about the aims and objectives of the research. The questionnaire was created in English and translated to Kazakh and Russian. To guarantee accuracy, a second, independent person reviewed and edited the translation for accuracy, natural flow in the target language, and adherence to the needs of the survey.

The data were collected in two periods, before and during the COVID-19 pandemic. The first wave of data collection ($n = 272$) was completed in a month period, between 10 December 2019 and 10 January 2020.

In March 2020, the first case of COVID-19 was reported in Kazakhstan and the government implemented a lockdown for two months, until May 2020. However, as soon as the restriction was eased, the number of cases of the disease increased sharply, reaching its peak in June–July 2020. Considering the situation and the government's measures to deal with it, in June 2020, we took the opportunity of exploring the effect of COVID-19 on respondents' WTP. Therefore, during the period of a month, between 13 June and 13 July 2020, 234 fully complete additional responses were collected, making a total of 506 observations.

The questionnaire consisted of five sections (awareness and support, CV, RAA, views on the past regime, sociodemographics, and location) and included a total 37 questions.

The aim and features of the governmental policy were delivered in the form of short informative text within the first section of the survey and respondents were asked to respond (a) if they have had information about co-operative creation and (b) if they agreed with the aim of the policy.

Within the CV section, respondents were asked to answer the WTP questions. During the pilot study in August 2018, we used open-ended questions allowing respondents to decide without giving options, then received an amount of money that was used to adjust main bids for WTP. Information from the pilot questionnaire was used to assign the prices for the WTP questions (KZT 10, 40, 70, 100, and 130). Thus, the amount of money Kazakh citizens are willing to pay for the transformation in the dairy sector was obtained by providing information about the governmental policy and asking them the following question: "Would you be willing to pay extra X amount of money for a litre of milk in order to support the government's policy?" where X amount of money was chosen randomly from the given bids.

If respondent answered "No", then the requested amount of money was decreased by KZT 15 (P^L) or it was increased to KZT 15 (P^H) if the answer was "Yes".

If a respondent ticked the fourth option and answers No–No, then further questioning was used to indicate the reasons. The third section of the questionnaire included questions on RAA in order to reveal psychological aspects underlying Kazakhs citizens' intention to pay an extra amount of money for a litre of milk. Salient beliefs of the respondents were defined during the pilot study in November 2019 by asking open-ended questions towards the support of the governmental policy aimed at co-operative creation; following that, the statements were identified and included in the survey. Respondents were asked to rate the RAA statements on a set of unipolar and bipolar evaluative adjective scales, with five places. To elicit attitude (A) toward paying an extra amount of money for a litre of milk in order to support the government policy, for instance, respondents were asked to score the strength of belief about a consequence of the behaviour from 1 to 5 (i.e., extremely unlikely–extremely likely), while evaluation of the belief was assessed from -2 on the negative side to $+2$ on the positive side. Thus, the higher the behavioural belief the more it was expected to have a positive influence on attitude. Consequently, the sum across all scales (since there are three behavioural outcomes, the possible range of the scale for A is from -30 to $+30$) was taken as a measure of a respondent's attitude towards co-operative production. The same procedure was applied to reveal SN and PBC with some differences on scoring, namely, (a) respondent's normative beliefs were scored from -2 to 2 (i.e., extremely unlikely–extremely likely), while the motivation to comply with a referent took on values from 1 to 5; (b) control beliefs were scored from 1 to 5, while the power (P) of the factor was scored from -2 to $+2$ on statements capturing facilitating factors (i.e., P1) and from 2 to -2 on statements capturing impeding factors (i.e., P2, P3, and P4) [36]. Hence, the scale for the SN and for the PBC ranged from -40 to $+40$.

Table 1 shows statements used to reveal Kazakh citizens' A, SN, and PBC. During the survey, prior to responding on RAA questions, respondents were informed about the aim and features of the governmental policy in the form of short informative text.

Table 1. Statements to reveal respondent's attitude, social norms, and perceived behavioural control towards the behaviour.

Item	Questionnaire Statements	Scale
	Attitude	
B1	Paying an extra amount of money for a litre of milk would improve the quality of milk	extremely unlikely–extremely likely
E1	For me improving of the quality of milk is	extremely bad–extremely good
B2	Paying an extra amount of money for a litre of milk would motivate farmers to produce better	extremely unlikely–extremely likely
E2	For me motivating farmers is	extremely bad–extremely good
B3	Paying an extra amount of money for a litre of milk would support domestic milk production	extremely unlikely–extremely likely
E3	For me increasing domestic milk production is	extremely bad–extremely good
	Social norms	
N1	My spouse/partner thinks that it would be good for me to pay an extra amount of money for a litre of milk	extremely unlikely–extremely likely
M1	With regards paying an extra amount of money for a litre of milk, I want to do what my spouse or partner thinks I should do	strongly disagree–strongly agree
N2	My close relatives think that it would be good for me to pay an extra amount of money for a litre of milk	extremely unlikely–extremely likely
M2	With regards paying an extra amount of money for a litre of milk, I want to do what my close relatives think I should do	strongly disagree–strongly agree
N3	My parents think that it would be good for me to pay an extra amount of money for a litre of milk	extremely unlikely–extremely likely
M3	With regards paying an extra amount of money for a litre of milk, I want to do what my parents think I should do	strongly disagree–strongly agree
N4	My close friend thinks that it would be good for me to pay an extra amount of money for a litre of milk	extremely unlikely–extremely likely
M4	With regards paying an extra amount of money for a litre of milk, I want to do what my close friend thinks I should do	strongly disagree–strongly agree

Table 1. *Cont.*

Item	Questionnaire Statements	Scale
	Perceived behavioural control	
C1	I have enough money to pay an extra amount of money for a litre of milk	extremely unlikely–extremely likely
P1	Having enough money would make it easier for me to pay an extra amount of money for a litre of milk	strongly disagree–strongly agree
C2	I don't trust dairy factories to pay an extra amount of money for a litre of milk	extremely unlikely–extremely likely
P2	The lack of trust in dairy factories would make it difficult for me to pay an extra amount of money for a litre of milk	strongly disagree–strongly agree
C3	I don't trust farmers (households) to pay an extra amount of money for a litre of milk	extremely unlikely–extremely likely
P3	The lack of trust in farmers (households) would make it difficult for me to pay an extra amount of money for a litre of milk	strongly disagree–strongly agree
C4	I don't trust the government's policy to pay an extra amount of money for a litre of milk	extremely unlikely–extremely likely
P4	The lack of trust in the government's policy would make it difficult for me to pay an extra amount of money for a litre of milk	strongly disagree–strongly agree

The statements "During the Soviet Union people had more healthy food"; "During the Soviet Union Kazakhstan's economy was better"; and "I like the idea of collective farming (kolkhozes) during the Soviet Union" in section 4 of the questionnaire were used to capture whether the respondent's views on the past regime are associated with their willingness to support the governmental policy.

Finally, age, education, gender, and income composed the sociodemographic part of the survey. Within this part, respondents were also asked to indicate the location where they reside.

2.5. Statistical Analysis

The analysis comprised a combination of quantitative methods including cluster analysis on the respondent's views on the Soviet Union (SU) and parameter model estimation using an interval regression model.

2.5.1. Cluster Analysis

Cluster analysis is used to group respondents according to their views on the past regime. Concisely, it involves a search through data for observations that have high similarity in comparison to one another but are very dissimilar with respect to objects in other clusters.

Two main approaches are known to cluster analysis: hierarchical and partitioning. Considering the hierarchical approach, which can also be interpreted as a top–down procedure, each observation represents its own cluster. At any following stage, similar and closer in characteristics clusters merge, creating a group and continue until cutting the tree at a suitable level. Otherwise, the procedure terminates when all members of a group are consistent, creating one common cluster at the top of a tree-like form, called a dendrogram [43–45].

In the partitioning (*k*-means) approach, a cluster can be formed by specifying the number of clusters prior to the analysis. Using this number as an input, the algorithm specifies an initial centre of the cluster (i.e., *k*), afterwards, observations are assigned to the cluster according to their nearest cluster centres (i.e., one of the *k* clusters). According to the *k*-means approach, the number of clusters is not known in advance [43–45]. Therefore, the choice of an initial configuration can be based on the results of hierarchical clustering [46]. Since *k*-means is stated as superior to the hierarchical methods due to its ease of implementation, simplicity, efficiency, and empirical success [44,46], we followed this approach. Thus, initially, the number of clusters was identified through the dendrogram, and then the *k*-means method was applied.

2.5.2. Interval Regression

An interval regression model, a generalisation of the Tobit model [47], was used to analyse factors underlying Kazakh citizens' WTP extra amount of money for a litre of milk in order to support the government policy aimed at dairy production and creating co-operatives. The singularity of this model is in the observed range of the dependent variable being censored, since the dependent variable y_i^* (i.e., respondent's WTP an extra amount of money for a litre of milk) is unobserved [48]. What is observed is an interval, which has lower m_i and upper M_i bounds,

$$m_i \leq y_i^* \leq M_i \tag{2}$$

where, basically, the data can be defined with three possible outcomes. In the case if the lower bound is known, but the upper is not, then "right-censored"; or vice versa, if the upper bound is known, but the lower is not, then "left-censored". If both lower and upper bound are known, then the data can be defined as an "interval" [49]. We can state that

$$y_i^* = x_i \beta + u_i, \ u_i | x_i \sim Normal \left(0, \sigma^2 \right) \tag{3}$$

where x_i is a vector of an explanatory variable of WTP of a respondent i and β is a parameter vector associated with explanatory variables x_i. These are the RAA variables (attitude, social norms, and perceived behavioural control), cluster variable accounting for respondents who like the past regime, policy awareness, sociodemographic variables (age, education and income), and location. The error term u_i is assumed to be normally distributed with mean zero and standard deviation σ [50,51].

3. Results and Discussions

3.1. Descriptive Statistics

Descriptive statistics of the explanatory variables are shown in Table 2. Lower and upper are dependent variables, which refer to left-censored and right-censored observations. A, SN, and PBC were generated following [36] (see Section 2.2). Two variables were created to indicate the awareness (i.e., infopolicy) and support (i.e., policyagree) of the considered policy, respectively. SU_likers is an explanatory variable obtained from the cluster analysis and captures respondent's views on the past regime, taking a value of 1 for those with a relatively positive view on the past regime and 0 otherwise. A dummy variable for COVID-19 was created with a value of 1 for respondents participating during the COVID-19 wave and 0 otherwise.

Finally, sociodemographic variables including age, education, gender, income, and location are the explanatory variables that refer to the sociodemographic and location part of the study. Almost 60% of the respondents were female. Nearly 50% belonged in the age band of 18–30, and up to 80% were aged below 50 years old. A quarter had education at school and college level, while undergraduate and postgraduate levels of education were 43% and 30%, respectively. Almost 40% of the respondents stated their income up to KZT 100,000, which can be defined as low income, about 25% indicated middle income (KZT 101,000–150,000), while the remaining 35% were respondents with high income. The majority of respondents reside in the capital (about 68%), while the rest were from different cities. Therefore, within the location variable, we treated the capital as a zero point and identified the distance to other cities in kilometres from the capital.

A comparison between the Kazakh population in 2019 and our survey sample is provided in Table 3.

The main difference is education at school and college level, and household income up to KZT 50,000 being underrepresented, while education at postgraduate degree and household income over KZT 100,000 are overrepresented. Education and level of income are highly correlated to one another, and since the survey was distributed mainly with the support of colleagues from national universities, the sample covered mostly educated and high-income earning respondents. Although most of the population hold the average per capita income of up to KZT 100,000, the sample household income was equally distributed amongst the 4 categories.

Table 2. Variable definitions and statistical descriptions.

Variable	Definition	Mean	Min	Max
Lower	Obs. (n = 284), lower bound	67.757	0	145
Upper	Obs. (n = 157), upper bound	66.382	0	145
A	Attitude of the respondents towards the co-operative creation policy	15.991	−13	30
SN	Perceived social norms of the respondents	7.126	−34	40
PBC	Perceived behavioural control of the respondents	−4.009	−40	24
SU_likers	cluster derived by the cluster analysis; dummy variable 1 = like the Soviet Union regime; 0 = otherwise	0.586	0	1
infopolicy	dummy variable, 0 = if otherwise; 1 = if the respondents received information about the government policy before;	0.233	0	1
policyagree	dummy variable, 0 = if otherwise; 1 = if the respondents agree with the aim of the policy	0.926	0	1
age	Age of the respondents 1 = 18–30; 2 = 31–49; 3 = 50 and older	1.733	1	3
education	The final completed education of the respondents 1 = school; 2 = college; 3 = undergraduate; 4 = postgraduate	2.932	1	4
gender	dummy variable, 0 = male, 1 = female	0.623	0	1
income	The respondent's monthly income 1 = KZT 0–50,000; 2 = KZT 51,000–100,000; 3 = KZT 101,000–150,000; 4 = KZT 151,000 and higher	2.797	1	4
location	the location of the respondents in kilometres from the capital	296.877	0	2600
covid	dummy variable, 0 = pre-COVID-19 period, 1 = COVID-19 period	0.592	0	1

Table 3. Socioeconomic characteristics of Kazakhstan population (2019), percentage of Kazakhstan population versus percentage of the sample.

	Number of Individuals	Kazakhstan Population (%)	Sample, n = 326 (%)
Total population	18,395,567	—	—
Female population	9,749,650	53	62
Male population	8,645,916	47	38
Age (15–34, Kazakhstan; 18–30, sample)	5,509,210	42	46
Age (35–54, Kazakhstan, 31–49, sample)	4,504,423	35	35
Age (55+)	3,034,521	23	19
School	117,204	28	10
College	144,333	34	17
Undergraduate	142,435	34	43
Postgraduate	22,765	5	30
Household income (<KZT 50,000)	n/a	50 *	15
Household income (KZT 51,000–100,000)	n/a	39 *	25
Household income (KZT 101,000–150,000)	n/a	8 *	25
Household income (>KZT 151,000)	n/a	3 *	35

Note: Figures for the level of education of the population are based on the number of individuals who finished each of the education categories during 2019; * distribution of population by average per capita income (by the number of the population is not available). An average nominal per capita income of the population was KZT 104,282 in 2019. The data were derived from the official website (www.stat.gov.kz, accessed on 10 January 2021) of the Statistics Committee of the Republic of Kazakhstan.

3.2. Cluster Analysis

Overall, three statements were used to define the views of respondents towards the past regime. Respondents were asked to evaluate these statements from strongly disagree to strongly agree on a 5-point Likert scale. Primarily, we conducted a hierarchical procedure for these variables to determine the number of clusters by using the dendrogram. Then, we checked the validation of the chosen number through Calinski and Harabasz's and Duda–Hart indices (i.e., cluster stopping rules). Both indices showed $n = 2$ cluster as appropriate.

Once the number of clusters was specified, a k-means procedure was carried out. Table 4 illustrates the summary statistics of the clusters by means. Cluster 2 was characterised by having higher mean rates, while cluster 1 had mean = 3 or less on the given statements. Therefore, cluster 2 is assumed that it captured the Soviet Union regime likers, while cluster 1 is not. We created a dummy variable with a value of 1 for SU_likers and a value of 0 otherwise (non-SU_likers).

Table 4. Summary statistics (by mean) of the clusters.

	During the Soviet Union People Had More Healthy Food	During the Soviet Union, Kazakhstan's Economy Was Better	I Like the Idea of Collective Farming (Kolkhozes) during the Soviet Union
0 = non-SU_likers (Cluster 1)	2.978	2.000	2.467
1 = SU_likers (Cluster 2)	4.654	3.702	3.974
Total	3.960	2.997	3.350

3.3. The Value of the Policy for Society

The average premium price of the respondents WTP for a litre of milk to support the policy was KZT 103. The average market price paid by respondents for a litre of milk in the period of the study was KZT 300. This means that on average respondents are prepared to pay 34% more than the market price to support the policy in production co-operative creation. However, this is possibly an overestimate given that our sample contains more respondents with relatively high levels of income. For the purpose of obtaining a WTP estimate that is more representative of the population, we looked at how the WTP varies according to sociodemographic characteristics (Table 5). Using the household income population information (Table 3), we weighted the estimated WTP by income group according to the population (%) in each income group. This gives the WTP of KZT 86.61 (i.e., a 29% premium price).

Table 5. The estimated average WTP according to sociodemographic characteristics of the respondents.

	Obs.	Mean	S.D.
Female population	203	100.32	40.60
Male population	123	106.88	41.24
Age (18–30, sample)	149	105.34	40.84
Age (31–49, sample)	115	100.34	39.62
Age (50+, sample)	62	101.24	43.64
School	32	109.15	47.05
College	56	100.46	44.83
Undergraduate	140	107.90	41.29
Postgraduate	98	94.75	34.59
Household income (<KZT 50,000)	50	77.49	32.17
Household income (KZT 51,000–100,000)	81	89.27	41.08
Household income (KZT 101,000–150,000)	80	121.89	40.02
Household income (>KZT 151,000)	115	110.04	36.21

The budget of the program, where the creation of co-operatives had been stated, was 2374.2 billion tenges (KZT) for five years (i.e., 2017–2021). We highlight that the program covered not only the support of small farmers through creating co-operatives but also other sectors, including (a) efficient use of water and land resources; (b) increasing the provision of agricultural producers with equipment and chemicals, and (c) scientific–technological, personnel and information–marketing support of the agroindustrial complex.

Once the individual average WTP for the policy is estimated, we can use it to estimate the economic value of the policy in a relatively simple way. Assuming that to evaluate the policy, a certain age needs to be reached, the total value of the policy was calculated by multiplying the number of Kazakh citizens at age 15 and over (13,000,000) (Table 3) by the corrected average WTP (i.e., KZT 86.61) times % Kazakh population consuming milk (approximately 90% of the population): kg milk/dairy consumed per month (22 kg) times 12 months. Then the estimate for the total economic value of the policy aiming at the creation of the co-operatives for the Kazakh citizens is KZT 267 billion per year, or KZT 1335 billion per five years (the five-year Program period), which is half of the total budget for the whole program. The economic value of the policy would equal the cost of the whole program after 10 years.

3.4. Drivers for WTP

Table 6 shows how elements of the RAA are associated with respondents' WTP. Namely, attitudes, social norms, and perceived behavioural control are associated with an increase in participants' WTP an extra amount of money for a litre of milk in order to support the government policy (p-values < 0.01). These results are in line with studies on consumer's willingness to purchase organic milk [52], to purchase pasture-raised livestock products [53], and to pay for meat from mobile slaughter units [54]. In other words, if the attitude towards the behaviour (i.e., paying a premium price for a litre of milk to support the policy) is more positive than negative, it is more likely that the behaviour will be performed. Furthermore, if other people (i.e., spouse/partner, close relatives, close friends, and parents) who are considered highly important by the individual are believed to approve rather than disapprove and also perform this behaviour, people are more likely to feel social pressure to engage in this behaviour. Additionally, following the model and the results of the study, if Kazakh citizens perceive more facilitating than inhibiting factors, perceived behavioural control should be high, consequently the behaviour will be performed.

Table 6. Results of the interval regression.

	Coefficient	z-Statistics
A	1.34 ***	2.59
SN	1.13 ***	3.19
PBC	1.22 ***	2.95
1. SU_likers	−33.90 ***	−3.60
1. Infopolicy	24.72 **	2.37
1. policyagree	9.88	0.63
Age (18–30, base category)		
31_49	−15.23	−1.51
50 and older	−28.81 **	−2.17
Education (School, base category)		
College	−9.50	−0.52
Undergraduate	−12.95	−0.73
Postgraduate	−32.59 *	−1.71
1. female	1.86	0.19
Income (<KZT 50,000, base category)		
KZT 51,000–100,000	6.78	0.49
KZT 101,000–150,000	49.76 ***	3.35
>KZT 151,000	34.62 **	2.40
Location	0.02 **	2.17
1. COVID-19	−26.20 ***	−2.79
_cons	94.30 ***	3.85
sigma	62.37	14.78
Number of observations	326	
Left-censored	42	
Right-censored	169	
Interval-censored	96	
Log-likelihood	−488.23	
LR chi2(17) = 83.93; Prob > chi2 = 0.0000		

Note: *, **, *** for 10, 5, and 1% of significance level, respectively.

The results also show that Kazakh citizens who like the Soviet Union regime were willing to pay KZT 33.90 (1 US dollar (USD) is equal to 426.84 tenges (KZT) as of 27 June 2021) less to support the policy on production co-operatives creation than citizens who do not like the Soviet Union regime (p-value < 0.01). Possible reasons for this result may relate to the possibility that individuals who like the Soviet Union (i.e., who perceive the past Communist as a better regime than the current regime) may also have a feeling of frustration with democracy [55]. Moreover, one of the reasons behind satisfaction with the past regime was its stability and guarantee of basic needs [55]. As pointed out by Toleubayev et al. [56], "Kazakhstani people express great nostalgia for their past lives in the Soviet era and their narratives express a strong appreciation for the level of social security, income stability, low food prices, and the sense of a more egalitarian communal life". This frustration present in post-communist countries may be consequence of a transition economy towards a "wild capitalism" characterized by "rapid and massive liberalization, by the lack or the inefficiency of the state intervention in the economy, by corruption, and significant social movements of protest", and not achieving the similar level of democracy such as in Western Europe [57,58].

The lower support for the policy on production co-operatives creation by Kazakh citizens who like the Soviet Union is reinforced by the finding that people aged over 50 are less supportive of the policy (Table 6). Hence, results suggest that Kazakh citizens with a positive attitude towards the old Soviet Union regime, and aged over 50, are more likely to perceive policies from the new regime (since independence) as unattractive and ineffective.

The results indicate that respondents' WTP is positively associated with having adequate information about the policy (p-value < 0.05). Kazakh citizens with relatively higher awareness about the policy are ready to pay about KZT 25 (1 US dollar (USD) is equal to 426.84 tenges (KZT) as of 27 June 2021) more than those who had no knowledge before. Undoubtedly, for a respondent receiving essential information about the product may be crucial for decision making. A similar finding was also reported by Stampa et al. [53] and Zhang et al. [14]. Moreover, Zhang et al. [59] found that increasing awareness of cultured meat influenced positively on Chinese consumer's acceptance of it. A similar effect was found by Roosen et al. [60], when investigating consumers' WTP for nanotechnology food differed according to the information provided.

The results showed an increase in income is associated with a higher WTP. Respondents with the income between KZT 101,000 and KZT 150,000, and more than KZT 151,000 are willing to pay KZT 50 and KZT 35 (1 US dollar (USD) is equal to 426.84 tenges (KZT) as of 27 June 2021) more, respectively, than respondents with monthly income up to KZT 50,000. This finding is expected and in line with [13,61], where a WTP was stated being increased with higher levels of income.

Although the respondents holding postgraduate level of education are less likely to support the policy (p-value < 0.10), the reason for this is unclear. However, it is noted that the share of highly educated respondents was higher in the sample of the study.

The location is found to be statistically significant (p-value < 0.05), and thus, individuals living apart from the capital are more inclined to pay a premium price for a litre of milk to support the policy. This is justified since the policy is oriented for the development of the rural areas and Kazakh citizens' living in regions (apart from the capital) perceive more the importance of the policy.

The parameter measuring the relationship between COVID-19 and respondents' WTP was found to be statistically significant (p-value < 0.01) suggesting that COVID-19 might have had some impact on individual's WTP. Kazakh citizens seem less likely to support the government policy on creating co-operatives under the COVID-19 situation. Results show that individuals average WTP for the government policy aimed at increasing the number of co-operatives was lower during the pandemic period compared to the pre-pandemic period. Thus, the average WTP to support the policy was KZT 118 (1 US dollar (USD) is equal to 426.84 tenges (KZT) as of 27 June 2021) prior to COVID-19 outbreak, whereas during the pandemic it decreased by 22% and was KZT 92. This can be due to the rise of unemployment [62], stated as one of the dramatic implications of the COVID-19, which touched Kazakhstan as well. According to the news agency "Khabar 24" [63], during the pandemic, the number of unemployed Kazakh citizens only in one city has increased by 3.5 times. Thousands of entrepreneurs were forced to pause their work; about 1.6 million employees were sent to leave without payment. Thus, widespread dissatisfaction with the measures taken by the government to stop the spread of the virus might cause decreased support of the current government by the general public.

3.5. Protest WTP Responds

Within $n = 506$ observations, $n = 180$ were labelled as protest bids and deleted, which is almost 35% of the sample.

Respondents were asked to state the reason for zero WTP, where the most common four reasons are found. Both "I am already paying tax and think that the government has to use that money to support" and "The prices of milk/dairy products are already expensive" were stated 67 times. Next was, "I am sceptical about that the money will go to the farmers" that was repeated in 52 places; 45 times protestors mentioned, "I will need to have more information about this policy". Although "I don't have enough income to pay extra money" was stated 56 times, this reason was labelled as true WTP of zero, therefore were not excluded from the sample.

3.6. Policy Implications

Our results show the readiness of the general public to support the government's plan in creating production co-operatives and the economic viability of the plan. However, it is important to acknowledge that the success of the policy also depends on the rural households' willingness to participate in the policy. Kaliyeva et al. [16] revealed the existing interest of rural households in joining and creating co-operatives in Kazakhstan. Hence, policies aimed at the creation of co-operatives can be a viable solution to increasing milk production in Kazakhstan. It is worth noting that the government could also take other approaches to increase dairy/milk production. For instance, policies such as promoting family farming by introducing tax relief and/or subsidies could also achieve the aim of increasing milk production, but farmers would not have the same level of access to information and technology that a co-operative would offer. The level of public support for policies promoting family farming is unknown, but this policy may find less opposition from individuals liking the SU.

The policy on co-operative creation might facilitate connection of farmers (rural households) with supply chains (dairy factories). Not only producers (farmers, dairy factories) might benefit from the policy, but also society. It is acknowledged that co-operatives can help developing local value chains as well as facilitate the access to local and global markets [64]. The structural changes in the dairy sector may enhance the production of domestic products, and as a result may positively affect the country's trade balance by reducing the demand on imported dairy products. Moreover, co-operatives are an acknowledged way of reducing poverty in rural areas and enhancing sustainable development [8–10].

Considering research findings in other countries, there are two points worth discussing: (a) what kind of co-operatives can help competitiveness in agriculture and (b) what has been the experience of policies supporting the creation of co-operatives. It is worth noting that research conducted in other countries on agricultural co-operatives is diverse and provides useful information to understand how regional characteristics/conditions may influence the potential effects of creating co-operatives on agricultural production and markets. The creation of co-operatives among enterprises in direct competition with each other allow producers to take advantage of synergies and reinforce bargaining power without major losses of freedom or flexibility [65]. This may be particularly important in developing countries where the size of the farming system is small. Li and Ito [66] show that agricultural co-operatives in regions where agricultural land size is relatively small (e.g., China) can help in developing other markets associated with agricultural production (e.g., development of land rental markets by reducing transaction costs). Liang [67] argues that producer co-operatives act as a competitive yardstick of markets leading to competitive markets. Liang [67] also shows that this yardstick effect resulted into higher farm gate prices for hog producers in China. In addition, the yardstick effect may lead to a reduction in production costs [67].

Co-operative and community-based forms of doing agriculture are common in most countries, especially in developed countries where "the access of small farmers to markets is usually facilitated by agricultural service co-operatives" [68]. According to recent research, 134 agricultural co-operatives in the US celebrated their 100th anniversary in 2014 [69]. Research on the longevity of agricultural co-operatives in developed countries listed several main reasons for that, such as the achievement of scale economic gains and the ability to adapt to dynamic situations. The success or failure of policies supporting the creation of co-operatives may depend on the existing institutional conditions as well as in the level of trust on the government by producers and the degree the regulatory policy with too regulatory policies being less likely to succeed, particularly in post-Soviet countries [70]. Research on the success and failure cases of agricultural co-operatives in developing countries revealed the lack of comprehensive support, including advice on best practices and monitoring co-operative activities as the main reasons for the failure of banana co-operatives in Rwanda [71]. Moreover, Moon [71] suggested that the success of the creation of co-operatives might be possible through the efforts of both the aid agency and the beneficiaries.

Although what share of the total budget was aimed to be used for co-operatives creation is not clear, the results of the study showed the importance of the policy for the Kazakh society. Extrapolating to the Kazakh population who consume milk/dairy products would mean that the economic value of the policy would be KZT 1335 bn for the length of the program at KZT 267 bn per year, which is approximately half the total program budget, and includes other interventions beyond the creation of co-operatives. The economic value of the policy would equal the cost of the program after 10 years. This indicates there is public support for this policy.

Our findings suggest that although there is general support for the policy, there are still parts of the population, i.e., individuals missing the SU regime, who may mistrust newly created organisational forms of the current government. Therefore, as a country with a transition economy, the Kazakhstan government may face nonacceptance of the policy by some of the population. The main reason is found to be the implications of the wild capitalism that Kazakh people faced after the transition from communism to a market economy. Public rejection of the policy might also be connected with COVID-19, which had dramatic damage to the economy of the country. Therefore, the government attempts for increasing its attractiveness will lead the policy to be more widely supported.

Provision of information about the policy (e.g., aims, implementation) was found to be important in respondents supporting the policy. We therefore recommend that policymakers need to resolve any unambiguity in definitions of the use of the term "co-operative" under the current policy, "that will prevent any possibility of misunderstanding or misinterpreting the strategic intentions" [68]. Hence, in order to gain policy support for increasing dairy/milk production by creating co-operatives, good communication of the policy seems key to building trust amongst Kazakh citizens. Finally, a "top–down" route to the creation of agricultural co-operatives has been widely criticized around the world due to its nonviability and noneffectiveness [68]. Survey results in this research showed that information on the policy aimed at creating co-operatives had neither been widely distributed nor explained to Kazakh citizens and rural households [16]. The majority of the participants only discovered the existence of the program from the researchers during the survey. However, in the developing world the "top–down" process can be a legitimate way of organising co-operatives [72]. For instance, the classic form of establishing co-operatives in China that involves the participation of the state and farmers has been regarded as widespread and effective. In post-socialist Vietnam, state involvement also played a crucial role in the development of agricultural co-operatives, where the sector suffered from low levels of initiative on the part of farmers [72,73]. Despite this, we believe that the initiative to create co-operatives should come from rural households. Moreover, dairy factories need to also be involved in such initiatives from the outset. Otherwise, the top–down process may not be implemented successfully.

4. Conclusions

We assessed the public support for a policy aimed at increasing milk production through co-operatives by estimating the monetary value for society of the policy. It was found that Kazakh citizens showed support for the government policy. The findings presented in this paper might also be relevant for post-communist countries, such as Russia, Ukraine, and Kyrgyzstan, the agricultural development of which has a similar pattern to Kazakhstan's.

Psychological factors played an important role in the success of the policy—namely, holding a positive attitude towards the behaviour, having positive endorsement regarding the behaviour (the support of the policy) from the social referent (e.g., family members and friends), and being in a position to control the behaviour, i.e., A, SN, and PBC, significantly influence Kazakh citizens' WTP support of the policy. Moreover, individual awareness of the policy was found to be important in supporting the policy. Therefore, good communication of the policy and its aims to the general public is key for policy support. Findings suggest that countries that have transitioned to new policy regimes can face difficulties in implementing policy programmes in cases where significant parts of the population miss characteristics of the past regime. We also found some evidence of reprioritisation of people's preferences under COVID-19, with relatively lower support for the policy. Therefore, to achieve the support of the general public, the government should take measures to increase its attractiveness and try to earn public acceptance.

In this study, we investigate the success of the Kazakh government policy aimed at increasing milk production through an increase in co-operative production. We mainly based our analysis on the opinion and reactions of the general public in Kazakhstan. However, other policy outcomes, such as an increase in the competitiveness of Kazakh's milk production in international markets, could also generate further benefits (e.g., extra government revenue). In addition, accounting for any environmental effects (e.g., landscape and habitat, biodiversity, soil) associated with a change from current production to co-operative production would also be needed in a cost–benefit analysis.

Additionally, it should be emphasized that this research considered only a single attribute, i.e., the value of the policy on creation of production co-operatives. However, there is a potential for exploring the general public's willingness to pay for co-operatives through including other specifications. These might include other attributes, including diversity of co-operatives such as service co-operatives. Alternatively, consumers' preferences can be explained by extending product attributes, e.g., quality and price of the milk from co-operatives. In such a case, a choice experiment approach can be utilized to investigate individuals' WTP for welfare changes by offering different attributes of goods/policies and choosing a preferred option across several sets [74,75].

Author Contributions: Conceptualization, S.K., F.J.A. and Y.G.; methodology, S.K., F.J.A. and Y.G.; software, S.K., F.J.A. and Y.G.; formal analysis, S.K., F.J.A. and Y.G.; investigation, S.K., F.J.A. and Y.G.; data curation, S.K.; writing—original draft preparation, S.K.; writing—review and editing, F.J.A. and Y.G.; visualization, S.K., F.J.A. and Y.G.; supervision, F.J.A. and Y.G. All authors have read and agreed to the published version of the manuscript.

Funding: This study was funded by the University of Reading and the Centre for International Programs (CIP) "Bolashak" of the Republic of Kazakhstan.

Institutional Review Board Statement: This study was approved by the Ethics Committee of the University of Reading (protocol code/ethical clearance application number 001151P, approved on 2 December 2019.

Informed Consent Statement: Informed consent was obtained from all subjects involved in the study.

Conflicts of Interest: The authors declare no conflict of interest.

References

1. Kinyakin, A. The Eurasian Economic Union: Between co-existence, confrontation and cooperation with the EU. *Rocz. Integr. Eur.* **2016**, 461–480. [CrossRef]
2. Legal Information System of Regulatory Legal Acts of the Republic of Kazakhstan. Available online: http://adilet.zan.kz/eng (accessed on 10 January 2021).
3. OECD. *Strengthening Agricultural Co-Operatives in Kazakhstan*; OECD: Paris, France, 2015.
4. Schmitz, A.; Meyers, W.H. *Transition to Agricultural Market Economies: The Future of Kazakhstan, Russia and Ukraine*; CABI: Wallingford, UK, 2015. [CrossRef]
5. Sheikin, D.; Kulbayeva, A. *Food Industry of the Republic of Kazakhstan*; JSC Rating Agency of the Regional Financial Center of Almaty: Almaty, Kazakhstan, 2015.
6. Csaki, C.; Gray, K.; Lerman, Z.; Thiesenhusen, W. Land Reform and the Restructuring of Kolkhozes and Sovkhozes. In *Food and Agricultural Policy Reforms in the Former USSR: An Agenda for Transition*; World Bank, Europe and Central Asia Region: Washington, DC, USA, 1992; p. WP3.1.
7. Ministry of Agriculture of the Republic of Kazakhstan. State Program for the Development of the Agro-Inductrial Complex of the Republic of Kazakhstan for 2017–2021. 2017. Available online: https://primeminister.kz/ru/page/view/razvitie_agropromishlennogo_kompleksa (accessed on 9 July 2018).
8. Ajates, R. An integrated conceptual framework for the study of agricultural cooperatives: From repolitisation to cooperative sustainability. *J. Rural. Stud.* **2020**, *78*, 467–479. [CrossRef]
9. Ishak, S.; Omar, A.R.C.; Sum, S.M.; Othman, A.S.; Jaafar, J. Smallholder Agriculture Co-operatives' Performance: What Is in the Minds of Management? *J. Co-Oper. Organ. Manag.* **2020**, *8*, 100110. [CrossRef]
10. Milovanovic, V.; Smutka, L. Cooperative rice farming within rural Bangladesh. *J. Co-Oper. Organ. Manag.* **2018**, *6*, 11–19. [CrossRef]
11. Sultana, M.; Ahmed, J.U.; Shiratake, Y. Sustainable conditions of agriculture cooperative with a case study of dairy cooperative of Sirajgonj District in Bangladesh. *J. Co-Oper. Organ. Manag.* **2020**, *8*, 100105. [CrossRef]
12. Price, C. Valuation of unpriced products: Contingent valuation, cost–benefit analysis and participatory democracy. *Land Use Policy* **2000**, *17*, 187–196. [CrossRef]
13. Tienhaara, A.; Ahtiainen, H.; Pouta, E. Consumer and citizen roles and motives in the valuation of agricultural genetic resources in Finland. *Ecol. Econ.* **2015**, *114*, 1–10. [CrossRef]
14. Zhang, B.; Fu, Z.; Huang, J.; Wang, J.; Xu, S.; Zhang, L. Consumers' perceptions, purchase intention, and willingness to pay a premium price for safe vegetables: A case study of Beijing, China. *J. Clean. Prod.* **2018**, *197*, 1498–1507. [CrossRef]
15. McGurk, E.; Hynes, S.; Thorne, F. Participation in agri-environmental schemes: A contingent valuation study of farmers in Ireland. *J. Environ. Manag.* **2020**, *262*, 110243. [CrossRef] [PubMed]
16. Kaliyeva, S.; Areal, F.J.; Gadanakis, Y. Attitudes of Kazakh Rural Households towards Joining and Creating Cooperatives. *Agriculture* **2020**, *10*, 568. [CrossRef]
17. Luo, J.; Han, H.; Jia, F.; Dong, H. Agricultural Co-operatives in the western world: A bibliometric analysis. *J. Clean. Prod.* **2020**, *273*, 122945. [CrossRef]
18. Fahad, S.; Jing, W. Evaluation of Pakistani farmers' willingness to pay for crop insurance using contingent valuation method: The case of Khyber Pakhtunkhwa province. *Land Use Policy* **2018**, *72*, 570–577. [CrossRef]
19. Elbakidze, L.; Nayga, R. The effects of information on willingness to pay for animal welfare in dairy production: Application of nonhypothetical valuation mechanisms. *J. Dairy Sci.* **2012**, *95*, 1099–1107. [CrossRef]
20. Bett, R.; Bett, H.; Kahi, A.; Peters, K. Evaluation and effectiveness of breeding and production services for dairy goat farmers in Kenya. *Ecol. Econ.* **2009**, *68*, 2451–2460. [CrossRef]
21. Mogas, J.; Riera, P.; Bennett, J. A comparison of contingent valuation and choice modelling with second-order interactions. *J. For. Econ.* **2006**, *12*, 5–30. [CrossRef]
22. Tuan, T.H.; Navrud, S. Valuing cultural heritage in developing countries: Comparing and pooling contingent valuation and choice modelling estimates. *Environ. Resour. Econ.* **2007**, *38*, 51–69. [CrossRef]
23. Bradford, W.D.; Kleit, A.N.; A Krousel-Wood, M.; Re, R.M. Willingness to pay for telemedicine assessed by the double-bounded dichotomous choice method. *J. Telemed. Telecare* **2004**, *10*, 325–330. [CrossRef] [PubMed]
24. Venkatachalam, L. The contingent valuation method: A review. *Environ. Impact Assess. Rev.* **2004**, *24*, 89–124. [CrossRef]
25. Bennett, R.; Balcombe, K. Farmers' Willingness to Pay for a Tuberculosis Cattle Vaccine. *J. Agric. Econ.* **2012**, *63*, 408–424. [CrossRef]
26. Hanemann, M.; Loomis, J.; Kanninen, B. Statistical Efficiency of Double-Bounded Dichotomous Choice Contingent Valuation. *Am. J. Agric. Econ.* **1991**, *73*, 1255–1263. [CrossRef]
27. Kanninen, B.J. Erratum: Optimal Experimental Design for Double-Bounded Dichotomous Choice Contingent Valuation. *Land Econ.* **1993**, *69*, 443. [CrossRef]

28. Kajale, D.B.; Becker, T. Willingness to Pay for Golden Rice in India: A Contingent Valuation Method Analysis. *J. Food Prod. Mark.* **2015**, *21*, 319–336. [CrossRef]

29. Hernandez, J.I.; Hunt, A.; Pazzona, M.; Lavín, F.V. Protest treatment and its impact on the WTP and WTA estimates for theft and robbery in the UK. *Oxf. Econ. Pap.* **2018**, *70*, 468–484. [CrossRef]

30. Frey, U.J.; Pirscher, F. Distinguishing protest responses in contingent valuation: A conceptualization of motivations and attitudes behind them. *PLoS ONE* **2019**, *14*, e0209872. [CrossRef]

31. Halstead, J.M.; Luloff, A.; Stevens, T.H. Protest Bidders in Contingent Valuation. *Northeast. J. Agric. Resour. Econ.* **1992**, *21*, 160–169. [CrossRef]

32. Johansson, P.-O.; Kriström, B. Why rational agents report zero or negative WTPs in valuation experiments. *J. Environ. Econ. Policy* **2021**, *10*, 22–27. [CrossRef]

33. Fishbein, M.; Ajzen, I. *Belief, Attitude, Intention, and Behavior: An Introduction to Theory and Research*; Adison-Wesley: Reading, MA, USA, 1975.

34. Armitage, C.J.; Conner, M. Efficacy of the Theory of Planned Behaviour: A meta-analytic review. *Br. J. Soc. Psychol.* **2001**, *40*, 471–499. [CrossRef]

35. Ajzen, I. The theory of planned behavior. *Organ. Behav. Hum. Decis. Process.* **1991**, *50*, 179–211. [CrossRef]

36. Fishbein, M.; Ajzen, I. *Predicting and Changing Behavior: The Reasoned Action Approach*; Psychology Press: London, UK, 2010. [CrossRef]

37. FAO. *Dairy Development in Kazakhstan*; FAO: Rome, Italy, 2011.

38. Kucherov, S. The Future of the Soviet Collective Farm. *Am. Slav. East Eur. Rev.* **1960**, *19*, 180. [CrossRef]

39. Dadabaev, T. Evaluations of perestroika in post-Soviet Central Asia: Public views in contemporary Uzbekistan, Kazakhstan and Kyrgyzstan. *Communist Post-Communist Stud.* **2016**, *49*, 179–192. [CrossRef]

40. Easterlin, R.A. Lost in transition: Life satisfaction on the road to capitalism. *J. Econ. Behav. Organ.* **2009**, *71*, 130–145. [CrossRef]

41. Hinks, T. Bribery, motivations for bribery and life satisfaction in transitional countries. *World Dev. Perspect.* **2020**, *17*, 100172. [CrossRef]

42. Valiyev, A.; Babayev, A.; Huseynova, H.; Jafarova, K. Do citizens of the former Soviet Union trust state institutions and why: The case of Azerbaijan. *Communist Post-Communist Stud.* **2017**, *50*, 221–231. [CrossRef]

43. Babu, S.C.; Sanyal, P. Classifying households on food security and poverty dimensions—Application of K-mean cluster analysis. *Food Secur. Poverty Nutr. Policy Anal.* **2009**, 265–277. [CrossRef]

44. Jain, A.K. Data clustering: 50 years beyond K-means. *Pattern Recognit. Lett.* **2010**, *31*, 651–666. [CrossRef]

45. Mooi, E.; Sarstedt, M.; Mooi-Reci, I. *Market Research: The Process, Data, and Methods Using Stata*; Springer: Singapore, 2018. [CrossRef]

46. Hyland, J.J.; Heanue, K.; McKillop, J.; Micha, E. Factors underlying farmers' intentions to adopt best practices: The case of paddock based grazing systems. *Agric. Syst.* **2018**, *162*, 97–106. [CrossRef]

47. Gustavsen, G.W.; Hegnes, A.W. Individuals' personality and consumption of organic food. *J. Clean. Prod.* **2020**, *245*, 118772. [CrossRef]

48. Bettin, G.; Lucchetti, R. Interval regression models with endogenous explanatory variables. *Empir. Econ.* **2012**, *43*, 475–498. [CrossRef]

49. Martinez-Ribaya, B.; Areal, F.J. Is there an opportunity for product differentiation between GM and non-GM soya-based products in Argentina? *Food Control* **2020**, *109*, 106895. [CrossRef]

50. Shen, J. Understanding the Determinants of Consumers' Willingness to Pay for Eco-Labeled Products: An Empirical Analysis of the China Environmental Label. *J. Serv. Sci. Manag.* **2012**, *5*, 87–94. [CrossRef]

51. Wooldridge, M.J. *Econometric Analysis of Cross Section and Panel Data*; The MIT Press: London, UK, 2010.

52. Carfora, V.; Cavallo, C.; Caso, D.; Del Giudice, T.; De Devitiis, B.; Viscecchia, R.; Nardone, G.; Cicia, G. Explaining consumer purchase behavior for organic milk: Including trust and green self-identity within the theory of planned behavior. *Food Qual. Prefer.* **2019**, *76*, 1–9. [CrossRef]

53. Stampa, E.; Schipmann-Schwarze, C.; Hamm, U. Consumer perceptions, preferences, and behavior regarding pasture-raised livestock products: A review. *Food Qual. Prefer.* **2020**, *82*, 103872. [CrossRef]

54. Hoeksma, D.L.; Gerritzen, M.A.; Lokhorst, A.M.; Poortvliet, P.M. An extended theory of planned behavior to predict consumers' willingness to buy mobile slaughter unit meat. *Meat Sci.* **2017**, *128*, 15–23. [CrossRef]

55. Klicperova-Baker, M.; Košťál, J. Post-communist democracy vs. totalitarianism: Contrasting patterns of need satisfaction and societal frustration. *Communist Post-Communist Stud.* **2017**, *50*, 99–111. [CrossRef]

56. Toleubayev, K.; Jansen, K.; Van Huis, A. Knowledge and agrarian de-collectivisation in Kazakhstan. *J. Peasant. Stud.* **2010**, *37*, 353–377. [CrossRef]

57. Dascălu, D.I. Individualism and Morality in the Post-communist Capitalism. *Procedia Soc. Behav. Sci.* **2014**, *149*, 280–285. [CrossRef]

58. Rabikowska, M. The ghosts of the past: 20 years after the fall of communism in Europe. *Communist Post-Communist Stud.* **2009**, *42*, 165–179. [CrossRef]

59. Zhang, M.; Li, L.; Bai, J. Consumer acceptance of cultured meat in urban areas of three cities in China. *Food Control* **2020**, *118*, 107390. [CrossRef]

60. Roosen, J.; Bieberstein, A.; Blanchemanche, S.; Goddard, E.; Marette, S.; Vandermoere, F. Trust and willingness to pay for nanotechnology food. *Food Policy* **2015**, *52*, 75–83. [CrossRef]
61. Yu, H.; Neal, J.A.; Sirsat, S.A. Consumers' food safety risk perceptions and willingness to pay for fresh-cut produce with lower risk of foodborne illness. *Food Control* **2018**, *86*, 83–89. [CrossRef]
62. Blustein, D.L.; Duffy, R.; Ferreira, J.A.; Cohen-Scali, V.; Cinamon, R.G.; Allan, B.A. Unemployment in the time of COVID-19: A research agenda. *J. Vocat. Behav.* **2020**, *119*, 103436. [CrossRef]
63. During the Dangerous Epidemic, the Number of Unemployed in Atyrau Increased by 3 and a Half Times. Available online: https://24.kz/kz/zha-aly-tar/o-am/item/420240-auipti-indet-kezinde-atyrauda-zh-myssyzdar-sany-3-zharym-esege-art-an (accessed on 18 August 2020). (In Kazakh)
64. Gava, O.; Ardakani, Z.; Delalić, A.; Azzi, N.; Bartolini, F. Agricultural cooperatives contributing to the alleviation of rural poverty. The case of Konjic (Bosnia and Herzegovina). *J. Rural. Stud.* **2021**, *82*, 328–339. [CrossRef]
65. Hannachi, M.; Fares, M.; Coleno, F.; Assens, C. The "new agricultural collectivism": How cooperatives horizontal coordination drive multi-stakeholders self-organization. *J. Co-Oper. Organ. Manag.* **2020**, *8*, 100111. [CrossRef]
66. Li, X.; Ito, J. An empirical study of land rental development in rural Gansu, China: The role of agricultural cooperatives and transaction costs. *Land Use Policy* **2021**, *109*, 105621. [CrossRef]
67. Liang, Q.; Wang, X. Cooperatives as competitive yardstick in the hog industry?—Evidence from China. *Agribusiness* **2020**, *36*, 127–145. [CrossRef]
68. Lerman, Z. *Co-Operative Development in Central Asia*; Policy Studies on Rural Transition No. 2013-4; FAO: Rome, Italy, 2013; 38p.
69. Iliopoulos, C.; Valentinov, V. Cooperative Longevity: Why Are So Many Cooperatives So Successful? *Sustainability* **2018**, *10*, 3449. [CrossRef]
70. Niyazmetov, D.; Soliev, I.; Theesfeld, I. Ordered to volunteer? Institutional compatibility assessment of establishing agricultural cooperatives in Uzbekistan. *Land Use Policy* **2021**, *108*, 105538. [CrossRef]
71. Moon, S.; Lee, S. A Strategy for Sustainable Development of Cooperatives in Developing Countries: The Success and Failure Case of Agricultural Cooperatives in Musambira Sector, Rwanda. *Sustainability* **2020**, *12*, 8632. [CrossRef]
72. Kurakin, A.; Visser, O. Post-socialist agricultural cooperatives in Russia: A case study of top-down cooperatives in the Belgorod region. *Post-Communist Econ.* **2017**, *29*, 158–181. [CrossRef]
73. Deng, H.; Huang, J.; Xu, Z.; Rozelle, S. Policy support and emerging farmer professional cooperatives in rural China. *China Econ. Rev.* **2010**, *21*, 495–507. [CrossRef]
74. Concu, G.B. Investigating distance effects on environmental values: A choice modelling approach. *Aust. J. Agric. Resour. Econ.* **2007**, *51*, 175–194. [CrossRef]
75. Schreiner, J.; Latacz-Lohmann, U. Farmers' valuation of incentives to produce genetically modified organism-free milk: Insights from a discrete choice experiment in Germany. *J. Dairy Sci.* **2015**, *98*, 7498–7509. [CrossRef]

 *agriculture*

Article

The Digitalization of the European Agri-Food Cooperative Sector. Determining Factors to Embrace Information and Communication Technologies

Javier Jorge-Vázquez [1,*], Mª Peana Chivite-Cebolla [1] and Francisco Salinas-Ramos [2]

[1] DEKIS Research Group, Department of Economics, Catholic University of Ávila, 05005 Avila, Spain; mpeana.chivite@ucavila.es
[2] Department of Economics, Catholic University of Ávila, 05005 Avila, Spain; francisco.salinas@ucavila.es
* Correspondence: javier.jorge@ucavila.es

Abstract: The digitization of the agri-food sector is a strategic priority in the political agenda of European institutions. The opportunity to improve the competitiveness and efficiency of the sector offered by new technologies comes together with its potential to face new economic and environmental challenges. This research aims to analyze the level of digitalization of the European agri-food cooperative sector from the construction of a composite synthetic index. Such an index is to be based on a diverse set of variables related to electronic commerce and the services offered through the internet. It also evaluates how European cooperatives influence the degree of technological adoption depending on their size or the wealth of the country where they carry out their activity. The empirical analytical method is thus used, through the analysis of frequencies and correlations. The results obtained reveal the existence of a suboptimal and heterogeneous degree of digitization of European agri-food cooperatives, clearly conditioned by their size and the wealth of the country where they operate. In this situation, it is recommended to promote public policies that guarantee high-performance digital connectivity, an improvement in training in digital skills and the promotion of cooperative integration processes.

Keywords: agroindustrial; agricultural cooperative; technology adoption; technology and competitiveness; information and communication technology; digital transformation; agri-food cooperatives

Citation: Jorge-Vázquez, J.; Chivite-Cebolla, M.P.; Salinas-Ramos, F. The Digitalization of the European Agri-Food Cooperative Sector. Determining Factors to Embrace Information and Communication Technologies. *Agriculture* **2021**, *11*, 514. https://doi.org/10.3390/agriculture11060514

Academic Editors: Adoración Mozas Moral and Domingo Fernandez Ucles

Received: 8 May 2021
Accepted: 28 May 2021
Published: 2 June 2021

1. Introduction

In the last few decades, society has digitalized in a generalized way in most of the developed countries and has also adopted a character of transversality that is encouraging reconsideration of the traditional forms and balances of economic and social organization. This phenomenon is inseparable from the vigorous and accelerated development of new digital technologies.

The vertiginous development of digital infrastructures together with the globalization of an increasingly agile and reliable network access and interconnection is causing a global digital ecosystem. Its configuration drives the concurrence of multiple disruptive processes, with a noticeable incidence in all productive sectors, pushing towards the transformation of business models and the change in economic growth patterns of developed countries. There is no doubt that the digital transformation comes today as a lever that drives development and economic growth while favoring profits in terms of competitiveness and business efficiency.

In this dynamic and highly competitive environment, the European agri-food cooperative sector must undertake a digitalization strategy that allows it to take advantage of the opportunities that arise from a hyperconnected global market such as the current one is. The access new technologies and the implementation of technical and organizational innovations must therefore be a priority for agri-food cooperatives that seek to obtain profits in terms of competitiveness and productivity and thus improve their market positioning.

This research intends to contribute to the study of the degree of implementation of new digital technologies in business organizations and, more particularly, in the European agri-food cooperative sector, which has unquestionable strategic importance for the European Union (EU) [1]. In this way, the main objectives set forth can be defined as two: (i) develop an exploratory study on the degree of digitalization of European agri-food cooperatives based on two specific dimensions of analysis: their presence on the Internet and the use of information and communication technology (ICT) for commercial purposes by evaluating the online sales channels and tools present on their websites; and (ii) identify the determining factors of the digitization index of European agri-food cooperatives, based on two variables: the size of the cooperative societies and the wealth of the country where they carry out their activity.

In order to achieve both objectives, we propose to build a composite index to allow for the development of a comparative analysis of the degree of technological integration of a representative sample composed of 454 EU (28) agri-food cooperatives. Additionally, the analysis of frequencies and correlations will allow for the determination of the degree of influence of the different factors on the digitization index of cooperatives.

The sequence of research is as follows: after defining the scope of research and, once the objectives of the study have been defined, we proceed to develop a brief review of the background of the research around the conditions of the issue that allow us to lay the groundwork to make research assumptions. Once the methodology has been thoroughly defined and the proposed assumptions presented, the analysis and discussion of the results obtained in the investigation, as well as the main conclusions reached, are presented by highlighting the implications of the findings and summarizing directions of ongoing research.

2. Theoretical Framework

Analyzing the economic effects associated with the process of digitalization of the economy has been a common object of interest for the scientific community in recent decades. These investigations have focused on the study of the economic impact of adding new technologies, mainly in three different areas: productivity gains, economic growth and the labor market.

Far from undertaking a systematic review of the literature on the current status, an objective that exceeds the scope and purpose of this research, we present below the main works that highlight the background of the investigation and the current state of knowledge.

The economic literature we reviewed evidences the existence of a large group of works that confirm the significant influence that implementing ICT has on profits in terms of total productivity of productive factors. In particular, Nordhaus [2] attributes the rebound we observed in the average productivity of the business sector since 1995 to the strong growth of productivity in sectors that are intensive in information and communication technologies. Along these lines, Besnaham et al. [3] also conclude that adopting information technologies causes positive effects on business productivity. However, they argue that such productivity increases when combined with certain organizational investments. Hernando et al. [4] also find evidence of a positive and relevant contribution of ICT to the growth of production and productivity in Spain in the period 1991–2000, while Astrostic et al. [5] assert that there is a clear link between information technology and productivity gains. For their part, Draca, M. [6] present a neoclassical framework to understand the role of ICT and productivity. In their study, they find that there is evidence of a strong association between information technologies and firm performance. For his part, Torrent [7] maintains that communication technologies, although they are not the only causal factor, "are consolidated as an essential instrument for the development of production, work and consumption in the network" (p. 19). Cardona et al. [8], after reviewing empirical literature, found that most studies point out the positive and significant effect of ICTs on productivity, although they argue that research in this field is still insufficient to better understand the externalities of ICTs in the economy. Other research has shown the significantly positive impact on agricultural productivity [9] and that such improvement in agricultural yields associated with the adoption of new technologies has contributed to reducing poverty and food insecurity rates in the most disadvantaged rural areas and, consequently, to economic growth [10]. On the contrary, some authors [11] have found certain limitations and a reduced impact of public programs for easy adoption of technology through extension programs based on ICT in the agricultural field. In any case, most studies coincide in pointing out the positive impact of investment in information technology (IT) on world economic growth, especially in the most industrialized economies and in developing Asian countries [12]. Other works [13,14] have also analyzed the impact of ICTs in Europe, concluding that the deployment and use of ICTs drives economic growth in developed European countries. Additionally, the use of ITC and, in particular, digital empowerment have positive economic effects on the labor market and on the inclusion of disadvantaged groups [15]. This positive impact on the economy responds largely to the improvement of the international competitiveness of companies [16] and the internal efficiency of companies [17]. In particular, some studies have analyzed the positive impact of ITCs on the economic efficiency of companies operating in the agri-food sector [18]. In most of these studies, the main benefits of the use of new technologies are the greater growth, development and economic efficiency of companies [19] by complementing other production factors and promoting innovation by significantly reducing transaction costs [20].

Along these same lines, some works [21] have confirmed that adopting ICT in the agri-food sector, along with other structural and organizational variables, constitutes a relevant factor to be considered in improving competitiveness, gains in economic efficiency and the development of the sector itself, while contributing to sustainability in agriculture and food systems [22].

On the other hand, the existence of conditioning factors to adopt new technologies in the agri-food sector has also been a common object matter of scientific interest.

Most studies on technological adoption in firms are based on the theories of the diffusion of innovation (DOI) and technology, organization and environment (TOE) [23]. In both models, the size of the firm measured by the number of employees is considered one of the determining factors in the adoption of innovation and technology in organizations [24,25]. Along these lines, some research such as that developed by [26] has explored the influence that social and demographic factors, commercial orientation or the size of farms generate in the adoption of information systems based on ICT.

Other studies have found that the level of ICT adoption has higher levels in the richest countries and that the return obtained from such implementation is also higher than in the poorest countries [9].

The studies on digital transformation developed in the field of the agri-food cooperative sector agree to point out the existence of some delay in adopting new technologies for business purposes [27]. Such a delay is conditioned by the size and subsector where cooperatives carry out their activity [28] and by the quality of their website [29]. Ultimately, this makes it difficult to include advanced functions on websites [30] or to take advantage of the opportunities offered by ICTs, such as traceability systems for the agri-food supply chain based on blockchain technology [31], among others.

In several areas and regions, there are still works that present digitization as a solution to the sustainability of agri-food systems around the world. In this regard, there are works that focus on studying the regions of the Middle East and North Africa [32]. In the Barents Region [33], digitalization can create conditions that are necessary to diversify organizational schemes and effectively monitor food processing operations that will help to promote food and nutrition security.

In Spain, for example, research indicates that, although the growing importance of digital communication stands out, Spanish cooperatives still do not invest or include this matter in their strategic plans [34]. For wine cooperatives to be competitive and improve the quality of their website, they will need to improve their digital communication [35]. Domestically, in the Catalonia region, it will be reported that cooperatives in this region continue to show very low levels due to the lack of presence of websites on the internet [36].

Based on the review of the research background and the current status of the issue raised, the following research assumptions are made for subsequent contrast:

Hypothesis 1. *Agri-food cooperatives in the EU (28) have a degree of digitalization below the average level observed in the European business sector as a whole.*

Hypothesis 2. *The size of the agri-food cooperative in the EU (28) constitutes a conditioning factor in the adoption of new technologies.*

Hypothesis 3. *The wealth of the country where the agri-food cooperative develops its activity exerts a significant influence on its degree of digital transformation.*

3. Materials and Methods

In order to comply with the scientific objectives formulated and proceed to contrast the research assumptions raised on the degree of digitalization of the European agri-food cooperative sector, it is proposed to apply the empirical analytical method, through the analysis of frequencies and correlations. To evaluate the website, we will choose the method of accounting and will apply content analysis techniques.

3.1. Population and Sample

The agri-food cooperatives that are active in the EU (28) make the study population of this work. The source used to obtain the European cooperatives operating in the agri-food sector is the Orbis database [37]. For this search, we obtained a total population of 35,384 cooperatives. By including the most updated availability criterion of the reported information as an indicator of business activity and taking cooperatives with data after 2016, the population is 16,184 registered cooperatives.

Once the population under study was identified, the sample size was determined through randomized stratified probabilistic sampling according to the country, for a 95% confidence level and a sampling error of 4.6%. That gave us a sample size of 441.52 cooperatives. Applying stratified random probabilistic sampling according to the country allows everyone to be represented, especially those with the largest number of cooperatives, according to that base, to a greater extent, which allows for an additional inter-territorial analysis.

To determine the sample size for each country, it is established that any countries that have the most cooperatives have up to 24, those that are average have 16, and those with the lowest number or least data availability (5 countries) have between 9 and 5, depending on said availability. Thus, the country with the lowest representation is Luxembourg, holding 5 cooperatives. The sample broken down by countries is distributed as shown in Table 1.

Although the sample came up to 442 cooperatives, 454 have finally been selected to allow greater representation.

Table 1. Sample of cooperatives by EU country (28). Source: Own development.

Country	Cooperative	Country	Cooperative
Austria	16	Italy	24
Belgium	16	Latvia	16
Bulgaria	24	Lithuania	16
Croatia	16	Luxembourg	5
Cyprus	6	Malta	8
CzechRepublic	16	Netherlands	16
Denmark	16	Poland	16
Estonia	16	Portugal	7
Finland	16	Romania	16
France	24	Slovakia	16
Germany	24	Slovenia	9
Greece	16	Spain	24
Hungary	24	Sweden	19
Ireland	16	UK	16
		TOTAL	**454**

3.2. Selection of Variables and Information Sources

The variables considered in this study, collected and described in Table 2, have been selected based on the recommendations provided by the European Parliament and the European Council for producing statistics on the Information Society and, in particular, to gather information related to the characteristics that must be collected from the companies that have a website. Following this recommendation, the taxonomy proposed by [28] has been adopted, insofar as it allows for the categorization of a broad set of parameters on the degree of digital transformation of cooperative societies by evaluating a series of indicators on the use of the internet and other electronic networks and, in particular, on the web services offered and electronic commerce. In turn, for better international comparison of the index of digital transformation of European agri-food cooperatives with other corporate legal forms, we have opted to select those variables included in the "Community survey on ICT usage and e-commerce in enterprises" that Eurostat publishes periodically and that allows one to perform said analysis based on a set of consistent and reliable data.

Table 2. Services offered on the internet and electronic commerce: selected variables. Source: own development based on the "Community survey on ICT usage and e-commerce in enterprises" (Eurostat, several years) and [28].

Category	Variable		Definition
Use of internet and other electronic networks by companies (electronic commerce)	B1	Cooperatives where the website provided online ordering or reservation or booking, e.g., shopping cart.	Regarding the existence of a sales channel through electronic commerce. It evaluates the existence of e-commerce platforms or platforms that allow for the reception of orders, the booking of goods or services through the internet or other telematic networks.
	B2	Cooperatives where the website provided description of goods or services, and price lists.	Refers to the possibility and ease of access, through the website, of catalogs of goods and/or services offered by the cooperative or publication of price rates for its products.
	B3	Cooperatives where the website provided possibilities for visitors to customize or design the products (webctm).	Is related to the inclusion of tools in the buying process that allow the user to personalize and/or take part in the design of the goods and services offered by the cooperative.
	B4	Cooperatives where the website provided order tracking available online.	Provision on the website of platforms or other telematic means that allow for real-time monitoring of the status of processing of the order, from the completion of the online purchase process to the effective delivery of the product to the customer.

Table 2. *Cont.*

Category	Variable		Definition
Quality of website and services offered on the internet	C1	Cooperatives with a website.	Includes the existence of a specific web portal of the cooperative company, as well as its positioning in the Google search engine.
	C2	Corporate presentation of the cooperative entity.	It is related to the publication of sufficient and adequate information about the cooperative entity and its activity.
	C3	Cooperatives with personalized content in the website for regular/repeated visitors (webper).	Is related to the adaptation of the contents and structure of the web based on the observed user's behavior, as well as its specific attributes (profile, location, etc.) in order to offer an improvement in the browsing experience.
	C4	Cooperatives where the website had links or references to the enterprise's social media profiles.	Presence in the corporate web portal of explicit references and links to the main communication platforms to allow interaction and exchange of content and information with suppliers, customers and other agents that are related to the activity of the cooperative.
	C5	Cooperatives where the website provided a private policy statement, a privacy seal or certification related to website safety.	Inclusion in the website of a specific section reserved for the description of the privacy and data protection policy, use of the page and limitations of use, use of cookies, security, etc.
	C6	Cooperatives where the website provided advertisement of open job positions or online job application.	Refers to the use of the website as an electronic means at the service of personnel recruitment processes. It includes elements such as the existence of a job offer board, the availability of a channel enabled for sending CV, etc.
	C7	Cooperatives where the website provided for the electronic submission of complaints.	Existence on the website of a specific channel enabled for the submission of claims or, failing that, the publication on the website of specific instructions for filing claims through other telematic means (for example: via email).
	C8	Adaptive web design.	The website has a "responsive" design, that is, it is optimized to be displayed according to the screen size of the device in use to visit it.

The search and data collection has been carried out in late 2018 and early 2019 through the direct analysis of the content and design of the Web pages corresponding to each of the 454 European agri-food cooperatives that make up the sample under study. In particular, 12 variables total have been verified. Said binary dichotomous variables are decided to be encoded so that they can take the value "1", should it have such attribute, or the value of "0" otherwise.

On the other hand, in order to identify determinants of the degree of digitalization of agri-food cooperatives and thus comply with the objectives formulated, a set of additional variables indicative of the size of the agri-food cooperative society are added: A.0 number of employees/members, A.1 ordinary results before taxes and A.2 total assets.

Finally, in order to determine the ability of the country's wealth to influence in the adoption of new technologies by the agri-food cooperatives under study, it is decided to consider as a measure of such wealth the gross domestic product (GDP) per capita, which is obtained from the statistics published by the World Bank for fiscal year 2018.

3.3. Method

To achieve the proposed scientific objectives and in order to proceed with the contrast of the formulated research assumptions, a combination of the following methods is applied: to evaluate the attributes of the website that are related to electronic commerce and the web services offered, we opted to apply the accounting method adopted in other research related to the evaluation of websites [38]. This method is based on the verification of a checklist made of a wide set of items that were verified through the application of web content analysis techniques [39,40].

In contrast to assumption 1, we mainly used the empirical analytical method, through the frequency analysis of the main variables shown in Table 2 and the construction of a composite synthetic index. Its purpose is to offer a synthetic and comparable view on the degree of digitalization of agri-food cooperative societies in the different Member States that make up the EU (28) as an equal measure of the different components that make up the following dimensions: electronic commerce, website quality and services offered on the internet.

Thus, in order to measure the degree of digitalization of the agri-food cooperatives that make up the sample, we created the aggregate variable "Level of digitization", defined as the sum of the set of variables *"B"* and *"C"*, according to the Formula (1), and shown in Table 2. This aggregate variable can take a maximum value of 12 and a minimum value of 0. This variable is additionally contrasted with another of the variables provided by the European Commission, specifically the digital intensity score for enterprises, as an aggregate of indices.

$$\sum_{i=1}^{4} \sum_{j=1}^{8} Bi\ Cj \tag{1}$$

In contrast to Assumption 2, we added a set of additional variables indicative of the size of the agri-food cooperative. In this phase of the research, the frequency analysis was combined with the correlation analysis between the selected variables.

To test the third assumption, we established the analysis of frequencies broken down by countries, and, in parallel, created a fictitious variable representing the "wealth of the country" measured as GDP per capita, based on data provided by the World Bank to 2018. It is considered that the country's wealth measured as its purchasing and productive capacity, GDP per capita, can be a determining factor in the level of digitalization of cooperatives. To measure this influence, two linear regressions are presented.

4. Discussion of Results

The data and specifications of the models and of the variables that allow for the contrast of the formulated assumptions are presented in this section. The results obtained are set out below in the order in which the assumptions were proposed.

4.1. Benchmarking of the Degree of Digitalization Existing between Agri-Food Cooperatives and All European Companies

Results obtained from the comparative analysis developed to contrast the existence of a greater delay in the digital transformation of agri-food cooperatives with respect to the entire business sector in the EU (28) are thus presented, as formulated in the first research assumption (H1).

The data collected for each of the selected variables as indicators of the degree of business digitalization are shown in Table 3.

The results obtained in this research confirm that, out of the 454 cooperatives that make up the sample under study, only 52.20% of them have an active website. The percentage reduces to 33.5% when excluding any websites that are not designed under a responsive design pattern. These results coincide with the estimates obtained in other studies, such as the study by [41], where it is quantified that on average, 53.41% of all olive oil producers had websites, or [42], which estimates that 43% of cooperatives in the second degree in Spain have a web page, or the research carried out by [43] that concludes that there are few cooperatives that have a web page in the region of the Canary Islands.

Table 3. EU agri-food cooperatives (28) 2018–2019: WEB services and electronic commerce (2019). Source: own development based on the data collected from the research, the community survey on ICT usage and e-commerce in enterprises (Eurostat, several years).

Category		Variable	Total Coop. *. (no.)	Total Coop. (%)	Total Coop. (%)	EU-28 (%) [1]	Differential (%)
Use of internet and other electronic networks by companies (electronic commerce)	B1	Cooperatives where the website provided online ordering or reservation or booking, e.g., shopping cart	36	15.19 [2]	7.93	19	−11.07
	B2	Cooperatives where the website provided description of goods or services, price lists	172	72.57 [2]	37.89	56	−18.11
	B3	Cooperatives where the website provided possibilities for visitors to customize or design the products (webctm)	4	1.69 [2]	0.88	18	−17.12
	B4	Cooperatives where the website provided order tracking available online	11	4.64 [2]	2.42	9	−6.58
Quality of website and services offered on the internet	C1	Cooperatives with a website	237	52.20 [1]	52.20	77	−24.80
	C2	Corporate presentation of the cooperative entity	228	96.20 [2]	50.22	56	−5.78
	C3	Cooperatives with personalized content in the website for regular/repeated visitors (webper)	71	29.96 [2]	15.64	58	−42.36
	C4	Cooperatives where the website had links or references to the enterprise's social media profiles	105	44.30 [2]	23.13	38	−14.87
	C5	Cooperatives where the website provided a private policy statement, a privacy seal or certification related to website safety	129	54.43 [2]	28.41	31 [3]	−2.59
	C6	Cooperatives where the website provided advertisement of open job positions or online job application	49	20.68 [2]	10.79	27 [4]	−16.21
	C7	Cooperatives where the website provided for the electronic submission of complaints	6	2.53 [2]	1.32	30 [3]	−28.68
	C8	adaptive web design	152	64.14 [2]	33.48	n.d.	n.d.

[1] Data on the cooperative companies analyzed total. [2] Data on cooperative companies with webpage total. (*) All enterprises, without financial sector (10 persons employed or more). [3] Latest available data 2014. [4] Latest available data 2016.

Regarding the quality of the website and the services offered, within the cooperative societies having a website, 96.2% prioritize their corporate presentation, whereas 29.6% offer the possibility of website personalization and 44.3% make reference to corporate profiles in social media. Only 20.7% of the agri-food cooperatives use the web as a staff recruitment channel, while few communication channels enabled to file claims are observed. Regarding the dimension of electronic commerce, 72.6% of the cooperatives that have websites offer access to a catalog of products or price lists, while only 15.2% allow for the formalization of online orders through their website. The possibility of product customization and online tracking of orders is barely available on the websites analyzed, confirming the difficulties of the agri-food cooperative sector in the digitalization of sales channels.

If we use the survey on the use of ICTs in companies published annually by Eurostat (several years) and take the values in the selected variables, shown in Table 4, we can see a relatively heterogeneous degree of digitalization between the different countries that make up the EU (28). Thus, the most developed European countries have better results in each of the items analyzed, and countries such as Netherlands or Finland stand out, compared to other Member States such as Romania or Bulgaria whose business sector has a much poorer level of digitalization.

Table 4. Digitalization of the European business sector (EU28): website functionalities and ecommerce (2018). Source: own development based on Eurostat (several years) and of the data collected in the research.

Country	Selected Variables **										
	B1	B2	B3	B4	C1	C1 Coop	C3	C4	C5	C6	C7
European Union—(UE-28)	19	56	18	9	77	52	58	38	31 [4]	27 [2]	30 [4]
Belgium	23	66	23	13	84	81	68	45	28 [5]	41 [2]	32 [4]
Bulgaria	14	41	14	9	51	8	42	18	17 [3]	9 [2]	14 [4]
Czechia	28	54 [1]	23	8	83	88	42	32	21 [4]	23 [5]	39 [4]
Denmark	33	66	30	10	96	88	68	59	19 [3]	47 [2]	35 [4]
Germany	16	74	16	7	87	54	75	35	56 [5]	41 [2]	41 [4]
Estonia	17	76	17	7	78	25	76	32	15 [3]	20 [2]	19 [4]
Ireland	29	62	27	12	79	69	66	50	43 [3]	28 [2]	28 [4]
Greece	14	42	13	6	65	38	44	42	20 [3]	17 [2]	24 [4]
Spain	15	37	14	8	76	54	39	37	51 [3]	17 [2]	19 [4]
France	18	58	18	11	69	58	60	33	26 [4]	22 [2]	26 [4]
Croatia	14	38	13	7	73	25	41	34	29 [3]	17 [2]	46 [4]
Italy	15	32	13	8	71	50	35	37	43 [3]	10 [2]	20 [4]
Cyprus	12	71	12	3	71	33	71	45	28 [3]	23 [2]	39 [4]
Latvia	9	59	5	3	63	31	59	26	13 [3]	16 [2]	15 [4]
Lithuania	20	54	19	13	78	50	57	30	29 [3]	21 [2]	30 [4]
Luxembourg	19	64	19	9	83	80	66	42	28 [3]	35 [2]	25 [4]
Hungary	20	56	19	9	66	29	58	25	14 [3]	20 [2]	27 [4]
Malta	37	78	36	14	82	38	80	61	38 [5]	35 [2]	46 [4]
Netherlands	36	79	34	13	94	81	82	62	36 [3]	57 [2]	45 [4]
Austria	22	60	21	5	88	88	61	42	31 [5]	29 [5]	36 [4]
Poland	14	61	14	9	67	75	62	22	32 [3]	18 [2]	20 [4]
Portugal	10	43	10	7	63	100	47	32	28 [3]	16 [2]	22 [4]
Romania	19	42	18	10	44	25	43	17	7 [5]	10 [5]	14 [4]
Slovenia	16	81	16	6	84	38	81	34	31 [3]	27 [2]	32 [4]
Slovakia	23	68	23	9	76	67	69	24	24 [3]	26 [2]	25 [4]
Finland	26	85	25	10	96	69	86	68	22 [3]	42 [2]	53 [4]
Sweden	36	48	32	9	92	21	51	54	24 [5]	na	60 [4]
United Kingdom	21	58	20	9	82	69	59	51	38 [5]	na	33 [4]

All enterprises, without financial sector (10 persons employed or more) ** See correspondence of variables (Table 2); na: not available; [1] data relating to the year 2017; [2] data relating to the year 2016; [3] data relating to the year 2015; [4] data relating to the year 2014; and [5] data relating to the year 2013. B1 Cooperatives where the website provided online ordering or reservation or booking. B2 Cooperatives where the website provided description of goods or services, and price lists. B3 Cooperatives where the website provided possibilities for visitors to customize or design the products (webctm). B4 Cooperatives where the website provided order tracking available online. C1 Cooperatives with a website. C2 Corporate presentation of the cooperative entity. C3 Cooperatives with personalized content on the website for regular/repeated visitors. C4 Cooperatives where the website had links or references to the enterprise's social media profiles. C5 Cooperatives where the website provided a private policy statement, a privacy seal or certification related to website safety. C6 Cooperatives where the website provided advertisement of open job positions or online job application. C7 Cooperatives where the website provided for the electronic submission of complaints.

Additionally, the variable "C1coop" has been included in Table 4. It is noteworthy that the percentage of cooperatives with a website is, in general, lower than that in the business group (C1) for 90% of European countries. If we exclude Portugal, which is atypical in the selection of the sample, it is worth highlighting the cases of Poland and the Czech Republic as the only countries that have a higher percentage of website availability in cooperatives compared to the business sector in their country.

To complete the information and in order to develop a benchmarking that allows for a contrast to assumption 1, the data provided by Eurostat (several years) is used in the "Community survey on ICT usage and e-commerce in enterprises", from which we extracted the data that are most directly related to the variables selected and analyzed for the particular case of agri-food cooperatives in Europe. The results of this benchmarking are presented synthesized in Table 3 and clearly confirm assumption 1—that is, the degree of digitalization of the European agri-food cooperative sector is much lower than that observed in all European companies, which indicates the existence of certain delay in the adoption of ICT by the cooperative societies analyzed. This finding is consistent with results in the literature on the delay with which cooperatives embrace ICTs [28,29,41].

This statement is proven by verifying that the agri-food cooperatives have worse results in all the indicators on the level of digital transformation selected. What is especially striking is the differential in parameters such as the possibility of personalization and availability of the website or in the dimension of electronic commerce in the access to product catalogs or price lists.

On the other hand, in order to build a composite synthetic index that allows for the characterization of the degree of digitalization achieved by agri-food cooperative societies, we have created the aggregate variable "Level of digitization", defined as the sum of the set of variables "B, C" listed in Table 2. This aggregate variable can take a maximum value of 12 and a minimum value of 0. Table 5 shows the results obtained, globally and itemized by countries. Each column indicates the score that can be obtained, from 0 to 12, and for each country the cooperatives that have reached those scores. The highest score, 11, is obtained by a cooperative in Denmark.

Table 5. "Level of digitization" * for European agri-food cooperatives EU (28). 2018–2019. Source: Own development.

Country	0	1	2	3	4	5	6	7	8	9	10	11	12	Average	Total
Austria	2	0	0	1	1	5	4	3	0	0	0	0	0	4.81	16
Belgium	3	0	2	1	1	2	2	3	1	1	0	0	0	4.44	16
Bulgaria	22	0	0	0	1	0	1	0	0	0	0	0	0	0.42	24
Croatia	12	0	0	2	1	1	0	0	0	0	0	0	0	0.94	16
Cyprus	4	0	1	1	0	0	0	0	0	0	0	0	0	0.83	6
Czechia	2	1	6	3	3	0	0	0	1	0	0	0	0	2.63	16
Denmark	2	0	4	1	3	0	2	2	0	1	0	1	0	4.31	16
Estonia	12	0	0	2	1	0	1	0	0	0	0	0	0	1.00	16
Finland	5	0	1	2	2	1	1	2	2	0	0	0	0	3.56	16
France	10	0	0	1	2	4	3	2	2	0	0	0	0	3.29	24
Germany	11	0	0	0	2	2	3	2	2	2	0	0	0	3.50	24
Greece	10	0	0	0	3	1	1	1	0	0	0	0	0	1.88	16
Hungary	17	0	0	2	3	1	1	0	0	0	0	0	0	1.21	24
Ireland	5	0	0	2	2	2	2	3	0	0	0	0	0	3.56	16
Italy	12	0	0	1	0	2	5	2	2	0	0	0	0	3.04	24
Latvia	11	0	0	1	1	1	2	0	0	0	0	0	0	1.50	16
Lithuania	8	0	1	1	2	2	1	1	0	0	0	0	0	2.25	16
Luxembourg	1	0	1	0	1	1	0	0	0	0	1	0	0	4.20	5
Malta	5	0	2	1	0	0	0	0	0	0	0	0	0	0.88	8
Netherlands	3	0	0	2	1	2	2	2	2	2	0	0	0	5.00	16
Poland	4	0	0	4	2	5	0	0	0	0	1	0	0	3.44	16
Portugal	0	0	1	0	1	1	0	1	3	0	0	0	0	6.00	7
Romania	12	1	0	0	0	2	1	0	0	0	0	0	0	1.06	16
Slovakia	10	1	2	0	1	0	0	0	2	0	0	0	0	1.56	16
Slovenia	3	0	2	0	1	2	0	1	0	0	0	0	0	2.78	9
Spain	11	0	1	2	5	3	1	0	0	1	0	0	0	2.42	24
Sweden	15	0	0	1	1	0	0	1	0	1	0	0	0	1.21	19
United Kingdom	5	0	0	0	1	3	1	2	3	0	1	0	0	4.56	16
European Union (EU28)	217	3	24	31	42	43	34	28	20	8	3	1	0	2.64	454

* "Level of digitization", defined as the sum of the set of variables "B, C" shown in Table 2.

On the other hand, it is interesting to verify that only two countries would reach an "approved" digitization index, with at least a 5-point average rating. It should be noted that as for one of them, the case of Portugal is atypical, since we only considered the 7 cooperatives reported by the database consulted, and all of them had a web page. The average for the EU is 2.64.

If we turn to the European Commission (EC) and, in particular, the index on digitalization that it designs to measure such transformation (DESI), it brings together the results achieved according to 4 different levels, as shown in Table 6.

Table 6. Digital intensity score for enterprises [1] (2018) EU (28) and aggregate variable "Level of digitization" for agri-food cooperatives EU (28) clustered (2018–2019). Measure: percentage (%). Source: Own development based on data from the study and the EC Economy and Digital Society index available at: https://digital-agenda-data.eu/datasets/desi/visualizations (accessed on 20 August 2019).

Country		Very Low (0–3)		Low (4–6)		High (7–9)		Very High (10–12)	
		Enter [2,3]	COOP	Enter [2,3]	COOP	Enter [2,3]	COOP	Enter [2,3]	COOP
Austria	AT	40.73	18.75	42.52	62.50	14.52	18.75	2.23	0.00
Belgium	BE	32.81	37.50	39.64	31.25	22.07	31.25	5.48	0.00
Bulgaria	BG	66.62	91.67	24.82	8.33	7.81	0.00	0.74	0.00
Croatia	HR	50.92	87.50	33.55	12.50	13.64	0.00	1.89	0.00
Cyprus	CY	44.47	100.00	40.97	0.00	13.26	0.00	1.3	0.00
Czechia	CZ	48.03	75.00	34.94	18.75	14.27	6.25	2.76	0.00
Denmark	DK	13.51	43.75	37.01	31.25	38.29	18.75	11.19	6.25
Estonia	EE	41.88	87.50	37.67	12.50	17.23	0.00	3.22	0.00
European Union	EU	**45.84**	**60.57**	**36.2**	**26.21**	**15.88**	**12.33**	**2.08**	**0.88**
Finland	FI	11.12	50.00	39.58	25.00	37.57	25.00	11.73	0.00
France	FR	50.28	45.83	34.88	37.50	13.4	16.67	1.44	0.00
Germany	DE	41.36	45.83	42.29	29.17	15.2	25.00	1.16	0.00
Greece	EL	59.77	62.50	30.56	31.25	8.76	6.25	0.91	0.00
Hungary	HU	54.77	79.17	30.21	20.83	13.06	0.00	1.95	0.00
Ireland	IE	33.89	43.75	37.57	37.50	25.44	18.75	3.1	0.00
Italy	IT	54.6	54.17	31.48	29.17	12.55	16.67	1.36	0.00
Latvia	LV	58.26	75.00	32.18	25.00	9.33	0.00	0.23	0.00
Lithuania	LT	32.7	62.50	40.87	31.25	21.16	6.25	5.27	0.00
Luxembourg	LU	38.21	40.00	41.43	40.00	19.04	0.00	1.32	20.00
Malta	MT	28.86	100.00	39.19	0.00	26.83	0.00	5.12	0.00
Netherlands	NL	21.18	31.25	41.75	31.25	32.4	37.50	4.67	0.00
Poland	PL	56.25	50.00	31.34	43.75	10.96	0.00	1.45	6.25
Portugal	PT	51.13	14.29	33.1	28.57	14.4	57.14	1.38	0.00
Romania	RO	60.52	81.25	28	18.75	10.29	0.00	1.18	0.00
Slovakia	SK	51.59	81.25	35.39	6.25	11.84	12.50	1.18	0.00
Slovenia	SI	31.62	55.56	41.64	33.33	23.42	11.11	3.32	0.00
Spain	ES	56.81	58.33	30.16	37.50	11.96	4.17	1.08	0.00
Sweden	SE	21.84	84.21	37.13	5.26	33.04	10.53	7.99	0.00
United Kingdom	UK	38.38	31.25	39.91	31.25	19.35	31.25	2.35	6.25

[1] The digital intensity score is based on counting how many out of 12 technologies are used by each enterprise. Then they are divided into four clusters of digital intensity: Very Low (scores 0–3), Low (score 4–6), High (score 7–9) and Very High (score 10–12). [2] "The 2015 list of technologies includes: usage of internet by a majority of the workers; access to ICT specialist skills; fixed broadband speed >30 Mbps; mobile devices used by more than 20% of employed persons; has a website; has some sophisticated functions on the website; presence on social media; does e-sales for at least 1% of turnover; exploit the B2C opportunities of web sales; use an ERP software; use a CRM software; share electronically supply chain management information." [3] Percentage of enterprises (all sectors).

Similarly, as the EC did, the cooperatives were classified according to the scores obtained but according to the index created for this research. Although it is true that the number of variables included in the DESI index was greater, it can be verified that the trend analyzed was maintained. Thus, whereas companies in general within the EU (28) had a Very Low level at 45.84% of the companies, cooperatives had a higher level at 60.57%. For the Low level, it was 36.20% compared to 26.21% in cooperatives, and in High and Very

High, compared to 15.88% and 2.08%, respectively, in cooperatives, it remained at 12.33% and 0.88%.

4.2. Influence of the Variables Size and Wealth of the Country on the Digital Transformation of Agri-Food Cooperatives

In order to contrast the second and third assumptions and verify whether the variables size and wealth of the country exert some influence on the degree of digital transformation of the agri-food cooperatives, we carried out the corresponding correlation analysis and collected it in the following tables. Additionally, an inter-territorial analysis was included to complete the analysis. However, given that the sample by country was not high in this aspect, Table 1 was left for future research to elaborate on this line.

As Table 7 shows, there is a high correlation between the aggregate variable, which measured the level of digitization of European agri-food cooperatives, and the variables proposed to measure the size of the cooperative, such as the number of employees/members, the ordinary results before taxes, or total assets, which leads one to confirm the existence of a correlation between the size of the cooperative and the level of digitalization thereof. Along the same lines, other studies [44] have assessed the influence of firm size, corporate website quality and outsourcing of ICT management on organizational performance in the agri-food cooperative sector measured in terms of efficiency. The results obtained also point to the existence of a direct relationship.

Table 7. Correlations of Spearman Agri-food Coop EU (28). Source: own development.

		V ADDED (1)	GDP PER CAPT (2)	C1 Cooperatives with a Website (3)	Number of Employees Last Year Available (4)	Ordinary Results before Taxes Thousand EUR Last Year Available (5)	Total Assets Thousand EUR Last Year Available (6)
(1)	Correlation coefficient Next (bilateral) N	1.000	0.326 ** 0.000 454	0.918 ** 0.000 454	0.505 ** 0.000 333	0.300 ** 0.000 353	0.570 ** 0.000 385
(2)	Correlation coefficient Next (bilateral) N		1.000	0.284 ** 0.000 454	−0.011 0.847 333	0.101 0.057 353	0.270 ** 0.000 385
(3)	Correlation coefficient Next (bilateral) N			1.000	0.461 ** 0.000 333	0.277 ** 0.000 353	0.534 ** 0.000 385
(4)	Correlation coefficient Next (bilateral) N				1.000	0.453 ** 0.000 272	0.778 ** 0.000 292
(5)	Correlation coefficient Next (bilateral) N					1.000	0.565 ** 0.000 353
(6)	Correlation coefficient						1.000

** Correlation is significant at the 0.01 level (bilateral).

On the other hand, it is also verified that there is a correlation between the country's GDP per capita and the level of digitalization, confirming in the same way that greater wealth meets greater digitalization.

To complete the study, it includes, in a complementary way, an assessment of the influence and significance level of the different variables analyzed on the level of digitalization of cooperatives through linear regression. Specifically, the following expressions are proposed:

$$\text{V ADDED} = \beta o + \beta_1 \, \text{GDP} + \beta_2 \, \text{Assets} + \varepsilon \; (\text{Model A}) \tag{2}$$

$$\text{V ADDED} = \beta o + \beta_1 \, \text{GDP} + \beta_2 \, \text{Ord Results} + \beta_3 \, \text{no. of employees} + \varepsilon \tag{3}$$
$$(\text{Model B})$$

Due to a very high correlation between assets, ordinary results and number of employees, they cannot be entered in the same regression. However, in order to see the influence on the aggregate variable, as a measure of the digital transformation of cooperatives, it may be of interest, and hence they are separated into two regressions.

Regarding the level of digitalization of agri-food cooperatives in the EU (28), Tables 8 and 9 show how the variables size of the cooperative, as well as the wealth of the country, measured as GDP per capita, influence their transformation. However, it cannot explain, to a large extent, (R2), such a transformation, but it certainly affects it, as it seemed when analyzing the correlations.

Table 8. Descriptive statistics (a). Source: own development.

	Average	Dev. Deviation	N
V ADDED	2.75	2.98	385
GDP PER CAPT	30,888	17,083.97	385
Total assets thousand EUR Last year available	58,625	287,367.38	385

Table 9. Model A coefficients Source: own development.

	Coef.	T	Next
(Constant)		3.00	0.00
GDP PER CAPT	0.33	6.94	0.00
Total assets thousand EUR Last year available	0.20	4.33	0.00
N		385	
R-sq (R2)		0.169	
Ad, RSq		0.165	
F		39.099	
(P-F)		0.000	

Should the second regression be checked, the results are similar, as shown in Tables 10 and 11, although the explanatory capacity of the model would increase somewhat.

Table 10. Descriptive statistics (b). Source: own development.

	Average	Dev. Deviation	N
V ADDED	3.01	3.051	272
GDP PER CAPT	30,087	16,192.51	272
Ordinary results before taxes thousand EUR Last year available	1420	9986.14	272
Number of employees Last year available	185.01	719.44	272

Table 11. Model B coefficients. Source: own development.

	Coef.	T	Next
(Constant)		1.03	0.30
GDP PER CAPT	0.44	8.31	0.00
Ordinary results before taxes thousand EUR Last year available	0.14	2.50	0.01
Number of employees Last year available	0.14	2.61	0.01
N		272.000	
R-sq (R2)		0.291	
Ad, RSq		0.283	
F		36.645	
(P-F)		0.000	

Again, one can check that the proposed model B shows again significance in the influence of the size of the cooperative, as well as of the country's wealth does in the level of digital transformation of the cooperative. In the latter case, the model has some more explanatory capacity.

5. Conclusions

The analysis of the processes of digitalization of business structures constitutes an indisputable element of interest as a catalyst phenomenon of a set of disruptive processes that lead to profit in terms of efficiency, productivity and business competitiveness. The agri-food sector, as a strategic sector of the European productive model, cannot be left out of this opportunity. On the contrary, the adoption of ICT offers competitive advantages by improving the productive yields of the sector while promoting the development of more sustainable, efficient and safe production models.

This research is based on three main research hypotheses that, after being contrasted, were all accepted. In the first place, considering Hypothesis 1, the degree of development of the level of digitization of European agri-food cooperatives is in general terms suboptimal, which entails the existence of a certain "technological backwardness". We can observe said deficiency in technological adoption by cooperatives applied to electronic commerce and services offered on the Internet. Such deficiency is even greater when compared with the data relating to the whole of the European business net. All the indicators analyzed on the degree of digital transformation show worse results in the cooperative agri-food sector. This is also confirmed by the synthetic index "digitization level" constructed in this research. The results obtained in this digitization index show an extremely low average score for European agri-food cooperatives, in particular 2.64 out of 12 points. Second, according to Hypothesis 2, the size of the cooperative is determining for the degree of digitization of European agri-food cooperatives. Thus, those cooperatives that have greater size or volume of resources clearly present a higher level of digitization in the two dimensions of analysis observed: electronic commerce and web services offered. We can also say that the level of digitization has a positive influence on size. There is a similar correlation between the benefit of the cooperative and digitization, and although it has been interpreted in one sense, the analysis could be done in the opposite direction, concluding that the greater the digitization, the greater the benefit, size and therefore growth. In this context, we need to adopt policies that promote cooperative integration processes to allow cooperatives to increase their size and thus improve the conditions for better adopting technology. Third, the contrast of hypothesis 3 allows us to affirm the significant influence of the country's wealth on the degree of digital transformation of the cooperatives under study. From a territorial point of view, the analysis carried out confirms that there has been a very uneven digital transformation among the EU Member States (28) and, in particular, that new technologies are more frequently adopted by cooperatives whose activity develops in territories with greater wealth per capita.

This research has revealed the deficient degree of digital transformation of the European agri-food cooperative sector. Additionally, there is an urgent need to promote public policies that encourage greater adoption of technology in the sector to improve levels of competitiveness, productivity and efficiency. To this end, European public administrations are encouraged to guarantee high-performance digital connectivity in rural areas where the agri-food industry is mostly located. Additionally, promoting training programs in digital skills and information on existing technologies that could be applied to production processes and marketing channels is important. This would allow for a greater dynamism of electronic commerce and an increase in the number of services offered on the internet by agri-food cooperatives. It would also make it possible to face new challenges such as the digitization of the value chain or the integration of new technologies such as artificial intelligence (AI), blockchain, robotics or the internet of things (IoT). In short, the digitization of the agri-food cooperative sector offers a real opportunity to reshape the functioning of the agri-food markets and respond to the economic and environmental challenges facing the sector.

However, we are aware of the limitations of the study, since there are factors that have not been studied in depth. Among them, it is recommended, for future work, to analyze the type of cooperatives, the different subsectors and the greatest need or convenience of digitization, according to the specific circumstances of each cooperative. These efforts could help focus the efforts of institutions on more efficient digitization.

Author Contributions: Conceptualization, J.J.-V. and M.P.C.-C.; methodology, J.J.-V. and M.P.C.-C.; software, M.P.C.-C.; validation, J.J.-V., M.P.C.-C. and F.S-.R.; formal analysis, J.J.-V. and M.P.C.-C.; investigation, J.J.-V. and M.P.C.-C.; resources, J.J.-V. and M.P.C.-C.; data curation, M.P.C.-C.; writing—original draft preparation, J.J.-V. and M.P.C.-C.; writing—review and editing, J.J.-V., M.P.C.-C. and F.S.-R.; visualization, J.J.-V. and F.S.-R.; supervision, J.J.-V.; project administration, J.J.-V.; funding acquisition, J.J.-V. All authors have read and agreed to the published version of the manuscript.

Funding: This research was funded by AAUCAV, grant number BI1920 and The APC was partially funded by the incentive granted to the authors by the Catholic University of Ávila.

Institutional Review Board Statement: Not applicable.

Informed Consent Statement: Not applicable.

Data Availability Statement: Not applicable.

Conflicts of Interest: The authors declare no conflict of interest.

References

1. Colom Gorgues, A.; Cos Sánchez, P.; Florensa Guiu, R.M. Cooperativismo agroalimentario en Europa. Dimensión, gobernanza y análisis BCG de las sociedades cooperativas TOP25 de la UE-28 y TOP10 en España. *REVESCO. Rev. Estud. Coop.* **2019**, *130*, 73–98. [CrossRef]
2. Nordhaus, W.D. Productivity growth and the new economy. *Brook. Pap. Econ. Act.* **2002**, *33*, 211–265. [CrossRef]
3. Bresnahan, T.F.; Brynjolfsson, E.; Hitt, L.M. Information Technology, Workplace Organization, and the Demand for Skilled Labor: Firm-Level Evidence. *Q. J. Econ.* **2002**, *117*, 339–376. [CrossRef]
4. Hernando, I.; Nuñez, S. The Contribution of ICT to Economic Activity: A Growth Accounting Exercise with Spanish Firm-level Data. *Investig. Econ.* **2004**, *28*, 315–348.
5. Atrostic, B.K.; Nguyen, S.V. IT and Productivity in U.S. Manufacturing: Do Computer Networks Matter? *Econ. Inq.* **2005**, *43*, 493–506. [CrossRef]
6. Draca, M.; Sadun, R.; Van Reenen, J. Productivity and ICTs: A Review of the Evidence. In *The Oxford Handbook of Information and Communication Technologies*; Mansell, R., Ed.; Oxford University Press: Oxford, UK, 2007; pp. 100–147.
7. Torrent, J. *La Empresa red. Tecnologías de la Información y la Comunicación, Productividad y Competitividad*; Ariel: Barcelona, Spain, 2008.
8. Cardona, M.; Kretschmer, T.; Strobel, T. ICT and productivity: Conclusions from the empirical literature. *Inf. Econ. Policy* **2013**, *25*, 109–125. [CrossRef]
9. Lio, M.; Liu, M.-C. ICT and agricultural productivity: Evidence from cross-country data. *Agric. Econ.* **2006**, *34*, 221–228. [CrossRef]
10. Minten, B.; Barrett, C.B. Agricultural Technology, Productivity, and Poverty in Madagascar. *World Dev.* **2008**, *36*, 797–822. [CrossRef]
11. Aker, J.C. Dial "A" for Agriculture: A Review of Information and Communication Technologies for Agricultural Extension in Developing Countries. *Agric. Econ.* **2011**, *42*, 631–647. [CrossRef]
12. Jorgenson, D.W.; Vu, K. Information Technology and the World Growth Resurgence. *Ger. Econ. Rev.* **2007**, *8*, 125–145. [CrossRef]
13. Fernández-Portillo, A.; Almodóvar-González, M.; Hernández-Mogollón, R. Impact of ICT development on economic growth. A study of OECD European union countries. *Technol. Soc.* **2020**, *63*, 101420. [CrossRef]
14. Hanclova, J.; Doucek, P.; Fischer, J.; Vltavska, K. Does ict capital affect economic growth in the EU-15 and EU-12 countries? *J. Bus. Econ. Manag.* **2014**, *16*, 387–406. [CrossRef]
15. Evangelista, R.; Guerrieri, P.; Meliciani, V. The economic impact of digital technologies in Europe. *Econ. Innov. New Technol.* **2014**, *23*, 802–824. [CrossRef]
16. Peña-Vinces, J.C.; Cepeda-Carrión, G.; Chin, W.W. Effect of ITC on the international competitiveness of firms. *Manag. Decis.* **2012**, *50*, 1045–1061. [CrossRef]
17. García, E.; Rialp, A.; Rialp, J. Tecnologías de la información y comunicación (TIC) y crecimiento de la empresa. Información Comercial Española. *ICE. Rev. de Econ.* **2007**, *838*, 125–145.
18. Bernal-Jurado, E.; Mozas-Moral, A.; Fernández-Uclés, D.; Medina-Viruel, M. Determining Factors for Economic Efficiency in the Organic Olive Oil Sector. *Sustainability* **2017**, *9*, 784. [CrossRef]

19. Fernández-Uclés, D.; Elfkih, S.; Mozas-Moral, A.; Bernal-Jurado, E.; Medina-Viruel, M.J.; Ben Abdallah, S. Economic Efficiency in the Tunisian Olive Oil Sector. *Agriculture* **2020**, *10*, 391. [CrossRef]
20. Deichmann, U.; Goyal, A.; Mishra, D. Will digital technologies transform agriculture in developing countries? *Agric. Econ.* **2016**, *47*, 21–33. [CrossRef]
21. Bernal-Jurado, E.; Mozas-Moral, A.; Medina-Viruel, M.; Fernández-Uclés, D. Evaluation of Corporate Websites and Their Influence on the Performance of Olive Oil Companies. *Sustainability* **2018**, *10*, 1274. [CrossRef]
22. El Bilali, H.; Allahyari, M.S. Transition towards sustainability in agriculture and food systems: Role of information and communication technologies. *Inf. Process. Agric.* **2018**, *5*, 456–464. [CrossRef]
23. Kossaï, M.; De Souza, M.L.L.; Ben Zaied, Y.; Nguyen, P. Determinants of the Adoption of Information and Communication Technologies (ICTs): The Case of Tunisian Electrical and Electronics Sector. *J. Knowl. Econ.* **2019**, *11*, 845–864. [CrossRef]
24. Drazin, R. The processes of technological innovation. *J. Technol. Transfer.* **1991**, *16*, 45–46. [CrossRef]
25. Rogers, E.M. *Diffusion of Innovations*, 4th ed.; EEUU; Free Press: New York, NY, USA, 1995.
26. Ali, J. Factors Affecting the Adoption of Information and Communication Technologies (ICTs) for Farming Decisions. *J. Agric. Food Inf.* **2012**, *13*, 78–96. [CrossRef]
27. Cristobal-Fransi, E.; Montegut-Salla, Y.; Ferrer-Rosell, B.; Daries, N. Rural cooperatives in the digital age: An analysis of the Internet presence and degree of maturity of agri-food cooperatives' e-commerce. *J. Rural. Stud.* **2020**, *74*, 55–66. [CrossRef]
28. Jorge-Vázquez, J.; Chivite-Cebolla, M.P.; Salinas-Ramos, F. La transformación digital en el sector cooperativo agroalimentario español: Situación y perspectivas. *CIRIEC-Esp. Rev. Econ. Pública Soc. Coop.* **2019**, *95*, 39–70. [CrossRef]
29. Bernal-Jurado, E.; Mozas-Moral, A.; Fernández-Uclés, D.; Medina-Viruel, M.; Poyatos, R.P. Calidad de los sitios web en el sector agroalimentario ecológico y sus factores explicativos: El papel del cooperativismo. *CIRIEC-Esp. Rev. Econ. Pública Soc. Coop.* **2019**, *95*, 95–118. [CrossRef]
30. Fernández-Uclés, D.; Bernal-Jurado, E.; Medina-Viruel, M.; Mozas-Moral, A. El sector cooperativo oleícolas y el uso de las TIC: Un estudio comparativo respecto a otras formas jurídicas. *REVESCO. Rev. Estud. Coop.* **2016**, *120*, 53–75. [CrossRef]
31. Borrero, J.D. Sistema de trazabilidad de la cadena de suministro agroalimentario para cooperativas de frutas y hortalizas basado en la tecnología Blockchain. *CIRIEC-Esp. Rev. Econ. Pública Soc. Coop.* **2019**, *95*, 71–94. [CrossRef]
32. Bahn, R.; Yehya, A.; Zurayk, R. Digitalization for Sustainable Agri-Food Systems: Potential, Status, and Risks for the MENA Region. *Sustainability* **2021**, *13*, 3223. [CrossRef]
33. Raheem, D.; Shishaev, M.; Dikovitsky, V. Food System Digitalization as a Means to Promote Food and Nutrition Security in the Barents Region. *Agriculture* **2019**, *9*, 168. [CrossRef]
34. Dueñas, P.P.M.; Carmona, D.G. La gestión de la comunicación digital en las cooperativas españolas. *CIRIEC-Esp. Rev. Econ. Pública Soc. Coop.* **2021**, *101*, 193–225. [CrossRef]
35. Bernal-Jurado, E.; Mozas-Moral, A.; Fernández-Uclés, D.; Medina-Viruel, M.J. Online popularity as a development factor for cooperatives in the winegrowing sector. *J. Bus. Res.* **2021**, *123*, 79–85. [CrossRef]
36. Araujo, A.; Serrano, E.; Jordan, V. La presencia de las cooperativas de Catalunya en Internet. *CIRIEC-Esp. Rev. Econ. Pública Soc. Coop.* **2020**, *99*, 37–56. [CrossRef]
37. ORBIS. Detailed Information on Companies, Several Years. *BvD (Bureau van Dijk).* Available online: https://www.informa.es/en/business-risk/sabi/orbis (accessed on 23 March 2019).
38. Law, R.; Qi, S.; Buhalis, D. Progress in tourism management: A review of website evaluation in tourism research. *Tour. Manag.* **2010**, *31*, 297–313. [CrossRef]
39. McMillan, S.J. The Microscope and the Moving Target: The Challenge of Applying Content Analysis to the World Wide Web. *J. Mass Commun. Q.* **2000**, *77*, 80–98. [CrossRef]
40. Weare, C.; Lin, W.-Y. Content Analysis of the World Wide Web: Opportunities and Challenges. *Soc. Sci. Comput. Rev.* **2000**, *18*, 272–292. [CrossRef]
41. Fernández-Uclés, D.; Bernal-Jurado, E.; Mozas-Moral, A.; Medina-Viruel, M.J. The importance of websites for organic agri-food producers. *Econ. Res. Ekon. Istraž.* **2019**, *33*, 2867–2880. [CrossRef]
42. Bernal-Jurado, E.; Mozas-Moral, A. Evaluación del uso comercial de la World Wide Web por parte de las cooperativas de segundo grado españolas. *Rev. Esp. Estud. Agrosoc. Pesq.* **2008**, *219*, 181–200.
43. González-Aponcio, Z. Determinantes de la calidad de la información divulgada vía Web por las pequeñas y medianas cooperativas de Canarias. *REVESCO. Rev. Estud. Coop.* **2020**, *133*, 1–15. [CrossRef]
44. Cristóba-Fransi, E.; Montegut-Salla, Y.; Daries-Ramon, N. Cooperativismo 2.0: Presencia en Internet y desarrollo del comercio electrónico en las cooperativas oleícolas de Cataluña. *REVESCO. Rev. Estud. Coop.* **2017**, *124*, 47–73. [CrossRef]

MDPI

St. Alban-Anlage 66

4052 Basel

Switzerland

Tel. +41 61 683 77 34

Fax +41 61 302 89 18

www.mdpi.com

Agriculture Editorial Office

E-mail: agriculture@mdpi.com

www.mdpi.com/journal/agriculture